Automotive Brake Systems

By Chek-Chart,
a Division of
The H.M. Gousha Company

Michael T. Calkins, CMAT, *Editor*
Richard K. DuPuy, *Contributing Editor*

HarperCollins*Publishers*

Acknowledgments

In producing this series of textbooks for automobile mechanics and technicians, Chek-Chart has drawn extensively on the technical and editorial knowledge of the nation's carmakers and component suppliers. Automotive design is a technical, fast-changing field, and we gratefully acknowledge the help of the following companies in allowing us to present the most up-to-date information and illustrations possible:

Alfred Teves Technologies, Inc.
American Motors Corporation
Ammco Tools, Inc.
Benwil Industries, Inc.
Bendix Aftermarket Brake Division, Allied Automotive
Chrysler Motors Corporation
DMC Technical Products/Diemolding Corporation
Easco/K-D Tools
Earl's Performance Products
ESI Brake Parts, Division of Standard Motor Products, Inc.
Ford Motor Company
FMC Corporation, Automotive Service Equipment Division
Failure Analysis Associates
General Motors Corporation
 AC-Delco Division
 Buick Motor Division
 Cadillac Motor Car Division
 Chevrolet Motor Division
 Oldsmobile Division
 Pontiac Division
Goodyear Tire & Rubber Company
Hako Minuteman
Kent-Moore Tool Group, Sealed Power Corporation
Kwik-Way Manufacturing Company
MAC Tools, Inc.
OTC Division, Sealed Power Corporation
Porsche Cars North America
Subaru of America, Inc.
Snap-On Tools Corporation
Stainless Steel Brakes Corporation
Vim Tools, Durston Manufacturing
The White Lung Association of New York

The comments, suggestions, and assistance of the following contributors were invaluable:

Ken Bullock, Benton Darda, and George Kurath; Ammco Tools Training Center, North Chicago, IL.
Scott Curtis; Curtis Circus Citröen/Maserati Service and Trihawk Sales
Robert Donaldson; Castrol, Inc., Edison, NJ.
Bernie Haren; Superior Clutch and Brake
Jay G. Osborn; Allied Automotive Aftermarket, Jackson, TN.
Gary Phillipi; MDC and Associates, Tapley Sales
Frank Skelton; Nuturn Corporation, Smithsville, TN.
Bill Benak, Mike Brandt, Steve Brown, Bob Leigh, Gary Lewis, Walt Merek, Jerry Mullen, Les Schwoob, Lee Spencer, Gary Toothman, Steve Walton; De Anza College Automotive Technology program

At Chek-Chart, Ray Lyons managed the production of this book. Original art and photographs were produced by John Badenhop, Jamey Dilbeck, Janet Jamieson, C. J. Hepworth, and F. J. Zienty. The project was directed by Roger L. Fennema, and was initiated under the auspices of Richard C. Thomson.

Library of Congress Cataloging and Publication Data:

Chek-Chart, 1987
 Automotive Brake Systems
 (Harper & Row/Chek-Chart Automotive Series)
 v. 1. Classroom Manual. v. 2. Shop Manual.

ISBN: 0-06-454010-3 (set)
Library of Congress Catalog Card No.: 87-12060

Contents

On the Cover:
Front—A poster depicting the many components of a typical brake system, courtesy of EIS Brake Parts Division of Standard Motor Products.
Rear—A pair of brake lathes and a brake vacuum, courtesy of Bear Automotive Equipment Company and Hako Minuteman.

Introduction to Automotive Brake Systems

Automotive Brake Systems is part of the Harper & Row/Chek-Chart Automotive Series. The package for each course has two volumes, a *Classroom Manual* and a *Shop Manual*. Other titles in this series include:

- Automatic Transmissions and Transaxles
- Fuel Systems and Emission Controls
- Engine Performance Diagnosis and Tune-Up
- Automotive Electrical and Electronic Systems
- Automotive Engine Repair and Rebuilding

Each book is written to help the instructor teach students how to become excellent professional automobile mechanics. The two-manual texts are the core of a complete learning system that leads a student from basic theories to actual hands-on experience.

The entire series is job-oriented, especially designed for students who intend to work in the automotive service profession. A student will be able to use the knowledge gained from these books and from the instructor to get and keep a job. Learning the material and techniques in these volumes is a big step toward a satisfying and rewarding career.

The books are divided into *Classroom Manuals* and *Shop Manuals* for an improved presentation of the descriptive information and study lessons, along with the practical testing, repair, and overhaul procedures. The manuals are designed to be used together: the descriptive chapters in the *Classroom Manual* correspond to the application chapters in the *Shop Manual*.

Each book is divided into several parts, and each of these parts is complete by itself. Instructors will find the chapters to be complete, readable, and well thought-out. Students will benefit from the many learning aids included, as well as from the thoroughness of the presentation.

The series was researched and written by the editorial staff of Chek-Chart, and was produced by Harper & Row Publishers. For over 58 years, Chek-Chart has provided car and equipment manufacturer's service specifications to the automotive service field. Chek-Chart's complete, up-to-date automotive data bank was used extensively in preparing this textbook series.

Because of the comprehensive material, the hundreds of high-quality illustrations, and the inclusion of the latest automotive technology, instructors and students alike will find that these books keep their value over the years. In fact, they will form the core of the master mechanic's professional library.

How To Use This Book

Why Are There Two Manuals?

This two-volume text — **Automotive Brake Systems** — is not like most other textbooks. It is actually two books, a *Classroom Manual* and a *Shop Manual* that should be used together. The *Classroom Manual* will teach you what you need to know about brake system theory and the braking systems on cars. The *Shop Manual* will show you how to repair and adjust complete systems, and their individual components.

The *Classroom Manual* will be valuable for study and reference both in class and at home. It contains text and illustrations you can use for years to refresh your memory about the basics of automotive brake systems.

In the *Shop Manual,* you will learn how to troubleshoot and overhaul the systems and parts you are studying in the *Classroom Manual.* These techniques are explained in an easy-to-understand manner, and some are illustrated with step-by-step photo-sequences that guide you through the procedures. Use the two manuals together to fully understand how the parts work, and how to fix them when they don't work.

What's In These Manuals?

There are several aids in the *Classroom Manual* that will help you learn more:

1. The text is broken into short sections with subheads for easier understanding and review.
2. Each chapter is fully illustrated with line drawings and photographs.
3. Key words are printed in **boldface type** and are defined on the same or the facing page, and also in a glossary at the end of the manual.
4. Review questions following each chapter allow you to test your knowledge of the material covered.
5. A brief summary at the end of each chapter helps you review for exams.
6. Every few pages you will find short blocks of "nice to know" information that supplement the main text.
7. At the back of the *Classroom Manual* is a sample test similar to those given for National Institute for Automotive Service Excellence (NIASE) certification. Use it to study and prepare yourself when you are ready to become certified as an expert in the area of brake system repair.

In addition to detailed instructions on overhaul, test, and service procedures, this is what you'll find in the *Shop Manual:*

1. Information on how to use and maintain shop tools and test equipment.
2. Detailed safety precautions.

3. Troubleshooting charts and tables to help you locate brake system problems.
4. Professional repair tips to help you perform repairs more rapidly and accurately.
5. A complete index to help you quickly find what you need.

Where Should I Begin?

If you already know something about automotive brake systems and how to repair them, you may find that parts of this book are a helpful review, while others address new technologies you are not yet familiar with. If you are just starting in car repair, the subjects covered in these manuals may be all new to you.

Your instructor will design a course to take advantage of what you already know, and what facilities and equipment are available to work with. You may be asked to take certain chapters of these manuals out of order. That's fine; the important thing is to fully understand each subject before you move on to the next. Study the vocabulary words, and use the review questions to help you comprehend the material.

While reading in the *Classroom Manual*, refer to your *Shop Manual* and relate the descriptive text to the service procedures. When working on actual car brake systems, look back at the *Classroom Manual* to keep basic information fresh in your mind. Working on a complicated modern brake system isn't always easy. Take advantage of the information in the *Classroom Manual*, the procedures of the *Shop Manual*, and the knowledge of your instructor to help you.

Remember that the *Shop Manual* is a good book for work, not just a good workbook. Keep it on hand while you're working on a brake system. For ease of use, the *Shop Manual* will fold flat on the workbench or under the car, and it can withstand quite a bit of rough handling.

To perform many brake repair and overhaul procedures, you will also need an accurate source of brake system specifications. Most shops have either automakers' service manuals, which lists these specifications, or an independent book such as the **Chek-Chart Car Care Guide**. This unique book, which is updated yearly, provides service instructions, brake system maintenance intervals, and brake bleeding sequence information you need to work on specific cars.

PART ONE

General Brake Service Operations

Chapter One
Shop Practices
and Special Tools,
Cleaners, and Lubricants

Chapter Two
Brake System Diagnosis

Chapter

1

Shop Practices and Special Tools, Cleaners, and Lubricants

Automotive brake systems are basically easy to service and overhaul, once you understand their construction. If you have a good grasp of the fundamentals presented in the *Classroom Manual*, you will find brake system service to be easy, interesting, and satisfying.

This chapter introduces a number of recommended shop practices for brake repair. Among the most important of these are the procedures used to jack cars and properly support them before you begin work. If you always follow good shop practices, your work will be both easier and safer.

This chapter also covers the unique tools used by the brake technician. Like any specialized area of automotive repair, brake work requires certain special tools in addition to the normal assortment of basic hand tools. These include tools designed to clean, adjust, disassemble, assemble, and measure wheel friction assemblies. Special tools are also required to refinish the brake hydraulic and friction components, and to work with the seamless, double-wall, steel tubing used for brake lines.

Finally, this chapter covers the various chemicals used to clean brake parts, and the unique lubricants that smooth the operation of the brake system and extend component life. These cleaners and lubricants are necessary to do accurate and professional brake work.

SHOP PRACTICES

The professional brake technician makes good shop practices a normal part of his or her on-the-job activities. Good shop practices are concerned primarily with three areas: convenience, safety, and cleanliness. Various recommended practices are listed below. As you read them, you may note that they are difficult to separate from each other — a clean, well-organized work area is also a safe one. The practices recommended below are both basic and common sense. This list is by no means complete; when you have finished reading it, see if you can think of other practices which will help you do a first-class job.

1. Organize your work area. This keeps unneeded movements to a minimum, which increases your efficiency and makes the repair procedure less tiring.
2. Keep your work bench clean and free of unnecessary tools, old parts, and other clutter. This provides more useful space to work in, and reduces the possibility of losing small parts.
3. Set aside enough bench space for any component overhauls that will be necessary.

Figure 1-1. Raising the car on a hoist makes it easier to service the wheel friction assemblies.

4. Equip your bench with a container suitable for draining brake fluid from hydraulic system components. This makes it easier to keep the work area clean and free from brake fluid that can damage painted surfaces.
5. Keep all tools clean and dry. A clean tool will not slip off the part on which you are working or out of your hand.
6. Replace tools in their proper storage area after each use so you will always know exactly where they can be found when needed.
7. Keep a supply of clean shop towels near your work area to wipe up spills.
8. Keep the shop floor and your work area clean and free from brake dust and fluid. To prevent asbestos particles from getting into the air when cleaning, wet sweep the floor or use a vacuum with a high efficiency particulate air (HEPA) filter.
9. Brake cleaning fluid fumes can be harmful. Provide adequate ventilation regardless of the weather.
10. Shop lighting should be adequate to prevent eyestrain. When using extension lights, make sure the bulbs are covered with safety shields to prevent burns. Keep electrical cords off the floor to prevent tripping on them.
11. Keep your hands as clean and dry as possible. A certain amount of dirt and dust is normal during disassembly, but should be washed away before eating or using the restroom. Never handle clean parts with dirty hands during reassembly.
12. Do not smoke while working, especially when working near brake dust and cleaning fluids. Inhalation of these items can cause serious diseases, and cleaning fluid vapors may be volatile and can cause major injuries if ignited.
13. Always wear safety glasses or goggles when there is a chance of foreign liquids or objects getting into your eyes. This can happen when cleaning friction assemblies, working under the car, or using almost any type of power tool.
14. Have plastic bags or other suitable containers available for saving old parts to be returned to the customer. Never dispose of old parts until the repairs are finished and the customer has been given the option of taking the parts home.
15. Keep up-to-date shop manuals, technical service bulletins, and other specification sources available for use. Refer to them whenever you are uncertain; *never guess.*
16. Maintain a professional personal appearance in keeping with your responsibilities. Customers tend to judge between a "grease monkey" and a "technician" solely on looks. If you are dirtier than the car on which you are working, you will inspire little confidence in your ability.

Lifts and Hoists

The easiest way to service the wheel brakes is to raise the vehicle off the ground so the friction assemblies are at a convenient working height. A lift or hoist, figure 1-1 is the safest and most efficient way to do this. There are several types of single- and double-post hoists available, and most of them are suitable for brake service. Naturally, hoists that support the car under the wheels, like those commonly used for exhaust system service, cannot be used when servicing the wheel brakes.

When raising a car on a lift or a hoist, you must position the lifting pads *only* at the locations designated by the vehicle manufacturer for this purpose. These lifting points are shown in the owner's manual, shop manual, and certain aftermarket service publications. Even very minor damage caused when the car is lifted in other locations can significantly weaken the body structure and reduce the crashworthiness of the vehicle; this is particularly true with unit-body cars.

Jacks and Safety Stands

When a hoist is unavailable, the car can be raised with a hydraulic floor jack and supported on safety stands, figure 1-2. Be sure to use a jack with adequate lifting capacity for the vehicle being serviced. Never use the jack that came with the car to lift or support the vehicle for brake service.

Compared to hoists, jacks and safety stands have a number of disadvantages. They limit the height to which the vehicle can be raised, and force you to work in an awkward and uncomfortable position which increases the amount of time the job will take. Safety stands are also more dangerous than hoists. If a stand is defective or extended beyond its safe working

Figure 1-2. A hydraulic jack and safety stands are an alternative to raising the car on a hoist.

Figure 1-3. Be absolutely sure safety stands are secure before you remove the wheels and begin work.

height, or if a careless person bumps into the car or stand, the vehicle can slip off and injure a person working underneath. Always check the stability of the stands before you begin work, figure 1-3.

TRICKS OF THE TRADE

There are no shortcuts in the proper diagnosis and repair of an automobile or light truck brake system. The safety concerns involved with vehicle brakes, and the legal liabilities that can result from improper repairs, make it foolish to do anything but a complete and professional repair in every case. However, even though factory shop manuals contain approved procedural sequences for every part of a particular job, no two brake technicians in the field approach a repair in exactly the same way. Experience has taught them different, and sometimes more efficient ways of doing the same job. These are often called "tricks of the trade."

The practices loosely defined by the expression "tricks of the trade" generally fall into two catagories, ways to organize the job, and alternative tools and procedures. Performing brake repairs in an organized manner requires less time and reduces the chances of mistakes. Every brake job can benefit from good organization. Alternative tools and procedures allow you to do some jobs without the special equipment that would otherwise be required, or allow you to save a great deal of time. However, you should never use an alternative tool or procedure if there is any chance it may cause damage or provide less than professional results.

There are literally hundreds of "tricks of the trade" and they cannot all be mentioned here. Some general examples are provided below to explain what we mean and to stimulate your thinking. Additional examples that apply to specific situations can be found throughout this text in the appropriate chapters.

Helpful Hints

1. When rebuilding hydraulic components, particularly master cylinders, it is often helpful to lay out the parts in the order they are removed, figure 1-4, or in the same sequence shown in an exploded drawing furnished by the factory. This helps prevent time-consuming mistakes when you overhaul a part with which you are unfamiliar.
2. When overhauling wheel friction assemblies with which you are unfamiliar, finish one wheel before starting on the other. It is easier to keep track of fewer parts, and you can use the other brake as a guide for reassembly. Taking apart the brakes at both wheels increases the possibility of mistakes.
3. Other than removing frozen pistons from hydraulic components, do not force brake parts apart or together. Virtually all parts in a brake system can be disassembled or assembled with hand pressure and the appropriate tools. If you cannot separate or install a part without the use of force, something has been put together wrong.

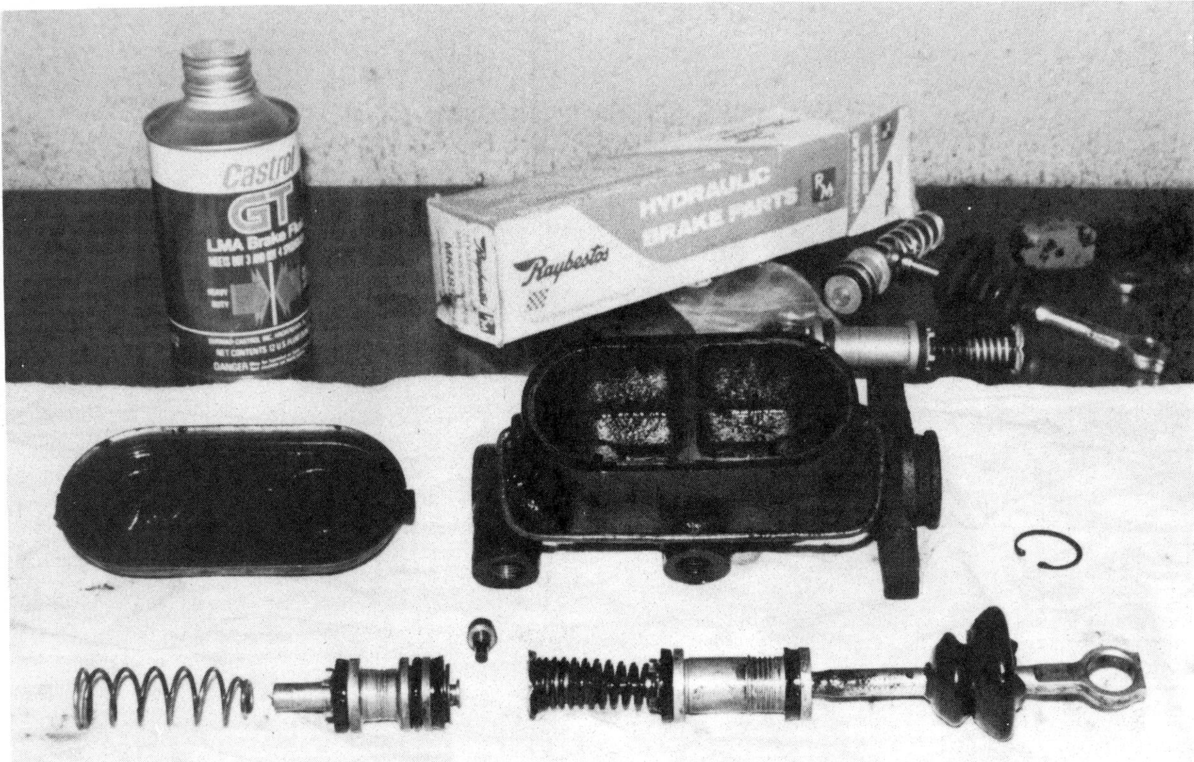

Figure 1-4. When disassembling a component such as a master cylinder, lay the parts out in the order in which they are removed.

4. Never wipe hydraulic parts with rags or shop towels. Lint and threads from rags and towels can collect on the parts. This debris will find its way into the hydraulic system and clog ports, causing seal leaks. Allow parts to air-dry, or dry them with clean, unlubricated compressed air after cleaning.
5. Lubricate all hydraulic pistons and seals with brake fluid or assembly lube before reassembly. This makes it easier to slide the parts together, helps prevent seal damage, and provides internal lubrication before the brake system is bled.
6. Certain brake repair operations are designed to be done with the aid of special tools available from the dealer. In many cases, aftermarket companies market similar tools for the same purpose. If you get stuck during a repair of an unfamiliar brake assembly, stop and consult a shop manual to determine if a special tool may be required. A number of common special tools are described in the following sections.

CLEANING TOOLS

Brake friction assemblies must be cleaned before a complete inspection or acceptable work can be performed. Road grime, wheel bearing grease, axle lubricants, brake fluid, and brake dust are some of the contaminants that collect on the brake assembly as a result of normal use or failures in the system. These contaminants disguise the condition of brake parts that may need to be replaced. And, if contaminants are not removed, they may lead to premature failure of new parts that are installed.

Another reason for cleaning the wheel friction assemblies is to avoid exposure to the asbestos fibers in brake dust and the health hazards they present. Special cleaning tools have been developed for this purpose. Under no conditions should an air hose be used to simply blow the brake dust off of a friction assembly into the shop air. The cleaning tools used to remove friction assembly contaminants and brake dust include:

- Brake vacuums
- Brake washers
- Brushes and scrapers
- Parts washers
- Air compressors.

Brake Vacuums

The special vacuums used to remove loose dust and dirt from the brake assembly are equipped with high efficiency particulate air (HEPA) filters designed to trap asbestos fibers. These

Figure 1-5. A brake vacuum is a safe and efficient means to remove dust containing asbestos fibers from a friction assembly.

fibers are small enough to pass through the conventional filters of general-use industrial vacuums, so a normal shop vacuum is *not* an acceptable substitute for a proper brake vacuum. Vacuums that collect asbestos fibers are also found on the brake shoe arcing equipment covered later in this chapter.

The brake vacuum shown in figure 1-5 is typical of those on the market. A shroud fits over the friction assembly to prevent the escape of fibers during the cleaning process. A vacuum is attached to the shroud to create a low pressure area around the friction assembly. A compressed-air nozzle built into the shroud is then used to dislodge dust and other contaminants which are drawn into the vacuum and trapped for later disposal.

Brake Washers

An alternative to the brake vacuum is the brake washer, figure 1-6. A brake washer connects to a compressed air hose and sprays a low-pressure stream of cleaning solution onto the brake. The solution is a mixture of water and a non-petroleum-base solvent. Brake washers are very effective at removing loose dirt and dust from the friction assembly, and the cleaning solution traps asbestos fibers so that they do not become airborne and present a health hazard. Because they use a solvent cleaning solution, brake washers are also able to wash away brake fluid, grease, or axle lubricant that has leaked onto the brake.

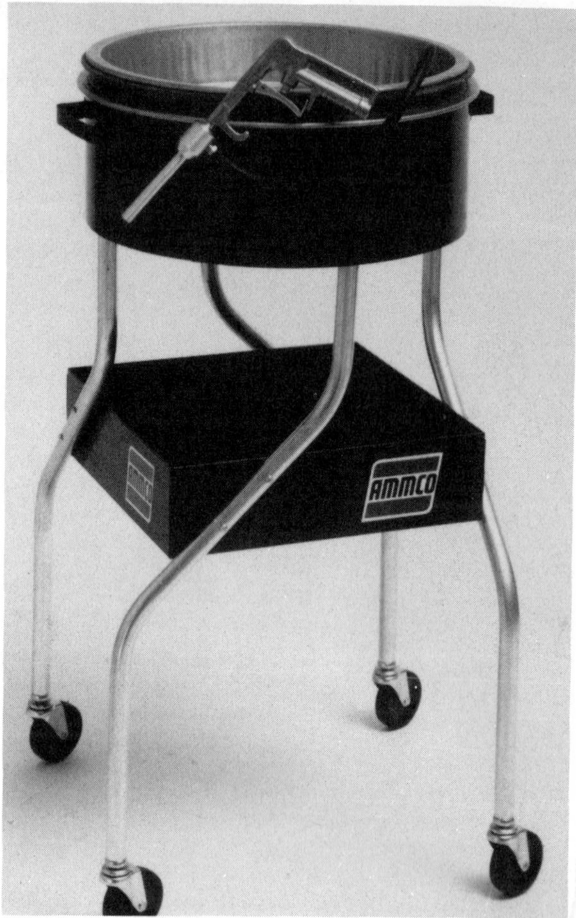

Figure 1-6. A brake washer cleans the friction assembly while it removes brake dust.

Figure 1-7. A wire brush used to clean brake friction assemblies.

Brushes and Scrapers

Once a brake vacuum or washer has been used to remove loose dust particles from a friction assembly, wire brushes and scrapers are often used to clean away baked-on contamination. To prevent dangerous asbestos particles from being released into the air, these tools should only be used when the part to be cleaned is wetted with brake cleaning solvent. A brush should *never* be used to remove asbestos dust from a dry friction assembly.

A wire brush, figure 1-7, consists of a wooden handle fitted with many stiff, mild-steel bristles. The bristles are able to clean hard-to-reach areas without damage to steel parts.

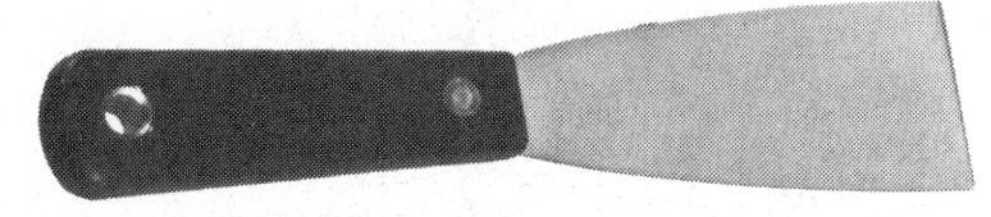

Figure 1-8. A scraper used to clean brake friction assemblies.

Figure 1-9. Parts washers have an electric pump that circulates a petroleum-base solvent.

Wire brushes should not be used on hydraulic cylinder bores or pistons, or on any aluminum part where the surface finish is important to proper brake operation.

Scrapers, figure 1-8, consist of a thin, unsharpened, steel blade fitted to a handle. Scrapers will remove stubborn buildup, and are most effective when cleaning flat surfaces.

Parts Washers

Virtually all automotive shops where brake work is performed have a parts washer, figure 1-9. The typical parts washer has a steel tank to hold the parts to be cleaned and a hose to direct a stream of solvent onto the parts in the tank. The solvent is circulated by a small electric pump, and filtered to remove larger particles of contamination.

Parts washers generally use petroleum-base solvents such as Stoddard solvent that make them unsuitable for cleaning most brake parts. Never clean brake system hydraulic parts in petroleum-base solvents. However, brake drums, rotors, and backing plates with a heavy buildup of contaminants may be cleaned in a parts washer if they are then cleaned with a non-petroleum-base brake cleaner to remove any oil left behind by the solvent.

Figure 1-10. An air compressor can be used to dry some brake system parts.

Air Compressors

An air compressor, figure 1-10, is a piece of equipment found in almost every automotive repair shop. Compressed air directed through a nozzle is a common method used to dry parts once they have been rinsed after cleaning in a solvent tank or other type of parts cleaner. For brake work, this practice is unacceptable in many cases because compressed air is usually contaminated with water and oil. Water lowers the boiling point of brake fluid, and *any* petroleum-base fluid will attack the rubber seals in the brake hydraulic system.

There are several ways compressed air becomes contaminated. First, condensation collects in the air storage tank as a result of temperature and pressure changes. In addition, moisture can be condensed out of the atmosphere as the high-pressure air goes through a pressure drop leaving the tip of the air nozzle.

Oil contamination can come from a couple of sources. A small amount of the lubricating oil in the compressor always leaks past the rings and into the air storage tank. Also, some compressor air outlets are fitted with automatic oilers that add oil to the airstream to keep impact wrenches and other pneumatic tools operating smoothly.

Compressed air should never be used to dry parts of the brake hydraulic system. In addition, it should not be used to dry brake drums and rotors because oil may contaminate the friction surfaces, and from them, the brake linings. Whenever compressed air is used to dry other parts of the brake system, you should be sure the compressor has been serviced regularly, and is not equipped with an automatic oiler. This will help ensure that the air supply is as free from contamination as possible.

CHEMICAL CLEANERS

Once the bulk of the dirt and grime is removed from a brake component, a final cleaning must be done to remove any trace of contamination; this is particularly true of hydraulic system parts. Final friction assembly cleaning, and all hydraulic component cleaning, must be done with a chemical cleaner that falls into one of three groups:

- Non-petroleum-base solvents
- Alcohol
- Detergents.

Non-Petroleum-Base Solvents

Petroleum- or oil-base solvents such as Stoddard solvent, kerosene, and gasoline cannot be used on most brake parts because they leave an oil film wherever they touch. Even if a film is not visible on the surface, oils from the solvents become trapped in the pores of the metal and emerge later to create problems in the brake system. Gasoline is highly flammable and its vapors are explosive; *never* use gasoline as a solvent to clean parts.

When hydraulic system parts that have been cleaned in petroleum-base solvents are reassembled, the oil film contaminates the brake fluid and causes the rubber seals and hoses in the system to shrink or swell. The results are fluid leaks, reduced strength, and possible brake system failure. The solvent film left on the friction surface of drums or rotors can reduce stopping ability when it contaminates the friction material on the brake shoes or pads.

Because conventional petroleum-base solvents are unsuitable for brake system service, non-petroleum-base solvents were developed. As discussed in the *Classroom Manual*, these solvents are made from chemicals like 1,1,1-trichlorethane, tetrachlorethane, and perchlorethylene or "perk." These solutions are powerful and effective in removing most dirt, grease, and carbon deposits, and they dry quickly without leaving harmful residue.

Perk-type solvents are available in bulk form or in convenient spray cans. The parts to be cleaned are either dipped in the bulk solvent, or sprayed with solvent from an aerosol can until all of the contaminants are removed. When a solvent spray is used to clean wheel friction assemblies that use asbestos-based friction materials, the runoff contains asbestos fibers and should be recovered in a catch basin and disposed of in a safe and approved manner.

Alcohol

Because of its chemical similarity to polyglycol brake fluid, denatured alcohol is another good cleaner for brake parts. Alcohol is not as powerful a solvent as perk-type cleaning solutions, but it is cheaper and also dries quickly without leaving harmful residue.

Denatured alcohol is also useful for flushing brake lines clean, and can be used as an assembly lubricant for hydraulic system components. In both these situations, however, it is essential that the alcohol be completely evaporated before the hydraulic system is sealed and filled with brake fluid. Alcohol has a very low boiling point that will significantly lower the brake fluid boiling point and lead to vapor lock if the two liquids are mixed.

Detergents

Mild dish-washing detergents are low cost alternatives to other types of brake cleaners in certain applications. Alcohol and other non-petroleum-base solvents evaporate readily and are used up quickly. For these reasons they tend to be fairly expensive. While not suitable for use on hydraulic system components, mild detergents do a good job of removing the oily film from parts such as drums and rotors that have been cleaned with petroleum-base solvents.

ADJUSTING TOOLS

Drum brakes without automatic adjusters can sometimes be adjusted using a standard screwdriver. However, special adjusting tools are available to make the job simpler and faster. These tools are also used to manually adjust drum brakes with automatic adjusters in certain

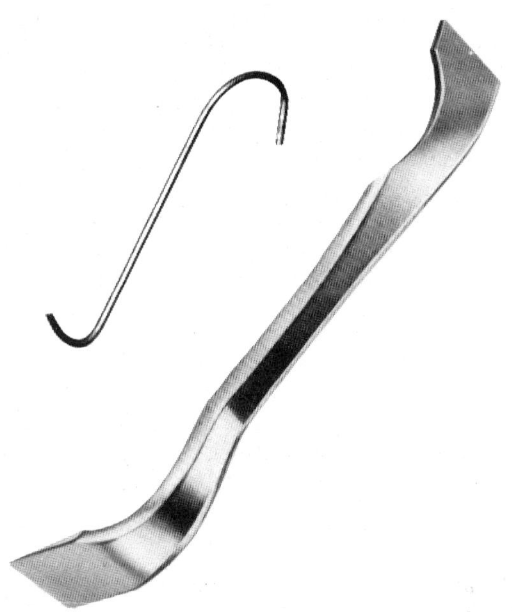

Figure 1-11. Brake spoons and wire hooks are used when rotating starwheel adjusting mechanisms.

situations. On some cars, a special tool is mandatory to adjust the brakes.

The two main types of brake adjusting tools are spoons and wire hooks, figure 1-11. Brake spoons are shaped to easily allow the technician to reach through a hole or slot in the backing plate or brake drum to engage the teeth on a starwheel adjuster and turn it. When the brakes on a vehicle with automatic starwheel adjusters are adjusted manually, a wire hook is used along with the brake spoon to hold the automatic adjuster lever away from the teeth of the starwheel. This allows the starwheel to be turned freely in both directions.

SPECIAL WRENCHES

Special wrenches used in brake service are designed to make certain jobs easier, to help avoid damaging special parts, or to tighten nuts and bolts to a specific degree of tightness. Their unique designs usually make these wrenches less suitable for general work, but perfect for their particular jobs. Special wrenches used in brake work include:

- Bleeder-screw wrenches
- Flare-nut wrenches
- Torque wrenches.

Figure 1-12. Bleeder screw wrenches help prevent damage to the bleeder screw.

Bleeder-Screw Wrenches

Bleeder-screw wrenches, figure 1-12, are box-end wrenches with special offsets that make it easier for the technician to loosen and tighten hard-to-reach bleeder screws. The box ends of the wrench are hexagonal, or six-point, like the bleeder screw itself. This design makes a bleeder-screw wrench less likely than a twelve-point box wrench, or an open end wrench, to round off the corners of the small hex on the bleeder screw.

Flare-Nut Wrenches

Flare-nut wrenches are used on brake tubing connections and fittings, figure 1-13; for this reason they are often called tubing wrenches. These tools are essentially box-end wrenches with one side of the box opening cut away so the wrench can be passed over the brake line.

Flare-nut wrenches are required because tubing fittings have relatively thin walls and can be easily damaged if an open-end wrench is used on them; even a small amount of distortion can cause a tubing fitting to leak. Open-end wrenches place all of the turning torque on only two flats of a hexagonal tubing fitting. Flare-nut wrenches circle the fitting to apply the force evenly and prevent distortion.

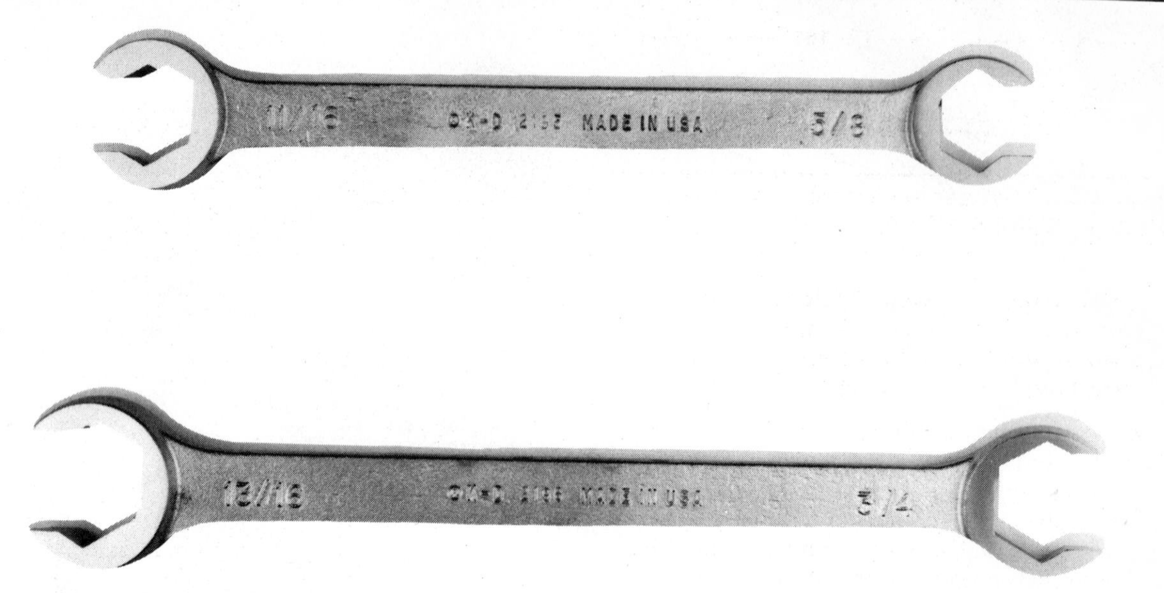

Figure 1-13. Flare-nut or tubing wrenches.

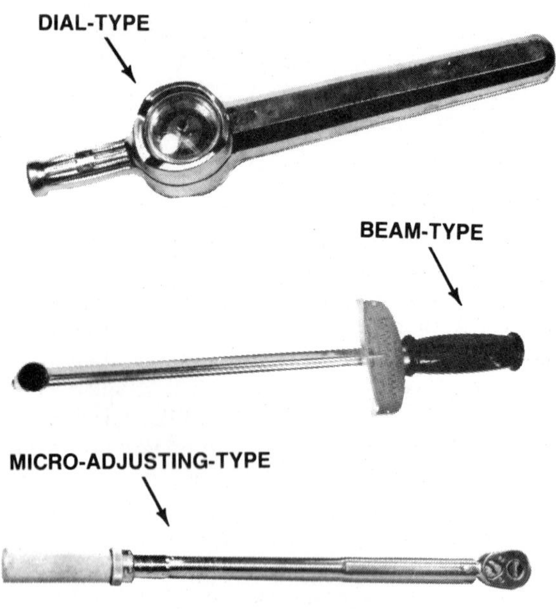

Figure 1-14. The three basic types of torque wrenches.

Flare-nut wrenches are available with either six- or twelve-point openings. The six-point type applies force over a larger area of the fitting than a twelve-point design and is less likely to round the corners of the fitting. However, the twelve-point type is easier to use where space to turn the wrench is limited.

Torque Wrenches

In brake repair, it is often essential that a nut or bolt be tightened a specific amount. This is true of *all* fasteners, but it is particularly important for fasteners in critical applications such as caliper bridge bolts or mounting bolts. On any part, tightening beyond the specifications, "for good measure," can stretch the fastener beyond its elastic limit, strip the threads, or distort the parts being fastened together. The end result is nut or bolt failure, or possibly fluid leaks.

To obtain proper fastener tightness, a special tool called a torque wrench is used. A torque wrench has a scale on it that indicates the amount of tightening force being applied to a nut or bolt. The scale may be graduated in inch-pounds, foot-pounds, Newton-meters, kilogram-meters, or a combination of these values depending on the wrench.

There are three basic types of torque wrenches: beam type, dial type, and micro-adjusting type, figure 1-14. With the beam-type wrench, a pointer indicates the torque on a scale near the handle. With the dial-type wrench, the torque is indicated by a needle and scale on a dial built into the wrench body. With the micro-adjusting-type wrench, the torque is preset by rotating the wrench handle until it aligns with a specific torque value marked on the body of the wrench; when the preset level of tightness is reached, the mechanism inside the wrench overcenters and produces a "click"

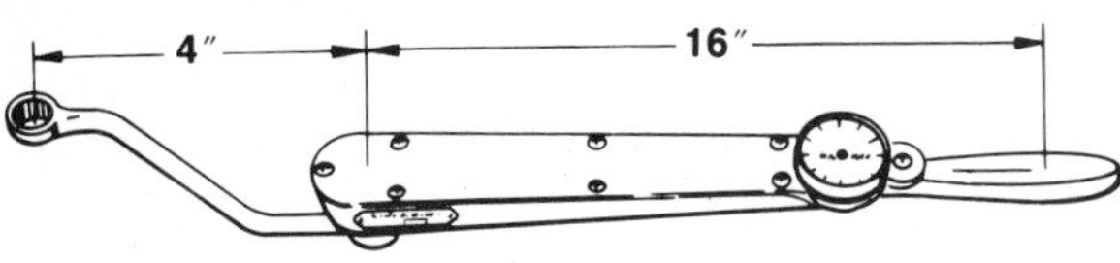

Figure 1-15. This adapter changes the effective length of the torque wrench.

that can be heard as well as felt through the wrench handle. For this reason, the micro-adjusting torque wrench is sometimes called a "click-type" torque wrench.

Torque wrench extensions

Torque wrenches can be used with various extensions and adapters that help get around obstructions or make it easier to use the wrench in tight quarters. Extensions on the handle end of the wrench do not affect the torque reading, although they make it easier to achieve because of the increased leverage. Extensions on the drive end of the wrench that increase the height of the wrench above the bolt or nut also will not affect the torque reading since the effective length of the wrench is not changed. However, adapters on the drive end of the wrench, figure 1-15, do increase the effective length of the wrench, and thus cause more torque to be applied to the fastener than is indicated by the scale on the wrench.

When using an adapter that lengthens the drive end of the wrench, the following formula is used to determine the proper wrench setting or reading for the torque required.

$$\text{wrench reading} = \frac{\text{torque at fastener} \times \text{wrench length}}{\text{wrench length} + \text{adapter}}$$

For example, if a technician needs to tighten a bolt to 30 foot-pounds using a 16-inch torque wrench with a 4-inch adapter, the formula would read:

$$\frac{30 \times 16}{16 + 4} = \frac{480}{20} = 24$$

As the equation shows, the bolt will be torqued to 30 foot-pounds when the torque wrench scale shows 24 foot-pounds.

TUBING TOOLS

Replacement steel brake lines are most often sold in fixed lengths and must be cut, bent, and flared to fit the individual application. These operations require three basic special tools:

- Tubing cutters
- Tubing benders
- Flaring tools.

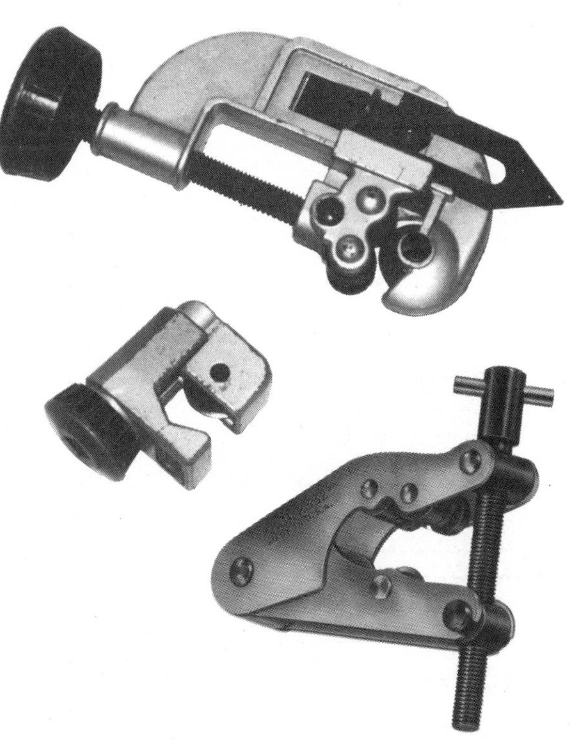

Figure 1-16. Typical brake tubing cutters.

Tubing Cutters

When a piece of brake line tubing is cut to length, the end must be left perfectly round and at a right angle to the body of the tubing so that fluid flow is not restricted and a good flare can be formed. This precision cut is achieved with a tubing cutter, figure 1-16, that supports the tubing on steel rollers perpendicular to a sharp, hardened cutting wheel. A screw forces the cutting wheel and tubing together, and the cutter is rotated around the tubing while the screw is slowly tightened until the cut is completed.

Many tubing cutters include a reamer bit used to remove flashing from the tubing bore after the cut is made. This prevents the flashing from breaking loose later and causing problems in the hydraulic system. Where the reamer is not a part of the tubing cutter, a similar separate tool must be used for the same purpose.

Tubing Benders

Although certain brands of brake tubing are designed to be bent by hand, most tubing must be bent with a special tubing bender that eliminates the chance of kinking and/or cracking the tubing. The two most common types of tubing benders are the lever type and the spring type.

Figure 1-17. Lever-type tubing benders are effective, but limited in the size of bend they can make.

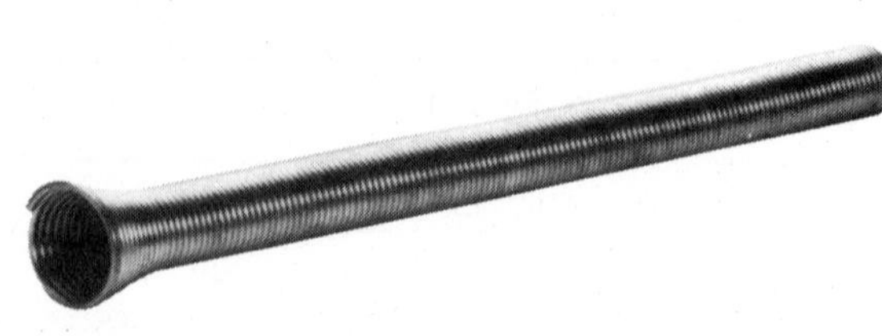

Figure 1-18. A spring-type tubing bender.

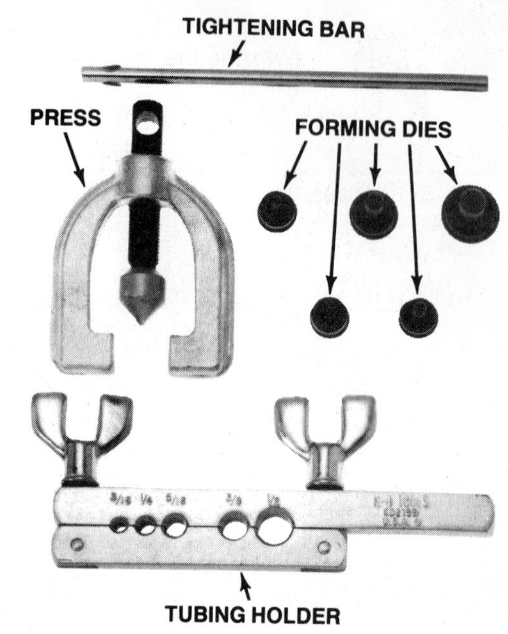

Figure 1-19. Double flaring tools have interchangeable forming dies for different tubing sizes.

Lever-type tubing benders, figure 1-17, use a pair of levers to form the tubing around a wheel or pulley that has grooves sized to match various tubing diameters. The walls of the grooves prevent distortion of the tubing that can cause a restriction in fluid flow. Some lever-type benders have interchangeable dies for different sizes of tubing, others have multiple grooves built into the tool itself. The lever-type bender has the advantage of being able to bend tubing whose ends are already flared.

Spring-type benders, figure 1-18, are tightly wound coil springs approximately 10 to 12 inches (254 to 305 mm) long. These springs have inside diameters from ⅛ to ⅜ inch to match the standard outside diameters of brake line tubing. To use a spring-type bender, the tubing is slipped inside the spring and bent into shape by hand; the strength of the spring coils prevents the tubing from becoming distorted. Spring-type benders can be used in tighter quarters than lever-type designs, and they make it possible to bend smaller and more intricate shapes. Spring-type benders cannot be used on pre-flared tubing, however, because the spring will not fit over the flare.

Flaring Tools

There are two types of flaring tools, those for making the SAE double flares used in most brake systems, and those for making the ISO flares that are becoming more common on newer brake systems. The SAE flare is formed in a two-step process with a double flaring tool, figure 1-19. This type of tool usually comes in a kit that includes a holder to clamp the tubing firmly in position, a press to shape the tubing, and flare forming dies for various tubing sizes.

The ISO flare is formed in a single-step process which requires only a tubing holder and a forming die, figure 1-20. However, a separate tool is required for each size of tubing since there is no way to change the clamp and forming die to accommodate different diameters of tubing.

ASSEMBLY/DISASSEMBLY TOOLS

A wide variety of special tools are available to aid in the assembly and disassembly of brake friction assemblies. Some of these tools are

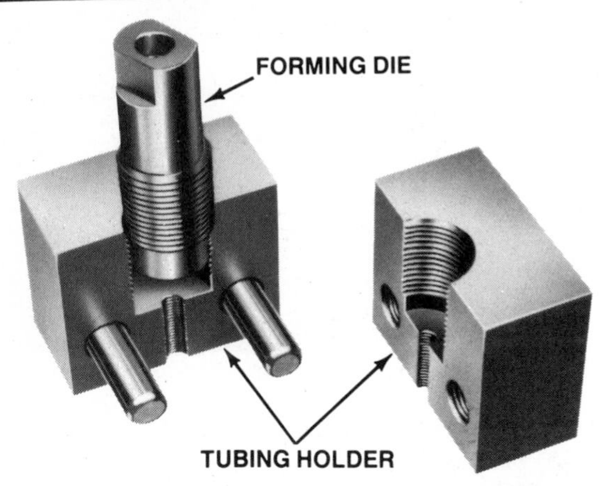

Figure 1-20. An ISO flaring tool for brake tubing.

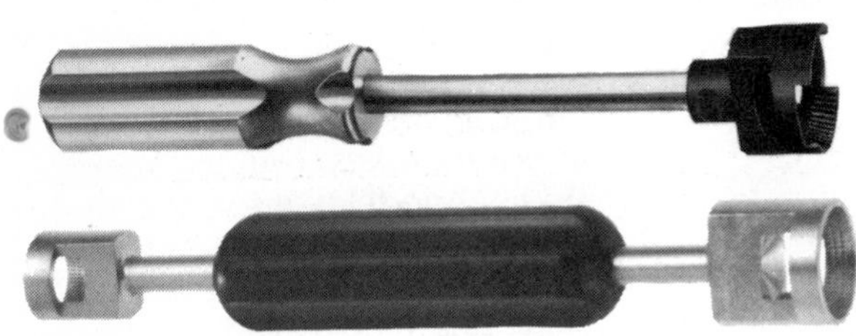

Figure 1-21. Typical holddown spring tools.

specific to certain car models, but others are fairly universal in their application. These tools include:

- Holddown spring tools
- Return spring tools
- Wheel cylinder removal tools
- Caliper piston removers
- Disc brake pad spreaders
- Wheel cylinder clamps
- Parking brake spring tools
- Dust boot tools
- C-clamps
- Swaging removal tools.

Holddown Spring Tools

The pin-and-spring holddowns that secure some brake shoes to the backing plate can often be serviced by hand. However, a special tool is available to speed their installation or removal in difficult situations. The holddown spring tool, figure 1-21, basically consists of a handle and steel shaft similar to a screwdriver. On the end of the shaft is a basket-shaped serrated socket which grips the retaining spring washer so it can be pushed down, rotated, and released from the holddown pin.

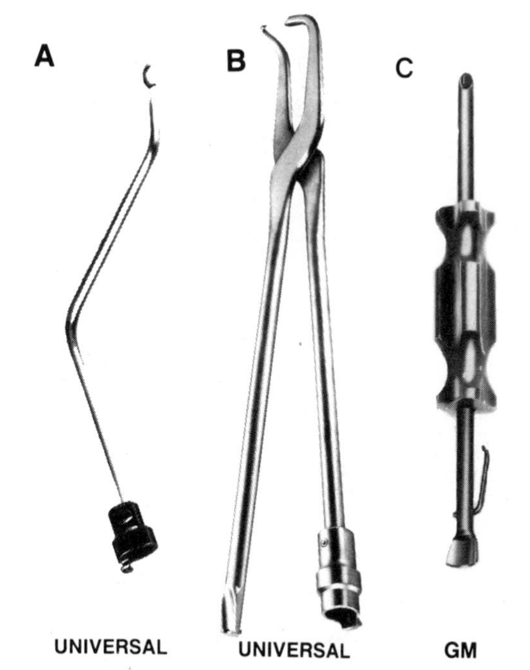

Figure 1-22. Brake return spring tools.

Another type of holddown spring tool is used to remove and install one-piece "beehive" coil-spring-type shoe holddowns. This tool consists of a handle and a straight shaft with a slot cut in the end. The shaft is inserted into the center of the holddown spring to engage the retaining hook, then the tool is depressed and rotated to release the holddown.

Return Spring Tools

Unlike holddown springs, brake shoe return springs are very strong; special removal and installation tools, figure 1-22, are mandatory to remove and install these springs. These tools usually have a special socket on one end and a cupped depression at the other, figure 1-22A. To remove a spring hooked over an anchor post, the socket end of the tool is placed over the post and rotated. A lip or protrusion on the rim of the socket opening then sweeps the return spring off of the anchor. To reinstall the spring, the cupped end of the tool is placed on the edge of the anchor post. The spring is then hooked over the tool shaft, and the tool is used to lever the spring into position on the post.

Another type of return spring tool, figure 1-22B is a special pliers used to remove and install springs that attach between two brake shoes. One end of the pliers rests against a rivet in the shoe lining, or on bonded linings,

Figure 1-23. A GM wheel cylinder retainer clip tool.

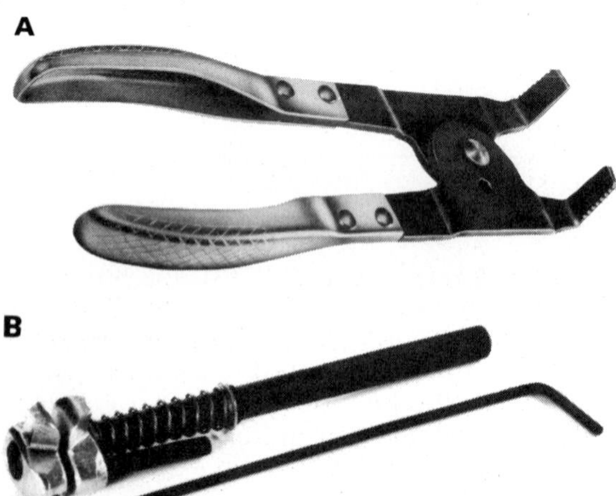

Figure 1-24. These special tools are used to remove brake caliper pistons.

against the friction material itself. The other end of the pliers has a hook that grabs the end of the spring. As the pliers are tightened, the spring is pulled toward the lining table until it can be removed from, or hooked into, its mounting hole in the shoe web. Regular pliers should never be used on brake shoe return springs because they can nick or otherwise damage the springs and lead to premature failure.

Many newer General Motors cars have return springs that attach to metal hooks on the shoe anchor, rather than to a more conventional anchor post. The special spring installation and removal tool for these brakes, figure 1-22C, has a separate pivoting wire hook to make servicing of this design easier.

Figure 1-25. Frozen pistons can be removed hydraulically using a caliper bench.

Wheel Cylinder Retainer Removal Tools

In addition to having different return spring attachments than other brakes, the rear wheel cylinders on some late-model General Motors cars are held to the backing plate with a unique spring-steel retainer clip. The clip is most easily removed and installed with a special retainer removal tool, figure 1-23. The tool fits inside the two retainer clip prongs, and expands them outward to release the clip from the wheel cylinder.

Caliper Piston Removal Tools

Rust and corrosion can make it very difficult to remove brake caliper pistons from their bores. In extreme cases, hundreds of pounds of force may be required. To make this job easier, two types of special caliper piston removal tools are available, mechanical and hydraulic.

The mechanical piston remover shown in figure 1-24A, is basically a pair of pliers with expanding jaws that have serrated teeth to grip the inside of the piston. Once the piston is grasped with this tool, it can be worked back and forth until it is free enough to be removed from the bore. A similar tool, figure 1-24B, is tightened against the inside of the piston with a long allen wrench. Generally, these types of removal tools will only work when the piston is slightly stuck in the caliper.

In cases of severely frozen pistons, the best removal tool is a hydraulic caliper piston remover, figure 1-25. With the caliper off the car, a hydraulic line from a hand-operated master cylinder is connected to the fluid inlet. The hand lever is then pumped until sufficient hydraulic pressure is created to force the caliper piston from its bore. This type of tool can create more than 1,000 psi (6,900 kPa) of pressure to

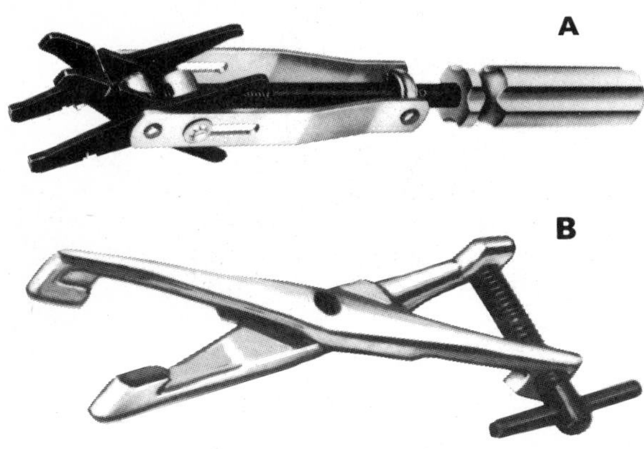

Figure 1-26. Disc brake pad spreaders.

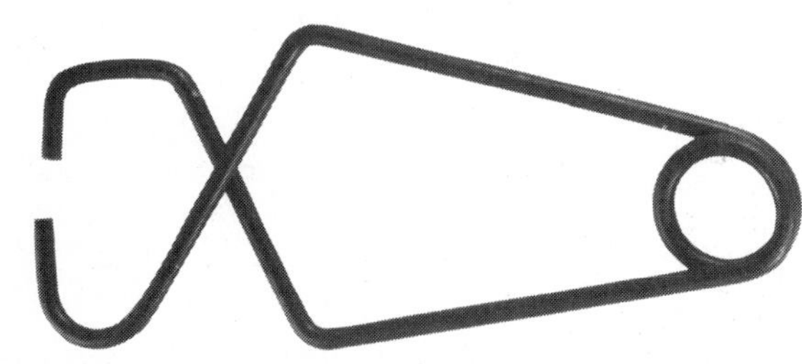

Figure 1-27. Wheel cylinder clamps hold the pistons in the cylinder bore during brake assembly.

free frozen pistons. To prevent injury from brake fluid spray, always wear eye protection when using this type of piston removal tool.

Disc Brake Pad Spreaders

Disc brake pad spreaders, figure 1-26, are used when assembling early four-piston brake calipers equipped with piston return springs. After the brake pads are installed in the caliper, the spreader is used to hold them apart against piston return spring pressure while the caliper is installed over the brake rotor. The pad spreader in figure 1-26A is operated by rotating its handle to expand the scissor-like legs. The spreader in figure 1-26B operates by rotating a threaded bolt at one end of the tool to spread the arms at the opposite end.

Wheel Cylinder Clamps

Wheel cylinder clamps, figure 1-27, are large spring-steel devices used to hold pistons in wheel cylinders bores against cup expander spring pressure. These clamps are used on brakes without piston stops to hold the wheel cylinders together during brake disassembly and assembly. The clamps prevent the wheel cylinder from coming apart unintentionally, and make it easier to install the brake shoes.

Figure 1-28. A typical parking brake cable tool.

Parking Brake Cable Tool

A parking brake cable tool, figure 1-28, is used when assembling drum brake friction assemblies that have a parking brake cable that is retained by a spring built into the end of the cable. The end of the tool fits between the spring and the beaded cable end to provide a pliers action that compresses the spring, making it easier to attach the cable to the parking brake lever.

Dust Boot Tools

Brake caliper dust boots attach tightly to the caliper piston and the caliper body to prevent contamination from entering between these two parts and causing rust, corrosion, or scuffing that can lead to fluid leaks and stuck pistons. Depending on the caliper and dust boot design, there are three particular special tools that may be required or helpful when assembling the caliper:

- Dust boot drivers
- Dust boot rings
- Dust boot pliers.

Dust boot drivers

Dust boot drivers, figure 1-29, are used when assembling calipers in which the outer edge of the dust boot contains a metal retaining ring that provides a press fit in the caliper body. These types of calipers are common on General Motors cars. The driver is centered on the outer circumference of the boot, then struck with a hammer; this provides equal pressure at all points and prevents distortion of the metal retaining ring.

Dust boot drivers are made in various sizes to match different diameter dust boots. Some drivers are one-piece plastic castings, but others have interchangeable metal drivers that attach to a common shaft.

Figure 1-29. Caliper dust boot drivers.

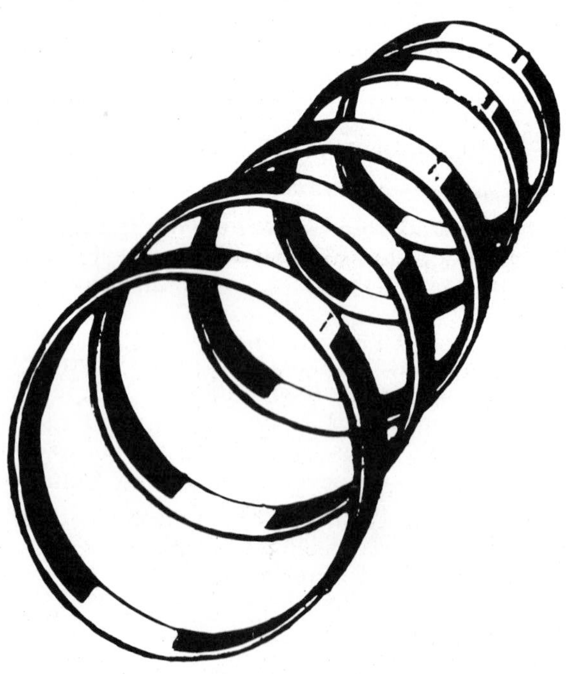

Figure 1-30. Dust boot installation rings.

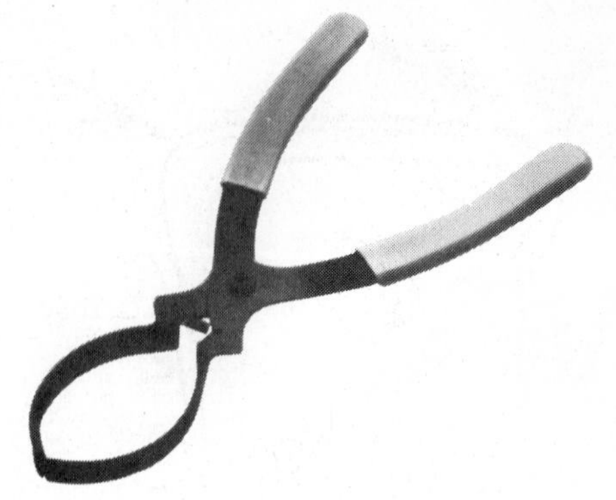

Figure 1-31. Dust boot installation pliers.

Dust boot rings and pliers

Dust boot rings and pliers, figures 1-30 and 1-31, are used to assemble calipers in which the dust boot installs into a groove in the caliper bore before the piston is installed. These types of calipers are common on Chrysler and Ford vehicles. The ring or pliers is used to expand the piston opening in the dust boot so the caliper piston can be installed through it.

Dust boot rings are usually made of plastic, and come in sets of five. Each ring has an inside diameter slightly larger than a common piston diameter. Dust boot pliers come in a single size that can be expanded or contracted as necessary. Because of this, a single pair of dust boot pliers can do the same job as a set of several size rings.

C-Clamps

Large C-clamps, figure 1-32, are often used on single-piston calipers to bottom the piston in its bore. When disassembling brakes, this makes it easier to remove the caliper from over the rotor, particularly if the rotor has a large ridge at its outer edge, or is scored to the point where grooves in the brake pads are interlocking with grooves in the rotor.

When assembling the brakes, a bottomed piston is necessary to install new brake pads. When used for this purpose, the C-clamp is positioned across the caliper with its threaded rod inside the caliper piston. The rod is then turned until the piston is fully seated in its bore.

Swaging Removal Tools

Swaging removal tools, figure 1-33, are required to remove brake drums that are swaged to the hub. The removal tool is placed over the stud and turned by a drill motor. The hardened

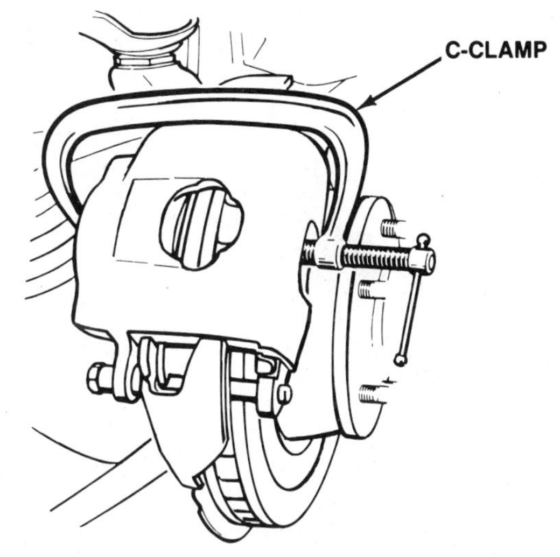

Figure 1-32. C-clamps can be used to bottom caliper pistons in their bores.

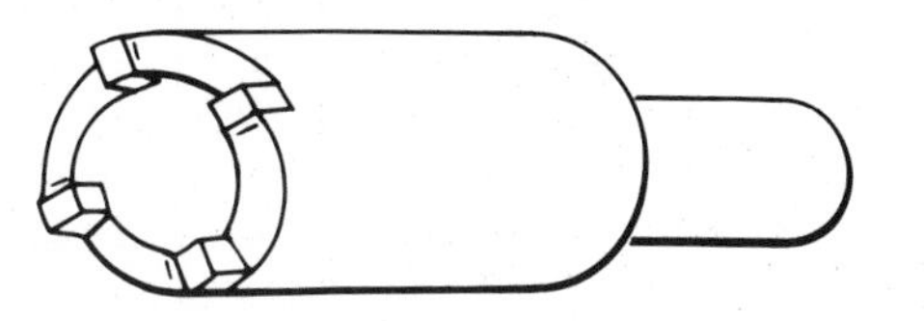

Figure 1-33. A swaging removal tool.

cutting teeth of the tool then mill away the swaged metal from around the stud, allowing the drum and hub to be parted.

MEASURING TOOLS

In order to properly inspect and refinish certain brake parts, it is important to know their exact dimensions. In some cases, brake adjustment can also benefit from a precise knowledge of component size. Truly accurate measurement is only possible when done with proper measuring tools. Tools of this type used in brake service include:

- Feeler gauges
- Micrometers
- Dial indicators
- Shoe setting calipers
- Drum micrometers
- Cylinder bore go/no-go gauges
- Pedal effort gauge
- Tapley brake testing meters.

Feeler Gauges

Feeler gauges, figure 1-34, are thin, flat pieces of metal manufactured to a precise thickness.

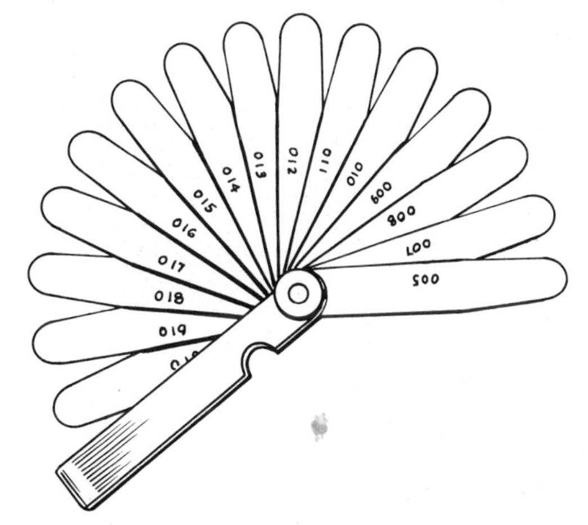

Figure 1-34. A typical set of feeler gauges.

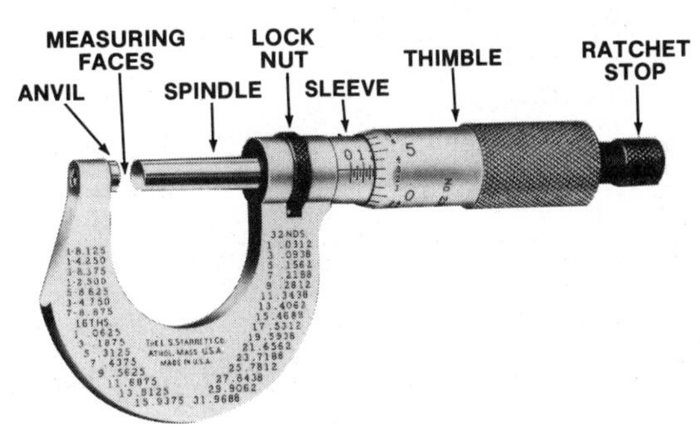

Figure 1-35. Outside micrometers are the type most often used in brake service.

Gauges of various sizes are inserted into a gap to be measured until one that just fits is found. The thickness of the gauge then indicates the size of the gap.

Feeler gauges that conform to the English system of measurement commonly range in thickness from .001″ to .025″ in .001″ increments. Metric feeler gauges typically range from 0.05 mm to 1.00 mm in 0.05-mm graduations. Thicker gauges of both types are available for special purposes, but in most instances, combinations of thinner gauges are stacked to obtain the desired measurement.

Micrometers

Micrometers are used to precisely measure the outside, inside, or depth of an object, opening, or depression. The micrometers used in automobile service are primarily outside micrometers that measure in thousandths of an inch or hundredths of a millimeter. An outside micrometer, figure 1-35, has a measuring range

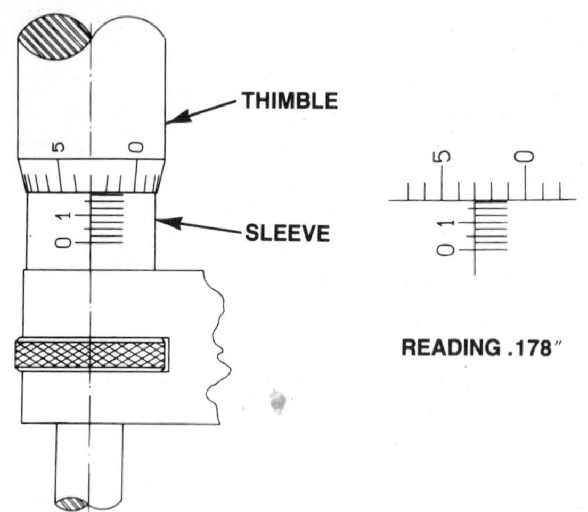

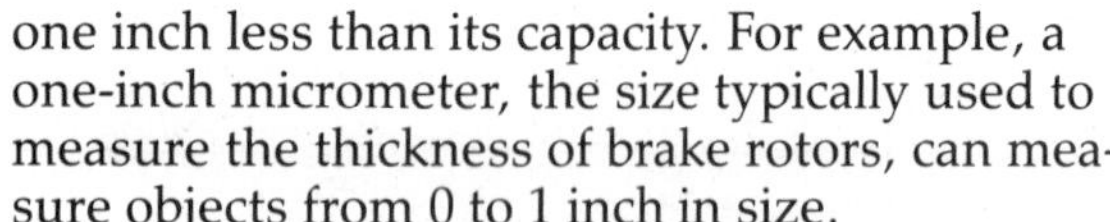

Figure 1-36. Reading the measurement on a standard micrometer.

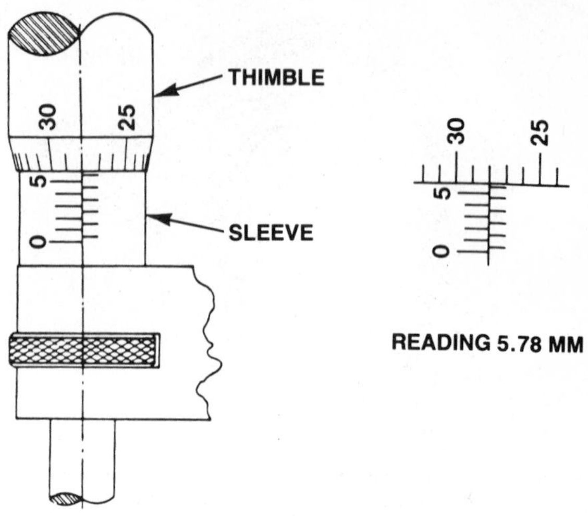

Figure 1-37. Reading the measurement on a metric micrometer.

one inch less than its capacity. For example, a one-inch micrometer, the size typically used to measure the thickness of brake rotors, can measure objects from 0 to 1 inch in size.

Most outside micrometers have anvils and spindles with flat surfaces. However, those designed specially for measuring brake rotors have a pointed anvil. This type of micrometer measures rotor thickness in the same manner as a standard micrometer, but the pointed anvil also permits measuring the depth of any scoring that may be present on the friction surfaces.

To use an outside micrometer, place the object to be measured between the anvil and spindle. Turn the thimble to move the spindle into contact with the object being measured until a slight amount of drag is felt when the micrometer is moved across the surface. Some micrometers, like the one shown in figure 1-35, have a ratchet built into them to provide the proper amount of drag; on these designs, turn the ratchet stop knob until you can feel it slip.

Once the proper tension is achieved, read the dimension of the object from the graduations on the sleeve and thimble. Each exposed line on the sleeve indicates .025″. Every fourth line of the sleeve is .100″, and is marked by a number to indicate this. Each revolution of the thimble moves the measuring face .025″, and the edge of the thimble is graduated with 25 lines, each representing .001″. Every fifth line on the thimble is marked with a number for easy identification.

To obtain the total reading, multiply the number of graduations exposed on the sleeve by .025″, then add that figure to the number of the graduation on the thimble that is aligned with the marker on the sleeve. In figure 1-36, there are seven graduations multiplied by .025″ giving .175″. The third graduation on the thimble is aligned with the mark on the sleeve giving .003″. Adding the two makes a total of .178″.

A metric micrometer, figure 1-37, is read in basically the same manner except for the difference in graduations. Each line on the top of the sleeve scale indicates 1 mm. Each line on the bottom of the sleeve scale indicates .5 mm. The lines on the thimble each indicate .01 mm. To obtain the reading shown, add the 5 mm indicated on the top sleeve scale and the .5 mm indicated on the bottom sleeve scale for a subtotal of 5.5 mm. To this, add the .28 mm indicated on the thimble scale, for a total of 5.78 mm.

Inside and depth micrometers are seldom used in brake service. Although they differ in their physical construction, the thimbles and sleeves of these micrometers operate, and are read, in the same way as those on outside micrometers.

Dial Indicators

Dial indicators, figure 1-38, measure the movement of a plunger and show the distance traveled on a dial. The measurement is generally calibrated in thousandths of an inch, or hundredths of a millimeter. Many kinds of holders, stands, and fixtures are available that allow the dial indicators to be used to measure a variety of components. In brake service, dial indicators

Figure 1-38. In brake service, dial indicators are used primarily to measure runout and out-of-round.

Figure 1-39. Shoe-setting calipers transfer the drum inside diameter measurement to the brake shoes.

are used primarily to measure rotor runout and brake drum out-of-round conditions. A dial indicator can also be used to check wheel bearing adjustment.

Shoe-Setting Calipers

Brake shoe-setting calipers, figure 1-39, are combination inside/outside measuring devices that consist of two interlocking pieces of steel that slide together and apart, and can be locked in position by tightening a knob. When new brake shoes are installed, the shoe-setting caliper is used to measure the inside diameter of the brake drum, and then locked in place. The opposite side of the caliper is then placed over the center of the brake shoes, allowing them to be quickly adjusted to the proper size to match the drum. The actual dimension of this measurement is not important for brake adjustment, so many shoe-setting calipers do not have inch or millimeter markings on them.

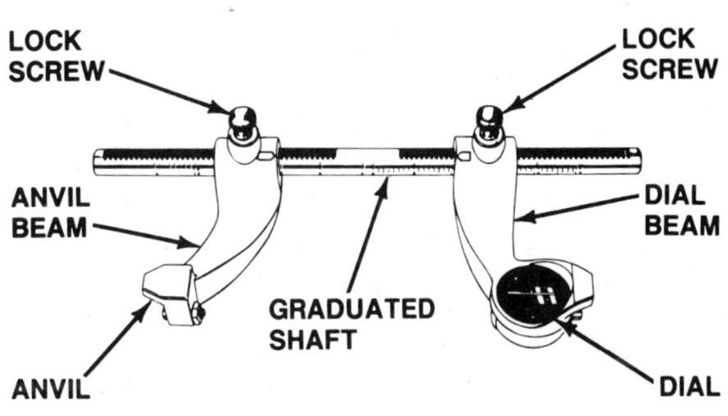

Figure 1-40. A typical brake drum micrometer.

Drum Micrometer

A brake drum micrometer is used to quickly and easily measure the inside diameter of brake drums. The typical drum micrometer, figure 1-40, consists of an anvil beam and a dial beam that are secured to a graduated shaft with lock screws. The shaft is marked in one-inch increments, and the lock screws for the beams fit into notches machined every .125″. The dial scale is graduated in .005″ increments.

To use a drum micrometer, loosen the lock screws and move the anvil and dial beams out along the shaft until they align with the graduations equivalent to the nominal size of the drum being measured. For example, to measure a drum with a standard inside diameter of 11.375 inches, figure 1-41, set one beam at the 11-inch marking on the shaft, and the other beam at the opposing 11-inch mark plus three .125″ graduations for a total of 11.375 inches.

Once the micrometer is set, place it inside the brake drum and hold the anvil steady against the friction surface. Slide the dial end of the micrometer back and forth until the highest reading is obtained on the dial scale. In this case the dial reads .015″ for a total drum diameter of 11.390 inches.

Drum micrometers graduated in metric measurements operate in much the same manner. The only exceptions are the dimension

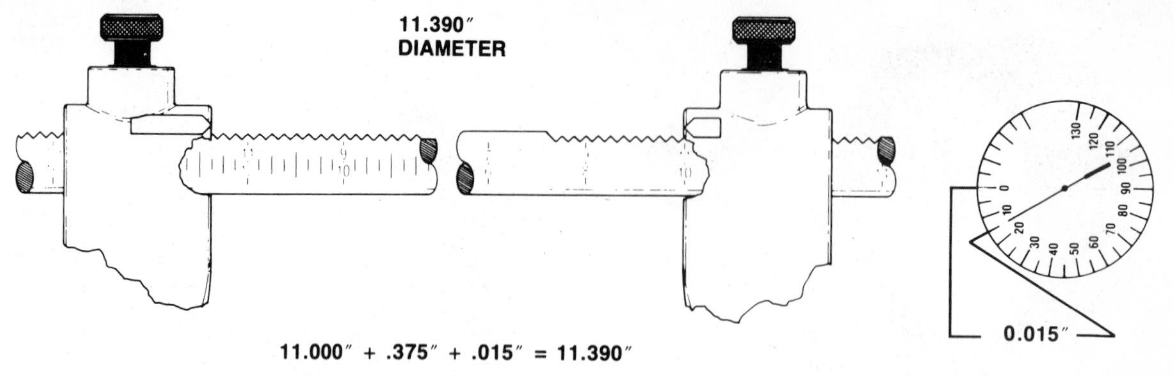

Figure 1-41. Reading the measurement on a standard brake drum micrometer.

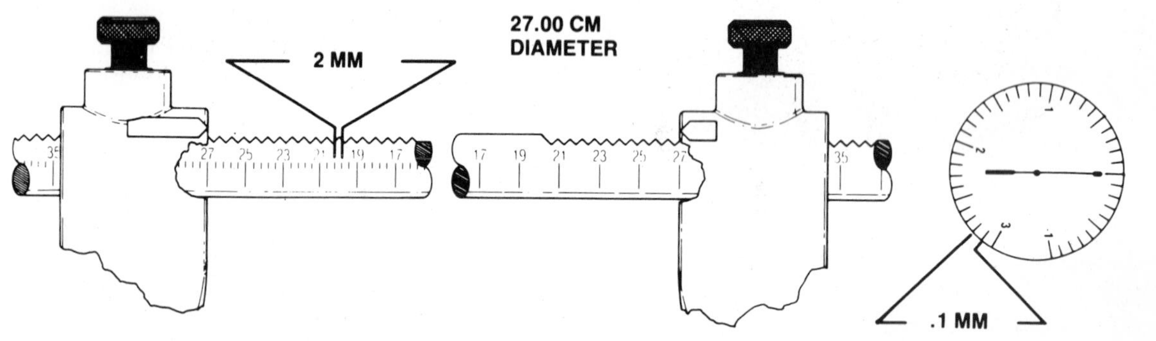

Figure 1-42. Reading the measurement on a metric brake drum micrometer.

Figure 1-43. Hydraulic cylinder bore go/no-go gauges are used to check for wear.

markings, figure 1-42. The shaft is marked in one-centimeter (cm) increments, and the lock screws for the two beams fit into notches machined every 2 mm. The dial scale is graduated in .1-mm increments.

Cylinder Bore Go/No-Go Gauges

Cylinder bore go/no-go gauges, figure 1-43, are used to quickly determine if a master cylinder or wheel cylinder bore is worn beyond service limits. If the bore of a hydraulic cylinder is too far oversize, the piston cup seals will not have adequate tension against the cylinder walls. This can cause the cylinder to leak fluid, enable air to enter the hydraulic system, or allow fluid under pressure to bypass the seals. A go/no-go gauge is generally used to check the size of the cylinder bore after honing to ensure that too much metal has not been removed.

A set of go/no-go gauges consists of a handle and various hardened steel plugs. The plugs are sized to match the diameters of typical cylinder bores, plus approximately .006 inch. To use the go/no-go gauge, select the size plug that corresponds to the nominal bore diameter and attach it to the handle. Attempt to insert the plug into the bore; if the plug fits, the bore is too far oversize and the master cylinder or wheel cylinder must be replaced.

Figure 1-44. A pedal effort gauge indicates the amount of force applied to the brake pedal.

Brake Pedal Effort Gauge

The brake pedal effort gauge, figure 1-44, measures the amount of force applied to the brake pedal by the operator's foot. Some manufacturers specify the use of this gauge when checking brake pedal travel or adjusting the parking brake. The pedal effort gauge is graduated in pounds or Newtons of force, and operates hydraulically through an attached pedal. The entire assembly is attached to a steel bracket that allows it to be installed over the vehicle brake pedal.

Tapley Brake Testing Meter

The Tapley brake testing meter, figure 1-45, is a decelerometer used to test the efficiency of braking systems. It does this by measuring braking force in gravities of deceleration. One gravity (g), is an acceleration or deceleration that changes at the rate of 32 feet per second, per second. Braking force is shown on the Tapley meter scale as a percentage of 1 g. For example, ½ g of braking force would show on the Tapley meter scale as 50 percent braking efficiency.

When the Tapley meter was introduced many years ago, only racing cars could decelerate at a rate anywhere near 1 g. However,

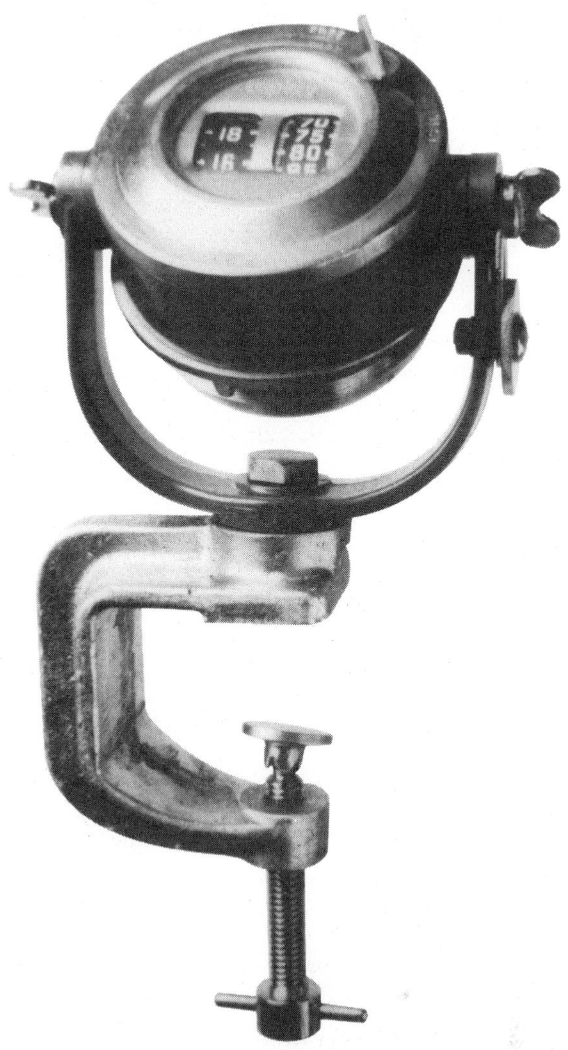

Figure 1-45. A Tapley brake testing meter.

because of advances in brake systems, suspensions, and tires, most modern cars can approach this figure, and some high-performance models can brake even harder than 1 g! Nevertheless, 1 g deceleration is still the practical maximum for most vehicle brake systems, and a measurement based on that value is useful when evaluating a car's braking performance.

To measure braking efficiency, mount the Tapley meter in the car in a level position using a clamp or the available 14-pound (6.4 Kg) floor mounting block. The meter must be free to swing forward and back along the lengthwise axis of the vehicle. Drive the car at approximately 20 mph (32 kph) on a smooth, dry, level road and apply the brakes smoothly and as hard as possible without locking the wheels.

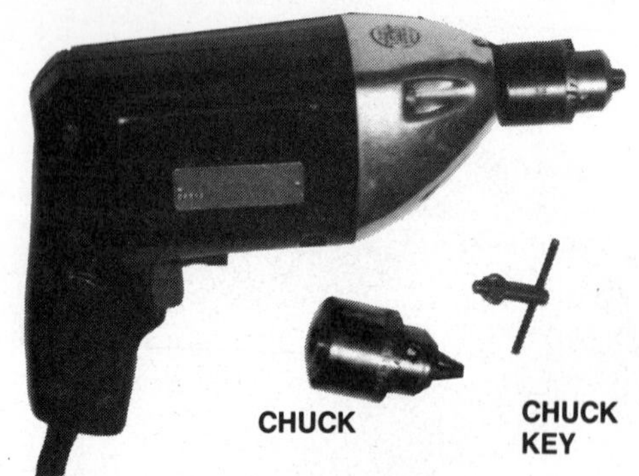

Figure 1-46. An electric drill motor, chuck, and chuck key.

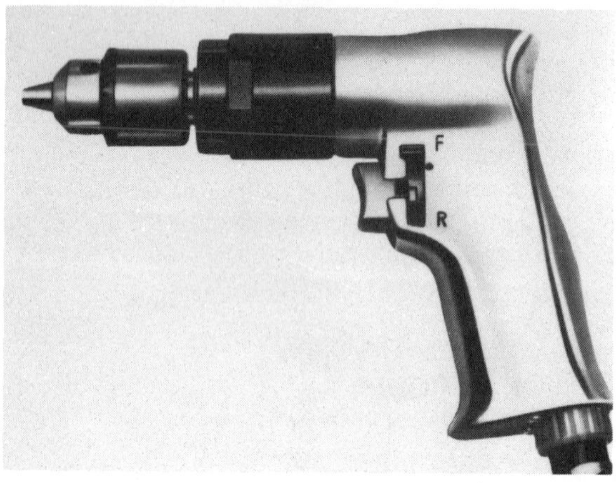

Figure 1-47. A pneumatic, or compressed-air-powered, drill motor.

The meter will swing forward under deceleration, and a reading of the vehicle's braking efficiency will be displayed on the readout until the meter is reset.

CYLINDER AND CALIPER REFINISHING TOOLS

To provide a proper sealing surface and prevent piston seizure, the bores of master cylinders, wheel cylinders, and brake calipers must be free from rust and corrosion, and have the proper finish. If the bore surface has only minor damage, special tools can be used to refinish it. These tools include:

- Drill motors
- Stone hones
- Ball hones.

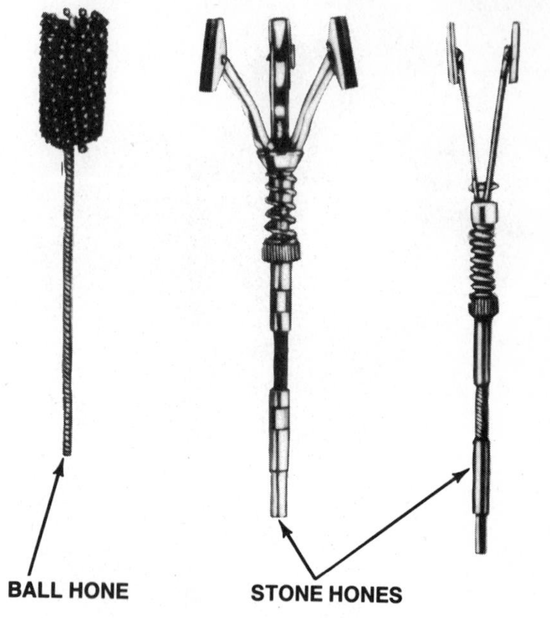

Figure 1-48. Typical brake cylinder hones.

Drill Motors

Electric drill motors and pneumatic drill motors, figures 1-46 and 1-47, commonly called electric and air drills, are hand-held motors with special clamps for holding other tools such as drill bits, stud removers, and hones. The clamp on a drill motor is called the chuck, and is tightened and loosened with a chuck key. The chuck key is the only tool that will properly tighten and loosen the chuck.

Hones

Cylinder hones, figure 1-48, are used to remove rust, corrosion, minor pits, and residue from the bores of master cylinders, wheel cylinders, and brake calipers. There are two basic types of cylinder hones, stone hones and ball hones.

Stone hones are available in both two-arm and three-arm models; the two-arm designs can fit into smaller bores than can the three-arm types. A replaceable abrasive stone is mounted to the end of each arm, and the arms are forced outward against the cylinder wall by spring pressure. The spring pressure is usually adjustable so the hone can apply proper tension to refinish a variety of bore diameters. The hone attaches to the drill motor through a flexible shaft that makes perfect alignment between the drill motor and the cylinder or caliper bore unnecessary.

Figure 1-49. A brake drum lathe.

Abrasive stones of various grits are available for master cylinder and wheel cylinder service. Medium-grit stones can be used to remove material rapidly, although care must be taken not to remove too much metal. Fine-grit stones are most common, and provide a smoother surface finish with less chance of honing the cylinder too far oversize. Coarse-grit stones are also available, but they are not recommended for master cylinder and wheel cylinder service. Coarse stones are normally used only to rapidly clean up the bores of brake calipers with fixed piston seals; the quality of the surface finish is less critical in these applications.

Ball hones consist of silicon carbide abrasive balls, with a medium-fine grit, attached to flexible metal bristles. The balls and bristles are connected to a flexible metal shaft that is inserted in the drill motor chuck. Ball hones are not adjustable but come a range of sizes to match different bore diameters. The nature of the abrasives on ball hones provides a better surface finish, and makes a ball hone less likely to remove excessive metal than a stone hone.

Figure 1-50. A brake rotor lathe.

FRICTION COMPONENT REFINISHING TOOLS

The friction components of the brake system are the drums and rotors, and shoes and pads. A proper fit between these parts is essential for maximum braking power. The tools for refinishing friction components include:

- Brake lathes
- Drum grinders
- Rotor resurfacers
- Shoe arcing machines.

Brake Lathes

As discussed in Chapter 11 of the *Classroom Manual,* brake lathes use an electric motor to turn a spindle that holds the drum or rotor to be machined. As the drum or rotor turns, a carbide steel tool bit, rigidly mounted on a boring bar, is passed over the friction surface to cut away a thin layer of metal and restore a smooth finish. Some lathes are designed to machine only drums, figure 1-49, while others will machine only rotors, figure 1-50. Still others can be adapted to machine drums or rotors. A typical lathe used for passenger-car brake service can machine drums from 6 to 28 inches (152 to 711 mm) in diameter, and rotors as large as 13 inches (330 mm) in diameter.

Most brake lathes require that the rotor be removed from the car for machining. However, compact, portable lathes are also available that bolt to the suspension and machine the rotor while it is still in place, figure 1-51. The lathe shown uses the car engine and drivetrain to turn the rotor during the machining operation; other designs have an electric motor built into them for this purpose.

Figure 1-51. A portable brake lathe that machines rotors on the car.

Figure 1-52. A drum grinding attachment for a brake lathe.

Figure 1-53. A rotor resurfacing attachment for a brake lathe.

Figure 1-54. A typical brake shoe arcing machine.

Drum Grinders

Drum grinders are attachments that mount to a brake lathe in place of the boring bar that holds the cutting tool bit. A drum grinder, figure 1-52, uses a stone grinding wheel to cut hard spots down flush with the drum friction surface.

Rotor Resurfacers

Rotor resurfacers, figure 1-53, use abrasive sanding discs to remove rust, corrosion, and lining deposits from the rotor, and give it a non-directional surface finish. Rotor resurfacers remove only very small amounts of metal to correct for minor rotor damage and distortion.

Shoe Arcing Machines

When brake shoes need to be ground or "arced" to properly match the curvature of the brake drum, a shoe arcing machine, figure 1-54, is used. The shoe to be arced is mounted in the machine and passed across an abrasive belt or drum that is spun at high speed by an electric motor. This removes friction material from the shoe lining to give it the proper shape to fit the drum. Modern shoe arcing equipment is fitted with a dust collection system consisting of a vacuum and a high efficiency particulate air (HEPA) filter to trap the asbestos fibers released from the linings during the arcing operation.

HYDRAULIC SERVICE TOOLS

The brake hydraulic system contains the fluid that transmits application force to the wheel friction assemblies. Much of the service on this system involves maintaining the proper fluid level, and bleeding trapped air from the brake lines, cylinders, and calipers. The two most common tools used in hydraulic system service are the:

- Brake fluid syringe
- Pressure bleeder.

Brake Fluid Syringe

The brake fluid syringe, figure 1-55, is a simple device used to withdraw fluids from reservoirs. This type of syringe consists of a hollow plastic

Figure 1-55. A brake fluid syringe.

tube attached to a rubber bulb. To use this tool, compress the bulb, place the open end of the tube into the fluid, then release pressure on the bulb. As the bulb returns to its relaxed state, it will draw a quantity of the fluid into the syringe. Brake fluid syringes are primarily used during brake pad replacement to remove fluid from the master cylinder reservoir before the caliper pistons are bottomed. A syringe is also used to remove old fluid from the reservoir when the hydraulic system is flushed and new brake fluid is installed.

Pressure Bleeder

A pressure bleeder, figure 1-56, uses compressed air to force brake fluid through the hydraulic system. This enables the system to be quickly and easily purged of any air that may be trapped in it.

A pressure bleeder consists of a sealed tank fitted with an air inlet valve and a fluid outlet valve. The tank is filled with a supply of brake fluid, then pressurized to approximately 30 psi (200 kPa); a relief valve prevents accidental overcharging. Some early, inexpensive pressure bleeders exposed the brake fluid directly to the air in the tank, but modern designs have a rubber diaphragm that separates the fluid from the air. This prevents the fluid from absorbing moisture out of the air which will greatly reduce its boiling point. It also prevents the fluid from becoming aerated which can make it impossible to properly bleed the brake system.

The pressurized brake fluid is routed from the tank through the fluid outlet valve into a rubber hose that ends in a quick release fitting. The fitting can be attached to a variety of adapters that fit various master cylinder fluid reservoirs, figure 1-57. To prevent injury from brake fluid spray, always wear eye protection when connecting or disconnecting a pressure bleeder.

Figure 1-56. Pressure bleeders are used to rapidly purge air from the brake hydraulic system.

LUBRICANTS

A variety of lubricants are used in brake work to aid assembly, preserve parts from corrosion, and prevent mechanical seizure. These lubricants include:

- Brake grease
- Assembly fluid
- Rubber grease
- Wheel bearing grease.

Brake Grease

Special high-temperature brake grease has a melting point of over 500°F (260°C) and contains solid lubricants such as zinc oxide. Even at extreme temperatures where the grease melts and runs off, the solid lubricants remain to provide protection. Like any grease, brake grease is designed to provide lubrication for moving parts, and thus prevent rust, corrosion and seizure.

Brake grease is generally applied to the shoe support pads on the backing plates of drum brakes, and to the sliding surfaces of sliding brake calipers. Several manufacturers also recommend that brake grease be applied to the locating pins, sleeves, and bushings of floating brake calipers.

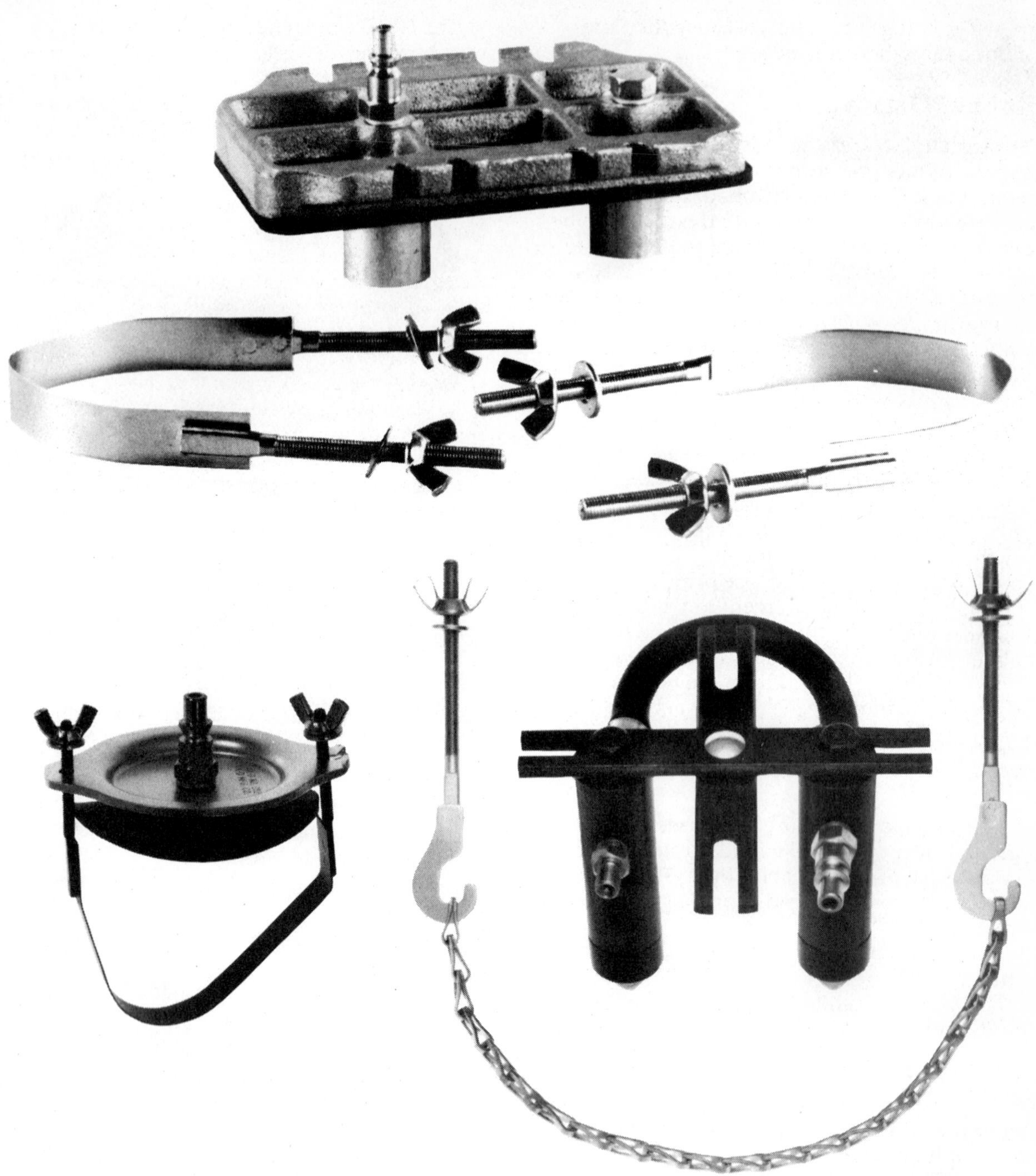

Figure 1-57. Typical master cylinder adapters for a pressure bleeder.

Assembly Fluid

When brake system hydraulic components are rebuilt, some type of lubricant is needed to help ease pistons past fixed seals, or piston seals into bores. Alcohol can be used for this purpose, but it evaporates quickly and leaves no residual lubrication. Many technicians use brake fluid to smooth the assembly of hydraulic components because it is readily available. However, the best product for this purpose is one of the special assembly fluids designed for the job. Assembly fluids are very similar to brake fluid, except that they have a much higher viscosity; some types have a consistency close to that of grease. The higher viscosity of assembly fluids provides better lubrication because the fluid will remain where it is needed.

Rubber Grease

Some manufacturers include a package of red or pink rubber grease in their master cylinder rebuild kits. These non-petroleum-base greases are intended to be packed into the master cylinder wiper boot where the pedal pushrod enters the cylinder. This provides lubrication between the boot and pushrod, and helps seal moisture away from the cylinder bore opening. A small amount of the grease between the pushrod and piston can also help prevent rust and corrosion. Rubber greases should never be mistaken for, or used in place of, brake assembly fluids.

Wheel Bearing Greases

All greases are made from oils that are thickened so the lubricant will cling to the surfaces where it is needed. As their name implies, wheel bearing greases are used to lubricate wheel bearings — although there are some wheel bearing greases that also make good chassis lubricants; these products are called multi-purpose greases. Greases are classified in two ways, by their National Lubricating Grease Institute (NLGI) number, and by the type of thickening agent used.

The NLGI number designates the viscosity or consistency of a grease; the higher the number, the thicker the grease. The NLGI number is only a means of comparison, it is not a measure of quality or performance. Virtually all wheel bearing greases have an NLGI #2 consistency number.

Greases are thickened with several substances including calcium, sodium, and lithium. The type and amount of thickening agent determine the melting point, appearance, texture, and water resistance of a grease. Today, most manufacturers recommend wheel bearing grease made with a lithium-based thickener because these greases have a high melting point and offer superior water resistance. It is important to recognize that different thickening agents do not intermix; always clean away all of the old grease when repacking wheel bearings.

Wheel bearing greases also contain additives that improve their performance. Sulphonates or amines are used to inhibit corrosion, and solid lubricants such as molybdenum disulphide (MS_2) improve the anti-seize properties of a grease. To increase the load-carrying ability of a grease under extreme pressure, sulphurized fatty oils are sometimes added. Always use grease that meets the vehicle manufacturer's specifications when you repack wheel bearings.

Chapter

2 Brake System Diagnosis

The professional brake technician knows that the most efficient way to identify a brake problem is to follow a thorough and logical diagnosis procedure. This makes it quick and easy to locate the cause of the complaint, and ensures that any other potential problems are located and can be repaired *before* a failure occurs. For reasons of safety, not to mention legal liability, vehicle brakes should *always* be serviced as a complete system. Never repair a brake problem without inspecting the entire system, and never allow a car to leave your shop unless the brakes are working perfectly. If a car owner decides not to have needed brake work done, note that fact on the work order along with the recommended repairs, and have the customer initial it.

Brake system diagnosis can be broken down into three steps. The first is to listen to the car owner's explanation of the problem, and ask questions to help isolate the defect. If the car can safely be driven, the second step is to perform a road test to check out the symptoms described by the car owner, and assess the overall performance of the brakes. The third and final step is to perform a complete brake system inspection to confirm your diagnosis and locate any additional problems.

The first three sections of this chapter discuss how to talk to the customer, perform the road test, and inspect the brake system. General guidelines are provided for each step, but it is not practical to list every detail of how these jobs are done. Professional brake system diagnosis requires a thorough knowledge of *all* the information in both volumes of this textbook set, as well as a certain "feel" for the job that only comes from experience. The concepts set forth in this chapter will serve as a framework on which you can build your own personal diagnosis procedure.

When you are just starting out in brake repair, you have only limited training and experience on which to base your diagnosis. And sometimes, even expert technicians are stumped by unusual or uncommon brake problems. In these situations, a diagnostic chart can be helpful in identifying possible causes of a brake problem. The final section of this chapter contains 26 diagnostic charts that describe typical brake system problems and their repair.

TALKING TO THE CUSTOMER

An important fact many technicians fail to realize is that brake repair is a business, not just a job. As in any successful business, the ultimate goal is not just to make a profit, but to satisfy the customer. While the actual repairs on the

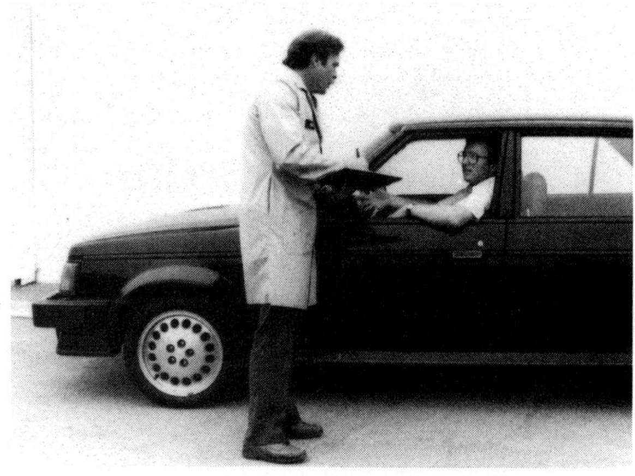

Figure 2-1. The customer can be of help when diagnosing many brake problems.

vehicle brake system are a part of achieving this goal, they are not the entire job. You must also treat customers with courtesy when they bring a car in for brake service, and pay attention to their description of the problem. This makes them feel they are receiving the service they pay for, and goes a long way toward ensuring repeat business for your shop.

When it comes to brake problem diagnosis, there is another good reason to pay attention to the customer. Although they may not realize it, car owners know a great deal about what is wrong with their cars. If you listen carefully to the customer's description of the problem, figure 2-1, then ask specific questions about likely problem areas, you can substantially shorten your diagnosis time. For example, if a customer complains of poor stopping power, you may ask if there is a grinding noise from the wheels, indicating that the brake pads are worn to the backing plates. Another question might be whether the brake pedal must be applied with greater force than in the past, possibly indicating a problem with the power booster.

The exact questions you need to ask will vary with the problem. However, there are some general questions that can be useful in many circumstances, for example: what are the symptoms of the brake system problem? When did the problem first occur? Is the problem present at all times, or does it occur only in certain situations? Have there been any other events, such as an accident or recent repairs, that took place about the same time the brake problem appeared?

When asking these and other questions, remember that most customers are unfamiliar with the expressions used by technicians to describe brake problems. Terms such as fade, pull, lockup, and others may not have any meaning to the car owner. Make sure you fully understand the condition being described, and if necessary, take the customer along on the test drive to identify the symptom causing the complaint. Sometimes a brake "problem," such as the hiss of air entering a vacuum booster as the brakes are applied, is just a normal part of brake operation that the customer does not understand.

BRAKE SYSTEM ROAD TESTING

In some cases, a skilled technician can diagnose a brake problem simply by questioning the customer without ever looking at the car. He is able to do this because he has repaired many similar brake problems in the past, knows the symptoms for most common brake problems, and may have knowledge of unique problems that are specific to certain vehicles. In these situations, a road test may not be required, although a complete brake system inspection should still be done to confirm the diagnosis and locate any other potential problems.

Experience is not always enough, however, when you have to diagnose unusual brake system problems, multiple problems occurring at the same time, or problems that are interrelated with other vehicle systems. In these cases, a road test can help better identify the problem before the brake system is inspected. As already mentioned, the road test allows you to confirm the symptoms described by the customer, but more importantly, it enables you to experience the condition first hand and determine if it is, in fact, a brake problem. Many tire, wheel, suspension, and driveline problems contribute to inefficient braking and display symptoms similar to those of brake problems.

To minimize the effect of the road surface on braking performance, a brake system road test should always be done on smooth, clean, dry, and level pavement, figure 2-2. Because braking force is limited by tire traction, a road that is bumpy, wet, greasy, or covered with loose dirt and leaves will not provide adequate traction for a proper road test. A road that is heavily crowned may cause the car to pull to one side. For safety, use a road that has only light traffic, and is safe for speeds up to 55 mph (88 kph).

ROAD TEST PROCEDURE

The road test focuses primarily on those symptoms that are best checked when the car is in motion. You use the road test to reduce the number of potential problem areas, then you use the more specific tests in the inspection

Figure 2-2. A road suitable for a brake system road test.

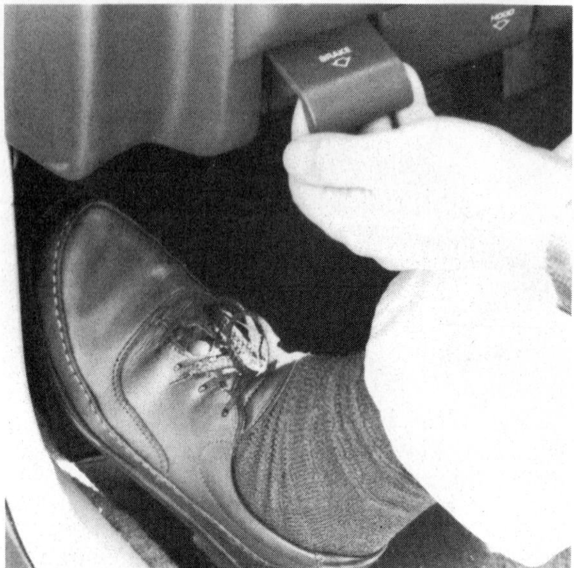

Figure 2-3. Deactivate the ratchet locking mechanism when using the parking brake to isolate a vibration.

procedure, and in later chapters of this *Shop Manual,* to isolate the exact problem. There are six basic areas you want to examine in a brake system road test:

- Driveline vibration
- Brake vibration
- Braking power
- Braking stability
- Brake pedal travel and feel
- Brake noise.

The procedure below describes how to isolate and identify brake problems related to these symptoms.

Before beginning the road test, make sure the tire pressures are set to the vehicle manufacturer's specifications. Check that the tire sizes are the same on each axle, and that all four tires are of the same construction. Both tires on each axle should have the same amount of tread wear, and you should inspect the tires for unusual wear patterns that may indicate a wheel alignment problem. Finally, take note of any obvious sagging of the suspension toward one corner, side, or end of the car. A problem in any of these areas will affect the car's braking behavior.

Driveline Vibration Test

The first part of the road test is a driveline vibration test. Driveline vibration is felt most strongly through the car body, and occurs regardless of whether the brakes are applied. To check for this condition, accelerate the car to 55 mph (88 kph), and coast down to 20 mph (30 kph) without using the brakes. Any vibrations that occur during this portion of the road test are most likely caused by bent, worn, or out-of-balance driveline parts; check for problems with the tires, wheels, axles, or driveshaft.

Brake Vibration Test

The second part of the road test is a brake vibration test. Brake vibration occurs *only* when the brakes are applied, and although it may shake the car body, it is usually felt most strongly through the brake pedal. To check for this condition, accelerate the car to 25 mph (40 kph), 40 mph (65 kph), and 55 mph (88 kph), and make stops with both light and heavy pedal pressure from each speed. Do not lock the wheels when making these stops. If you feel a body vibration or pedal pulsation when you apply the brakes, suspect distorted drums or rotors, or possibly loose wheel bearings.

Generally, front-brake vibration problems occur at speeds above 45 mph (70 kph), while rear-brake vibration problems occur below that speed. On cars where two of the service brake friction assemblies function as the parking brake, you can sometimes isolate a vibration by using the parking brake control to apply the brakes at only one axle. To do this, accelerate to speed, actuate the parking brake release lever or button to disengage the ratchet mechanism, figure 2-3, then slow the car using the parking brake. As the car decelerates, check for body vibration or a pulsation through the parking brake pedal, lever, or handle. Pulsations felt through the parking brake control are generally not as strong as those at the service brake pedal because parking brake cable stretch absorbs some of the variation in force.

Figure 2-4. Using a Tapley meter to measure braking efficiency.

If a vibration disappears when the parking brake is used to slow the car, and no pulsation can be felt through the parking brake control, the wheel brakes not used for the parking brake are the source of the problem. If a vibration remains at the same intensity when the parking brake is used to slow the car, and a pulsation can be felt through the parking brake control, the problem is in the parking brake friction assemblies. If a vibration continues at a reduced level when the parking brake is used to slow the car, and some pulsation can be felt through the parking brake control, the problem is most likely being caused by the brakes at all four wheels.

Braking Power

When you apply the brakes at the different speeds described above, note the amount of braking power available. The stopping distance from each speed should be appropriate for the type of vehicle being tested. Experience will give you a better idea of how hard particular types of vehicles can stop, but as a general rule the larger and heavier a vehicle is, the longer its stopping distances will be.

In some cases, a deceleration measuring device like the Tapley meter described in Chapter 1 can be helpful in assessing braking power, figure 2-4. All brake systems should provide enough stopping power to lock the wheels; unless, of course, the car is equipped with anti-lock brakes. If the stopping distances seem too long, or it is impossible to lock the wheels, suspect a problem with restricted brake lines, contaminated brake linings, or a defective power booster.

Don't overlook the parking brake when testing braking power; this system uses two of the service brakes on most cars, and can give you a good indication of the efficiency of those friction assemblies. Include a hill on your road test route, and apply the parking brake on the incline. The pedal, lever, or handle should apply smoothly without binding; if not, suspect cable problems. If more than two-thirds of the parking brake control travel is required to apply the brake, shoe and/or cable adjustments are called for. Finally, if the parking brake fails to hold the car in position on a good grade, inspect the friction assemblies for glazed linings or other problems that reduce braking power.

Braking Stability

As you apply the brakes at the different speeds described above, note the distribution of braking force to the wheels, and the overall stability of the vehicle. The car should stop smoothly with no tendency toward wheel lockup. During very heavy braking in ideal conditions, the front brakes should lock just before the rear brakes to ensure vehicle stability; however, not all brake systems behave in this manner when less than fully loaded.

If both brakes on the front or rear axle lock prematurely, suspect a problem with restricted brake lines, or the metering or proportioning valve. The master cylinder may also be at fault if the dual-circuit hydraulic system has a front/rear split. If only one wheel locks, the problem may be contaminated or misadjusted brake linings that are causing the friction assembly to grab. Sometimes, contaminated linings make a friction assembly slip, causing the brake in good condition on the opposite end of the axle to lock as more pedal pressure is applied to stop the car. You will determine the exact cause of the problem in the brake system inspection that follows the road test.

During the road test stops, the car should brake in a straight line with your hands resting lightly on the steering wheel. A side-to-side braking imbalance that is not severe enough to cause lockup will usually result in a pull to one side when you apply the brakes. The imbalance can be in the tires, wheels, or suspension just

Figure 2-5. An experienced foot on the brake pedal is an excellent diagnostic tool.

as easily as in the brake system. If your pre-road-test check did not indicate any unusual problems outside the brake system, suspect a restricted brake hose or a defective friction assembly with limited braking power. Once again, you will identify the precise cause of the problem in the brake system inspection that follows the road test.

Brake Pedal Travel and Feel Test

The travel and feel of the brake pedal while making the road test stops can tell you a great deal about the condition of the brake system, figure 2-5. In fact, almost every brake system problem eventually results in some type of pedal symptom. There are many possibilities, and these are all covered in the diagnostic charts at the end of the chapter. The most common symptoms and causes are described below.

A spongy pedal with longer than normal travel usually indicates a hydraulic problem, such as a fluid leak or air in the lines. Long pedal travel with a firm feel often indicates a mechanical problem such as a friction assembly in need of adjustment. However, if the brake pedal gradually sinks part way to the floor, then becomes firm, one circuit of the dual-circuit brake hydraulic system is probably at fault. If the pedal gradually sinks all the way to the floor, suspect a master cylinder that is bypassing internally. When excessive pedal pressure is required to stop the car, suspect a power booster problem, or defective wheel friction assemblies that are not providing full braking power.

Figure 2-6. Pad wear indicators are designed to create brake noise.

Brake Noise

Like pedal feel, brake noise is a significant aid in diagnosing brake problems. However, some noises are considered normal; certain semi-metallic pads cause a slight grinding sound, and the high-temperature organic friction materials used on many European import cars have a normal tendency to squeal. Fixed calipers are often noisier than other designs because their solid mountings transmit vibration directly into the chassis; the more flexible mountings of floating and sliding calipers damp many vibrations that might otherwise cause noise.

Although many different problems cause brake noise, the sounds are often similar, and in some cases resemble the noise of normal brake operation. Only experience can teach you which sounds are normal and which are not, and which of several problems is the source of a specific noise. During the test drive, listen for noises when the brakes are both released and applied. Common noises and their sources are described below.

When the brakes are released, rattles and scrapes are the most common noises. A rattle at low speeds indicates excessive clearance between the brake pads and rotor caused by pads that are not securely mounted as a result of worn, damaged, or missing parts. On many newer disc brakes, a light scraping or "chirping" noise when the brakes are not applied is caused by the pad wear indicator rubbing against the rotor to signal the need for pad replacement, figure 2-6. A louder scraping noise can be caused by a bent splash shield rubbing a rotor, or a bent backing plate contacting a brake drum.

Another noise that may be heard when the brakes are not applied is the growling sound

Figure 2-7. The master cylinder fluid reservoir should be free of residue.

that signals a worn wheel bearing. To confirm this problem, swerve the car back and forth to alternately load and unload the bearings on opposite sides of the car. The noise from the bad bearing will increase with the load on that side of the car, and decrease as the load is reduced. See Chapter 11 for complete wheel bearing diagnosis procedures.

Noises that occur when the brakes are applied include scrapes, grinding, squeaks, squeals, snaps, clicks, and thumps. A scraping or grinding sound usually indicates worn out brake linings that have resulted in metal-to-metal contact between the pad backing plate and rotor, or shoe lining table and drum. Squeaks and squeals commonly occur when the brakes are applied, and most often result from problems with the lining friction material, or worn, damaged, and missing vibration damping or anti-rattle parts. A snap or click on brake application is usually caused by loose, worn, or broken mechanical parts, or an improper rotor or drum surface finish. A thumping sound when the brakes are applied is usually associated with a cracked brake drum, defective brake shoes, or weak brake shoe return springs. The exact cause will be determined during the brake system inspection that follows the road test.

BRAKE INSPECTION

After you have discussed the brake problem with the customer and road tested the vehicle, you should have a fairly good idea of where the problem lies. The final step in the diagnosis procedure is to perform a complete brake system inspection to confirm your suspicions and locate any additional brake components in need of repair. Perform the basic inspection procedure described below on any vehicle that comes into your shop for brake work. As with the road test, it is not possible to cover every detail of brake inspection in this section; detailed checks and tests for each component are described in later chapters on each part or system.

If you were unable to road test the car, or a road test was unnecessary, perform the pre-road-test checks and inspections at this time. Make sure the tire pressures are set to the vehicle manufacturer's specifications. Check that the tire sizes are the same on each axle, and that all four tires are of the same construction. Both tires on each axle should have the same amount of tread wear, and you should inspect the tires for unusual wear patterns that may indicate a wheel alignment problem. Finally, take note of any obvious sagging of the suspension toward one corner, side, or end of the car. A problem in any of these areas will affect the car's braking behavior.

The first step in the inspection is to sit in the driver's seat, apply the brake pedal, and listen for noise. A squeak as the pedal is applied or released is usually caused by a lack of lubrication. If the noise is from under the dash, lubricate the pedal linkage. If the noise is from the wheel brakes, the caliper floating or sliding parts, or the shoe support pads on the drum brake backing plates, are in need of lubrication. As you apply the brake pedal for this test, make sure there is the proper amount of freeplay.

Next, if the brake system is so equipped, test the vacuum power booster. With the engine off, apply the brake pedal repeatedly with medium pressure until the booster reserve is depleted. At least two brake applications should have a power-assisted feel before the pedal hardens noticeably. If the pedal feels hard immediately, or after only one brake application, a vacuum leak or low level of engine vacuum is indicated. To test booster function once the reserve is depleted, hold moderate pressure on the brake pedal and start the engine. If the booster is working properly, the pedal will drop slightly toward the floor. During this part of the inspection, have an assistant verify that the stoplights come on when you apply the brake pedal.

Next, check the brake fluid level in the master cylinder reservoir. A low level indicates that either the brake linings are worn or there is a leak in the hydraulic system. Dip a clean finger into the fluid and wipe it across the bottom of the reservoir. Withdraw your finger and check it for signs of rust, corrosion, or other residue, figure 2-7. If there is any buildup, the hydraulic

Figure 2-8. In and out movement indicates wheel bearing wear or misadjustment.

Figure 2-9. Mark wheels and hubs so they can be reassembled in proper relation to one another.

system should be flushed, and the wheel cylinders and brake calipers may need to be rebuilt or replaced. The master cylinder is at the top of the hydraulic system; any contamination there is always found in larger quantities at the lower ends of the system.

Inspect the brake lines from the master cylinder to the point where they disappear under the car. Look for leaks and corrosion along the lines and at their connections. Heavily rusted or corroded brake lines, and any line with a kink in it, must be replaced. Also inspect any hydraulic control valves located in the engine compartment. Check them for leaks, and make sure that any electrical switches they contain are properly connected to the vehicle wiring harness.

Raise the vehicle on a lift, grasp the front tires at the top and bottom, then check for wheel bearing play by attempting to wobble the tires in and out, figure 2-8. You should feel little or no play. Experience will give you a better idea of how much play is acceptable, until then, you may want to use a dial indicator to check the bearing play as described in Chapter 11. More than a minimal amount of play indicates a need for bearing service.

Mark each wheel and one of its studs with chalk or a grease pencil so you can replace the wheels in their original positions, figure 2-9. This prevents problems if the wheels and tires have been balanced on the car. Remove the wheels from the car so you can inspect the brake friction assemblies.

On disc brakes, verify that the brake linings are at least 1/32 inch (.030″ or .75 mm) above the pad backing plate or rivet heads. Inspect the calipers for leaks, and make sure the mounting hardware is tight. The floating or sliding surfaces of the caliper should be properly lubricated and free of rust and corrosion. Inspect the rotors for scoring, cracks, heat checking, and hard spots. And finally, inspect the brake hoses for swelling, leaks, cracks, or abrasions.

Remove the drums from drum brake friction assemblies and verify that brake shoe linings are at least 1/32 inch (.030″ or .75 mm) above the lining table or rivet heads. Pull back the wheel cylinder boots and inspect for piston seal leakage; no more than a slight amount of dampness should be present. Inspect the drum friction surface for scoring, cracks, heat checking, or hard spots. And finally, inspect any brake hoses for swelling, leaks, cracks, or abrasions.

Continue your inspection of the brake lines under the car. Look for leaks along the lines and at the connections. Heavily rusted or corroded brake lines, and any line with a kink in it, must be replaced. While checking the fluid lines, inspect the parking brake cables for rust, corrosion, and fraying.

BRAKE SYSTEM DIAGNOSTIC CHARTS

The following diagnostic charts can help you determine the cause and cure of most brake problems. The charts are organized into four sections with a heading for each basic type of symptom: pedal, wheel brake, brake performance, and brake noise. Each symptom is then broken down into specific problems with a cause and a cure for each. On each chart, subheadings (All, Disc, and Drum) are used to indicate whether the problems listed apply to all brakes, only disc brakes, or only drum brakes.

I. PEDAL SYMPTOMS

1. Spongy Pedal — All

Cause	Remedy
1. Air in hydraulic system	1. Bleed brake system
2. Weak brake hose expanding under pressure	2. Replace hose
3. Water-contaminated brake fluid (low boiling point)	3. Flush brake hydraulic system
4. Master cylinder leaking internally	4. Rebuild or replace master cylinder
5. Master cylinder mounting bolts loose	5. Tighten bolts
6. Cracked firewall or power booster housing	6. Repair or replace as needed

• Spongy Pedal — Disc

Cause	Remedy
1. Caliper distortion from frozen piston (multi-piston calipers)	1. Rebuild or replace caliper

• Spongy Pedal — Drum

Cause	Remedy
1. Bent or warped brake shoes	1. Replace shoes in axle sets
2. Poor lining-to-drum contact	2. Arc or replace brake shoes
3. Thin or cracked brake drum	3. Replace drum
4. Faulty residual pressure check valve admitting air	4. Replace check valve and bleed brakes
5. Leaking wheel cylinder	5. Rebuild or replace wheel cylinder
6. Shoes not centered in drum (with adjustable anchors)	6. Adjust shoe anchors
7. Bent or broken backing plate	7. Replace backing plate

2. Excessive Pedal Travel — All

Cause	Correction
1. Air in hydraulic system	1. Bleed brake system
2. Water-contaminated brake fluid (low boiling point)	2. Flush brake hydraulic system
3. Master cylinder pushrod adjustment incorrect	3. Adjust pushrod
4. Brake booster output pushrod adjustment incorrect	4. Adjust pushrod
5. Master cylinder bypassing internally	5. Rebuild or replace master cylinder
6. Pressure loss in one half of split hydraulic system	6. Repair leak in hydraulic system
7. Fluid bypassing quick-take-up valve in master cylinder	7. Replace quick-take-up valve

• Excessive Pedal Travel — Disc

Cause	Correction
1. Loose, broken, or worn caliper attachment	1. Torque caliper mounting bolts to specifications or repair attachment
2. Excessive pad knockback from rotor runout	2. Turn or replace rotor
3. Damaged or worn caliper piston seals	3. Rebuild or replace caliper
4. Loose or worn wheel bearings	4. Service and adjust bearings
5. Warped or bent brake pad	5. Replace pads in axle sets

• Excessive Pedal Travel — Drum

Cause	Correction
1. Excessive clearance between linings and drums	1. Adjust brakes
2. Automatic adjusters not working	2. Service adjusters
3. Bent or warped brake shoes	3. Replace shoes in axle sets

3. Sinking Brake Pedal — All

Cause	Correction
1. Hydraulic system leaking externally	1. Locate and repair leak
2. Master cylinder leaking internally	2. Rebuild or replace master cylinder
3. Bleeder screw left open	3. Close screw and bleed brake system

4. Brake Pedal Vibration or Pulsation — All

Cause	Correction
1. Loose or worn wheel bearings	1. Service and adjust bearings

• Brake Pedal Vibration or Pulsation — Disc

Cause	Correction
1. Excessive rotor thickness variation or lateral runout	1. Turn or replace rotor
2. Rust on rotor surface	2. Resurface rotor

• Brake Pedal Vibration or Pulsation — Drum

Cause	Correction
1. Drum out of round or eccentric	1. Turn or replace drum

5. Excessive Pedal Effort — All	
1. Brake pedal linkage binding	1. Lubricate, repair, or replace linkage
2. Power booster failure	2. Repair vacuum, hydraulic, or electrical power supply to booster. Rebuild or replace booster if needed.
3. Friction material contaminated with grease or brake fluid	3. Replace leaking grease seal or repair brake fluid leak. Replace brake shoes or pads in axle sets
4. Glazed brake shoe or pad linings	4. Replace shoes or pads in axle sets
5. Restricted brake lines	5. Repair or replace lines
6. Frozen master cylinder piston	6. Rebuild or replace master cylinder
7. Center orifice in quick-take-up valve clogged	7. Replace quick-take-up valve
• Excessive Pedal Effort — Disc	
1. Frozen caliper pistons	1. Rebuild or replace caliper
2. Semi-metallic brake pads used in system designed for asbestos friction materials	2. Install pads with asbestos-based linings
• Excessive Pedal Effort — Drum	
1. Reversed primary and secondary brake shoes	1. Install shoes in proper locations — if badly worn, replace shoes in axle sets
2. Frozen wheel cylinder pistons	2. Rebuild or replace wheel cylinder

II. WHEEL BRAKE SYMPTOMS

6. One Brake Drags — All	
1. Restricted brake line preventing fluid return	1. Replace hose or tube
2. Loose or worn wheel bearings	2. Service and adjust bearings
• One Brake Drags — Disc	
1. Warped or bent brake pad	1. Replace pads in axle sets
2. Sticking caliper piston	2. Rebuild or replace caliper
3. Swollen caliper piston seal	3. Rebuild or replace caliper
• One Brake Drags — Drum	
1. Brake adjustment too tight	1. Adjust brakes
2. Weak or broken shoe return spring	2. Replace springs in axle sets
3. Shoes sticking on worn backing plate support pads	3. File support pads smooth and lubricate with brake grease
4. Warped or bent brake shoe	4. Replace shoes in axle sets
5. Sticking wheel cylinder piston	5. Rebuild or replace wheel cylinder
6. Adjustable shoe anchor loose	6. Tighten anchor

7. Front Brakes Drag — Disc	
1. Residual pressure check valve installed in master cylinder front brake outlet	1. Remove valve

8. Rear Brakes Drag — All

Cause	Remedy
1. Restricted brake line preventing fluid return	1. Replace hose or tube
2. Improper parking brake cable or shoe adjustment	2. Adjust cable and/or shoes
3. Sticking or frozen parking brake cables	3. Lubricate or replace cables

• Rear Brakes Drag — Drum

Cause	Remedy
1. Automatic adjusters damaged	1. Repair or replace adjusters
2. Brake adjustment too tight	2. Adjust brakes

9. All Brakes Drag — All

Cause	Remedy
1. Brake pedal linkage binding	1. Lubricate or repair linkage
2. Master cylinder pushrod adjustment incorrect	2. Adjust pushrod
3. Brake booster output pushrod adjustment incorrect	3. Adjust pushrod
4. Weak brake pedal return spring	4. Replace spring
5. Sticking master cylinder pistons	5. Rebuild or replace master cylinder
6. Swollen rubber parts from contaminated brake fluid	6. Replace all rubber parts, and flush brake hydraulic system

10. One Brake Locks — All

Cause	Remedy
1. Worn or mismatched tires	1. Make sure the tires on each axle have the same construction, size, and treadwear
2. Incorrect tire pressure	2. Adjust tire pressures to vehicle specifications
3. Failing wheel bearing	3. Replace and adjust bearing
4. Improper size or type of brake lining	4. Replace shoes or pads in axle sets
5. Brake linings contaminated by grease or brake fluid	5. Replace grease seal or repair fluid leak; replace shoes or pads in axle sets

11. Premature Front Brake Lockup — Disc

Cause	Remedy
1. Faulty metering valve allowing early application	1. Replace valve
2. Failure of master cylinder rear brake hydraulic circuit	2. Rebuild or replace master cylinder
3. Defective proportioning valve restricting fluid flow to rear brakes	3. Replace proportioning valve
4. Restricted brake line preventing fluid flow to rear brakes	4. Replace hose or tube

12. Premature Rear Brake Lockup — All

Cause	Correction
1. Failure of master cylinder front brake hydraulic circuit	1. Rebuild or replace master cylinder
2. Restricted brake line preventing fluid flow to front brakes	2. Replace hose or tube
3. Defective proportioning valve allowing excessive pressure to rear brakes	3. Replace proportioning valve

• Premature Rear Brake Lockup — Drum

Cause	Correction
1. Defective proportioning valve allowing full pressure to rear brakes	1. Replace proportioning valve

13. Uneven Lining Wear — Disc

Cause	Correction
1. Sticking or binding caliper mount (floating and sliding calipers)	1. Replace mounting hardware and lubricate properly
2. Sticking or frozen caliper pistons (fixed calipers)	2. Rebuild or replace caliper
3. Caliper loose on mount	3. Torque mounting bolts to specifications
4. Bent caliper mount	4. Replace spindle or anchor plate
5. Damaged or improperly machined caliper body	5. Replace caliper
6. Caliper not aligned over rotor (fixed calipers)	6. Align caliper

• Uneven Lining Wear — Drum

Cause	Correction
1. Frozen wheel cylinder piston	1. Rebuild or replace wheel cylinder
2. Shoes reversed in dual servo brake	2. Install shoes in correct positions
3. Bent backing plate	3. Replace backing plate
4. Shoes not centered in drum (with adjustable anchors)	4. Adjust anchors

14. Rapid Lining Wear — All

Cause	Correction
1. Incorrect type of brake linings	1. Replace with proper lining material
2. Inadequate type of brake linings (severe service)	2. Upgrade to semi-metallic or premium-quality lining
3. Rough drum or rotor surface finish	3. Resurface to proper finish
4. Parking brake adjusted too tight	4. Adjust parking brake
5. Master cylinder held in partially applied position	5. Adjust brake pedal freeplay

III. BRAKE PERFORMANCE SYMPTOMS

15. Pull During Braking — All

1. Unequal tire pressure side-to-side	1. Equalize pressures
2. Uneven tire size or treadwear side to side	2. Install matched tires with equal treadwear
3. Incorrect wheel alignment	3. Align front end
4. Worn suspension parts	4. Replace as necessary
5. Restricted brake line on one side of vehicle	5. Replace hose or tube
6. Brake linings contaminated with grease or brake fluid	6. Replace grease seal or repair fluid leak. Replace shoes or pads in axle sets
7. Water fade of brake linings	7. Apply brakes gently to dry linings

• Pull During Braking — Disc

1. Unequal brake action side to side.	1. Check for loose calipers; incorrect or broken parts; charred, glazed, contaminated, or worn pads; worn or damaged rotors; sticking caliper pistons; etc. Repair, replace, or adjust as necessary

• Pull During Braking — Drum

1. Unequal brake action side to side.	1. Check for unequal brake adjustment; incorrect or broken parts; charred, glazed, contaminated, or worn linings; worn or damaged drums; sticking wheel cylinder pistons; weak return springs; loose backing plate; etc. Repair, replace, or adjust as necessary

16. Brake Fade — All

1. Excessive repeated use of brakes without adequate cool-down time	1. Stop vehicle and allow brakes to cool
2. Water-contaminated brake fluid (low boiling point)	2. Flush brake hydraulic system
3. Low quality friction materials	3. Install premium brake shoes or pads in axle sets
4. Dragging brakes	4. See earlier chart
5. Drums or rotors worn beyond discard dimension	5. Replace drums or rotors
6. Glazed brake linings	6. Replace shoes or pads in axle sets

• Brake Fade — Drum

1. Poor lining to drum contact	1. Arc brake shoes or replace in axle sets

17. Steering Wheel Shimmy — All

Cause	Correction
1. Worn suspension and steering parts	1. Check and replace as needed
2. Damaged or out-of-balance tires and wheels	2. Check and replace as needed
• Steering Wheel Shimmy — Disc	
1. Excessive rotor runout	1. Replace and/or turn rotors
• Steering Wheel Shimmy — Drum	
1. Drum out of round	1. Replace and/or turn drum

18. Grabbiness or Sensitivity — All

Cause	Correction
1. Incorrect size or type of brake linings	1. Replace shoes or pads in axle sets
2. Brake linings contaminated with grease or brake fluid	2. Replace seal or repair fluid leak and replace shoes or pads in axle sets
3. Binding brake pedal linkage	3. Lubricate, repair, or replace linkage
4. Failing wheel bearings	4. Replace bearings
5. Power booster malfunction	5. Repair or replace booster
• Grabbiness or Sensitivity — Disc	
1. Caliper loose on mount	1. Torque mounting bolts to specifications
2. Defective metering valve allowing early application	2. Replace valve
3. Sticking or frozen caliper pistons	3. Rebuild or replace caliper
• Grabbiness or Sensitivity — Drum	
1. Brake adjustment too tight	1. Adjust brakes and repair automatic adjusters
2. Brake adjustment too loose (dual-servo brakes)	2. Adjust brakes and repair automatic adjusters
3. Lining loose on brake shoe	3. Replace shoes in axle sets
4. Excessive debris in drums	4. Clean friction assembly
5. Incorrect or warped brake shoes	5. Install proper shoes in axle sets
6. Shoes not centered in drum (with adjustable anchors)	6. Adjust anchors
7. Backing plate loose	7. Tighten mounting bolts
8. Scored or out-of-round drums	8. Replace and/or turn drums
9. Improper lining-to-drum contact	9. Arc linings or replace brake shoes in axle sets

19. Chatter or Shudder — All

Cause	Correction
1. Bent axle or axle flange	1. Replace axle
2. Incorrect lug nut/bolt torque or tightening sequence	2. Torque lug nuts/bolts to specification in correct sequence

• Chatter or Shudder — Drum

Cause	Correction
1. Incorrect brake adjustment	1. Adjust brakes and repair automatic adjusters
2. Backing plate, wheel cylinders, or anchors loose	2. Tighten mounting bolts
3. Weak or broken brake shoe return spring	3. Replace springs in axle sets
4. Brake lining contaminated with grease or brake fluid	4. Replace seal or repair fluid leak and replace shoes in axle sets.
5. Twisted or warped brake shoe	5. Replace shoes in axle sets
6. Out-of-round or damaged brake drum	6. Replace and/or turn drums

20. Unable to Fully Bleed System — All

Cause	Correction
1. Restricted brake lines	1. Replace hoses or tubes
2. Master cylinder ports plugged	2. Rebuild or replace master cylinder
3. Pressure differential switch not centered	3. Locate and repair leak in hydraulic system

• Unable to Fully Bleed System — Disc

Cause	Correction
1. Metering valve not disarmed	1. Disarm valve with proper tool
2. Fluid transfer passage between caliper halves blocked (fixed calipers)	2. Rebuild or replace caliper

IV. BRAKE NOISE

21. Squeak on Application at Rest — All

Cause	Correction
1. Brake pedal linkage dry	1. Lubricate linkage
2. Loose master cylinder or power booster mounting	2. Tighten mounting bolts

• Squeak on Application at Rest — Drum

Cause	Correction
1. Backing plate shoe support pads dry	1. Lubricate pads with brake grease
2. Cracked brake drum	2. Replace and turn drum

22. Click or Snap on Application — All

Cause	Correction
1. Loose or worn suspension parts	1. Tighten or replace as needed
• Click or Snap on Application — Disc	
1. Loose or missing caliper mounting bolts	1. Replace missing bolts and tighten to specifications
2. Spiral machining grooves on rotor friction surface	2. Resurface rotor
3. Worn brake pad backing plate creating excessive caliper end clearance	3. Replace pads in axle sets and check end clearance
• Click or Snap on Application — Drum	
1. Excessive lining-to-drum clearance	1. Adjust brakes
2. Loose backing plate, wheel cylinder, or shoe anchor	2. Tighten loose parts
3. Backing plate shoe support pads grooved	3. File pads smooth and lubricate
4. Spiral machining grooves on drum friction surface	4. Resurface drum
5. Shoe holddown pin bent	5. Replace pin
6. Weak holddown spring	6. Replace spring
7. Bent or warped brake shoe	7. Replace shoes in axle sets
8. Cracked drum	8. Replace drum

23. Thump on Application — Drum

Cause	Correction
1. Cracked drum	1. Replace drum
2. Shoe return springs weak or of unequal tension	2. Replace springs in axle sets

24. Squeal While Braking — All

Cause	Correction
1. Glazed friction material	1. Replace shoes or pads in axle sets
2. Incorrect type of friction material	2. Install linings with proper friction material
3. Poor-quality brake linings	3. Install premium quality shoes or pads in axle sets
4. Front wheel bearings worn or loose	4. Replace and/or adjust bearings
5. Loose or worn suspension parts	5. Tighten or replace parts
• Squeal While Braking — Disc	
1. Brake pad wear indicators contacting rotors	1. Replace pads in axle sets
2. Anti-rattle clips, springs, or shims are weak, missing, or incorrectly installed	2. Replace or reposition anti-rattle parts
3. Outboard brake pad retaining tabs not clinched tightly against caliper	3. Bend tabs to secure pad in caliper

Continued on next page

• Squeal While Braking — Drum

Cause	Correction
1. Poor brake shoe lining-to-drum contact	1. Arc or replace shoes
2. Debris imbedded in brake lining	2. Replace shoes if debris cannot be removed
3. Drum silencer spring missing or weak	3. Fit new spring
4. Riveted lining loose on brake shoe	4. Replace shoes in axle sets
5. Drum out-of-round	5. Turn or replace drum
6. Weak or broken shoe holddown springs	6. Replace springs in axle sets
7. Loose backing plate, wheel cylinder, or shoe anchor	7. Tighten loose parts
8. Bent backing plate, brake shoes, or anchors	8. Replace defective parts
9. Wrong size or type of brake shoe	9. Install correct shoes in axle sets
10. New linings installed in scored drum	10. Turn or replace drums
11. Debris in brake drum	11. Clean out drum

25. Scraping or Grinding — All

Cause	Correction
1. Worn brake linings causing metal to metal contact	1. Replace shoes or pads in axle sets and turn drums or rotors as needed
2. Debris imbedded in brake linings	2. Replace shoes or pads in axle sets and turn drums or rotors as needed

• Scraping or Grinding — Disc

Cause	Correction
1. Rotor rubbing against dust shield or caliper	1. Inspect alignment of parts and repair or adjust as needed
2. Caliper mounting bolts too long and contacting rotor	2. Install mounting bolts of the correct length

• Scraping or Grinding — Drum

Cause	Correction
1. Rough drum surface	1. Replace and/or turn drums
2. Brake shoe contacting drum web	2. Correct shoe installation or install correct shoes

26. Rattle When Brakes Unapplied — Disc

Cause	Correction
1. Excessive pad knockback from rotor runout	1. Replace and/or turn rotor
2. Anti-rattle clips, springs, or shims are weak, missing, or incorrectly installed	2. Replace or reposition anti-rattle parts

PART TWO

Brake Hydraulic System Service

Chapter

3

Fluid-Related Brake Service

Fluid-related service procedures are among the most basic, yet most important, brake system services. An adequate fluid reserve in the master cylinder reservoir, and a hydraulic system free of air and contamination, are critical to the performance and service life of the brake system. This chapter begins with simple fluid level checking procedures, then goes on to explain more involved fluid-related services such as brake bleeding, and brake fluid changing. The final section of the chapter explains how to reset the pressure differential switches used to trigger brake system warning lights.

FLUID LEVEL CHECKING

The brake fluid level in the master cylinder reservoir should be checked regularly as part of routine vehicle maintenance. Most manufacturers recommend that the fluid level be checked at least twice a year. A fluid level check should also be done both before and after any other brake work is performed. A low fluid level can allow air to enter the brake system; this causes an increase in brake pedal travel, and gives the pedal a spongy feel. If enough air enters the system, a total loss of braking power will result.

The most common reason for a low fluid level is brake lining wear that leaves the pistons in the brake calipers and wheel cylinder farther out in their bores. This increases the volume of the hydraulic system, and requires fluid from the master cylinder reservoir to keep the system filled. Because brake calipers have relatively large pistons that displace greater amounts of fluid than the smaller pistons in wheel cylinders, reservoirs that serve only disc brakes will show a significantly greater fluid level drop as the brake linings wear than will reservoirs that serve only drum brakes.

Brake lining wear, and the accompanying fluid level drop, occurs relatively slowly over a period of months. A rapid and complete loss of brake fluid in the reservoir is usually a sign of a leak in the hydraulic system. The leak may be caused by something as simple as a loose brake line fitting, but it is more likely to be the result of a leaking hydraulic system component.

Brake Fluid Inspection

Whenever the brake fluid level is checked, the fluid should also be inspected for dirt, moisture, or oil contamination. Fluid in good condition appears relatively clear, with only a slight tint; the exact color depends on the type of fluid. A cloudy appearance indicates fluid contaminated by moisture, while a dark appearance indicates contamination from dirt, rust,

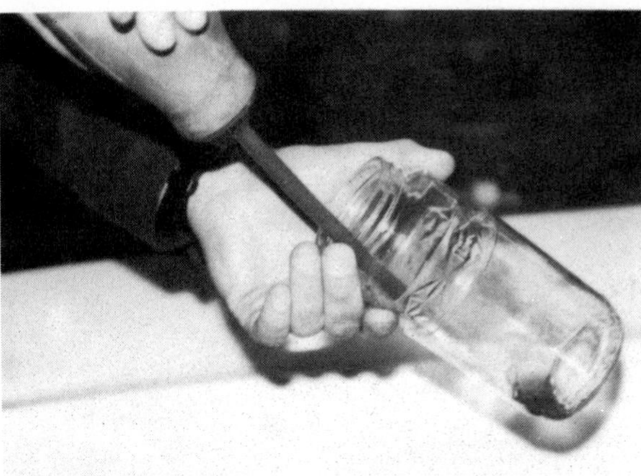
Figure 3-1. Checking a brake fluid sample for contamination.

corrosion, or brake dust. Oil contamination causes brake fluid to layer and separate because petroleum products are not compatible with polyglycol brake fluids. However, layering can also occur when silicone brake fluid is mixed with polyglycol fluid. The most common indications of oil-contaminated brake fluid are leaks caused by damage to the rubber seals and hoses in the hydraulic system.

If contamination is suspected, but the exact type is difficult to determine with the fluid in the reservoir, use a clean syringe to remove a sample of fluid and place it in a clear jar for closer examination, figure 3-1. Any visible contamination is sufficient reason to flush the system and install new fluid.

Brake Fluid Selection

When you top up the brake fluid level, always use the correct type and DOT grade of brake fluid recommended by the vehicle manufacturer. Many cars use polyglycol type DOT 3 grade fluids, however that is not always the correct choice. Ford Motor Company cars require a special DOT 3 fluid with an extremely high dry boiling point, and a number of imported cars require DOT 4 fluid. The use of DOT 5 silicone fluid as original equipment is quite rare, as are brake systems that require Hydraulic System Mineral Oil (HSMO). Consult the owner's manual or shop manual if you have any question about the type of fluid that should be used in a specific brake system.

Using a lower *grade* of fluid than is recommended, for example putting DOT 3 fluid in a system designed for DOT 4 fluid, will lower the boiling point of the fluid, and thus the temperature at which the system will vapor lock. If the boiling point drops below the highest temperature a fluid may reach under extreme braking conditions, a loss of braking power, and possibly an accident, will occur.

Using the wrong *type* of brake fluid can also create problems. While the three DOT grades of fluid are technically compatible, adding either DOT 3 or 4 polyglycol fluid to a system filled with DOT 5 silicone fluid will again reduce the boiling point. In addition, the heavier polyglycol fluid will sink to the lowest points in the system (the wheel cylinders and brake calipers) where any moisture it absorbs will lead to rust, corrosion, and leaks.

Another case where the wrong *type* of brake fluid will cause problems is if engine oil, HSMO, automatic transmission fluid, or any other petroleum-base product is used in a system designed for polyglycol fluid. Petroleum-base fluids cause the rubber parts in the system to soften and swell, resulting in fluid leaks, stuck pistons, and eventually, total hydraulic system failure. When this occurs, the entire system must be torn down, flushed of all contaminated fluid, and reassembled with all-new rubber parts. The same kind of damage will occur if polyglycol fluid is accidentally used in a system designed for Hydraulic System Mineral Oil (HSMO).

Polyglycol brake fluid used to top up a master cylinder should only be taken from a sealed container. If the container has been previously opened, it should have been stored for only a short time, and kept tightly capped in storage. Remember that polyglycol brake fluid is a powerful solvent that can cause injury and remove paint in seconds. Whenever you handle this type of fluid, wear safety glasses and take special care not to spill any on the vehicle finish.

If you should spill polyglycol brake fluid on a car's paint, do not attempt to wipe it off with a shop towel; this will only make the problem worse. Instead, immediately flush the area with large amounts of clean water. Once all traces of brake fluid are washed away, dry the area with a soft cloth. If the damage is not too great, you may be able to restore the finish with polishing compound and a good wax.

Specific Brake Fluid Level Checks

Checking the fluid level in most brake systems is a simple procedure. However, certain late-model cars have brake systems with electro-hydraulic power boosters that require special fluid level checking procedures. The following sections describe how to check the fluid levels in conventional brake systems, the GM Powermaster brake system, and two versions of the Teves anti-lock brake system.

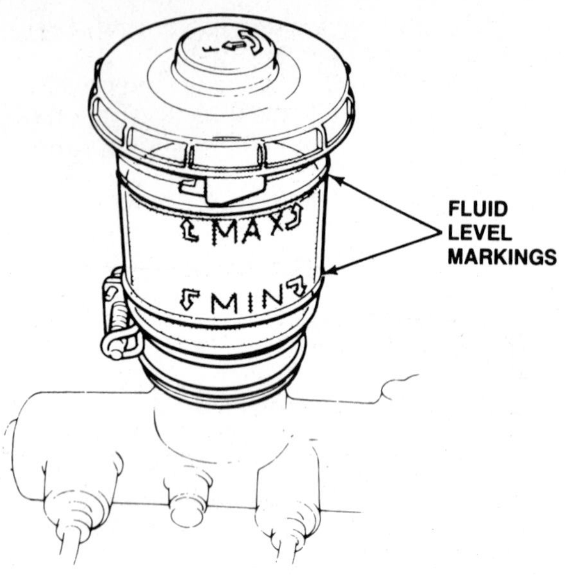

Figure 3-2. This translucent plastic reservoir has cast-in fluid level markings.

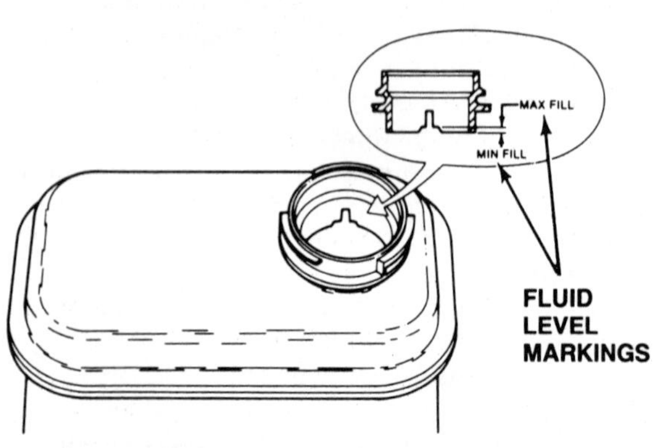

Figure 3-3. Opaque plastic reservoirs have internal fluid level markings.

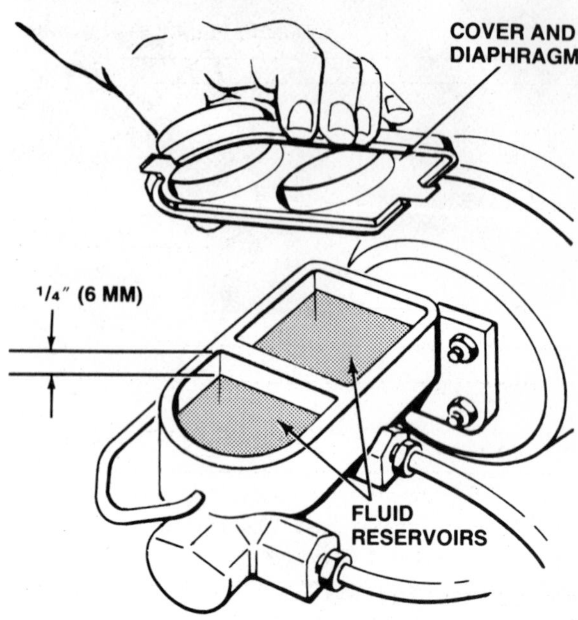

Figure 3-4. Unmarked master cylinders are typically filled to within ¼ inch (6 mm) of the top.

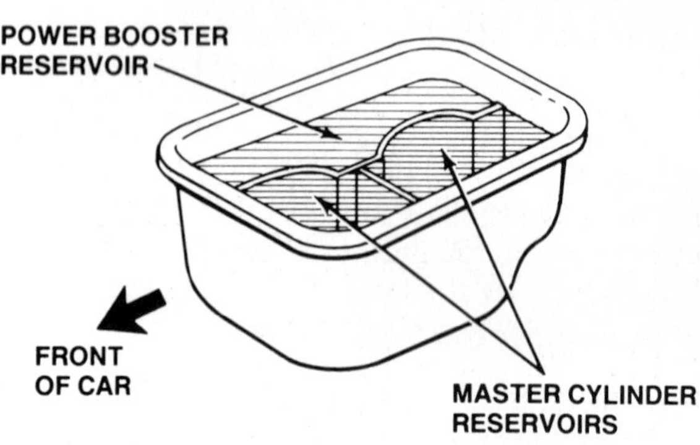

Figure 3-5. The GM Powermaster brake system fluid reservoir.

Conventional brake system level check

Brake system fluid reservoirs are made of plastic, or cast in metal as an integral part of the master cylinder body. Most translucent plastic reservoirs have marks or lines cast on the outside of them to indicate high and low fluid levels, figure 3-2. To check the level, simply sight through the reservoir. Some cars have opaque plastic reservoirs that are filled to a mark or plastic ring located inside the reservoir, figure 3-3. To check the fluid level in these master cylinders, remove the reservoir cover, or filler caps, to expose the markings.

The cover must also be removed to check the fluid level on a master cylinder with an integral metal reservoir. This type of reservoir is not marked to show the proper fluid level; instead, the maximum level is designed to be ¼ inch (6 mm) from the top, figure 3-4. To check and adjust the brake fluid to the proper level:

1. Wipe the fluid reservoir and cover clean.
2. Remove the reservoir cover and check the brake fluid level.
3. Add fluid to the reservoir, or draw fluid from it with a brake fluid syringe, to obtain the proper level.

Powermaster brake system level check

The General Motors Powermaster brake system combines an electro-hydraulic power booster and master cylinder into one unit. Because of this, the Powermaster fluid reservoir is divided into three chambers, figure 3-5. The left side of

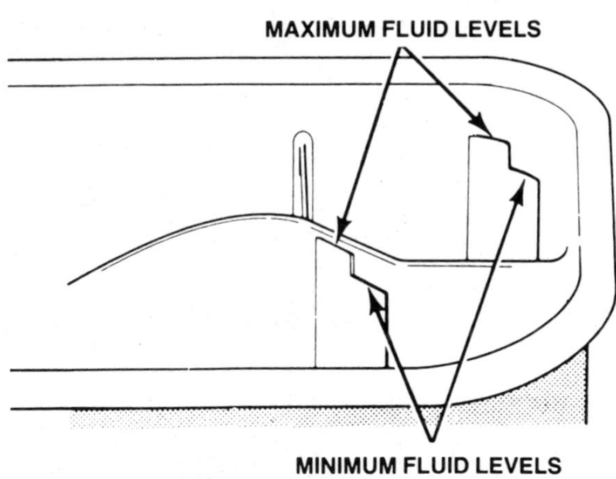

Figure 3-6. Powermaster reservoir fluid level markings.

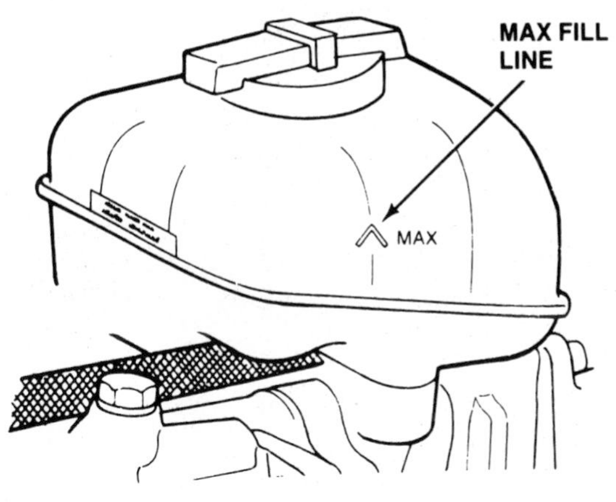

Figure 3-7. Check the fluid level in the Ford/Teves anti-lock system reservoir with the accumulator charged.

the reservoir has two small chambers that serve the master cylinder; the fluid level in these chambers is checked in basically the same manner as described above. There are minimum and maximum level markings cast into the walls of the reservoir chambers, figure 3-6.

The right side of the reservoir has a single large chamber that serves the power booster accumulator. Checking the level in this side of the reservoir requires a special procedure. In normal operation with the accumulator charged, the fluid level is very low, barely covering the pump ports in the bottom of the reservoir. When checking the fluid level, the accumulator must be discharged. To do this:

1. Leave the reservoir cover in place and make sure the ignition switch is OFF.
2. Pump the brake pedal at least 10 times using approximately 50 pounds (110 kg) of force. Pedal feel will become noticeably harder when the accumulator is discharged.
3. Remove the reservoir cover and check the fluid level; it should be between the maximum and minimum level markings cast into the wall of the reservoir, figure 3-6.

General Motors states that only new DOT 3 brake fluid should be used when adjusting the fluid level in the Powermaster brake system. Silicone DOT 5 fluids are specifically *not* recommended. As discussed in the *Classroom Manual*, never transfer fluid from the power booster chamber of the reservoir to the master cylinder chambers; the power booster chamber is not sealed off from the atmosphere so its fluid is free to absorb moisture.

Teves anti-lock brake system level check

The Teves anti-lock brake system used by both General Motors and Ford incorporates an electro-hydraulic power booster that shares a single fluid reservoir with the master cylinder. However, differences in the fluid level markings on the reservoirs dictate a unique checking procedure for each manufacturer's unit. The Ford/Teves system is checked with the accumulator charged; the GM/Teves system, like the Powermaster, is checked with the accumulator discharged.

To check the brake fluid level in the Teves anti-lock system used by Ford:

1. Turn the ignition switch ON.
2. Pump the brake pedal until the hydraulic pump motor begins to run. When the pump stops, the accumulator is fully charged.
3. Check that the fluid level is even with the "MAX" marking cast into the reservoir body, figure 3-7. Use DOT 3 fluid that meets Ford specifications when adjusting the fluid level.

To check the hydraulic fluid level in the Teves anti-lock system used by General Motors:

1. Turn the ignition switch OFF.
2. Pump the brake pedal a minimum of 20 times. Pedal feel will become noticeably harder when the accumulator is discharged.
3. Adjust the fluid level in the reservoir until it is even with the full mark cast into the reservoir body, figure 3-8.

BRAKE BLEEDING

Brake bleeding is a process that pushes new brake fluid through the brake system to force out contaminated fluid and trapped air. Air is present in the hydraulic system when the car

Figure 3-8. Check the fluid level in the GM/Teves anti-lock system reservoir with the accumulator discharged.

is first assembled, and enters the brake lines whenever the system is opened for service. Defective seals can also allow air to enter during normal brake operation.

A system free of air has a high, firm brake pedal. A system with air trapped in it has a spongy, low, or bottoming brake pedal. The low pedal will rise and become somewhat firmer if it is pumped rapidly. However, when the pedal is released for even a short period of time, it falls to its initial level on the next stroke. An air-entrapment test can be used to help determine if there is air in the brake system:

1. Remove the master cylinder cover and adjust the brake fluid to the proper level. Replace the master cylinder cover but do not secure it in place.
2. Have an assistant pump the brakes rapidly 10 to 20 times, then hold the pedal down firmly.
3. Remove the master cylinder cover and have the assistant release the pedal quickly. If a squirt of brake fluid occurs above the surface in either reservoir, air is trapped in the system.

The air entrapment test works because any air in the system is compressed when the brake pedal is pumped. When the pedal is released, the compressed air pushes the fluid back through the lines and the compensating ports. Note the side of the reservoir in which the squirt of brake fluid occurs; this indicates the side of the dual-circuit split brake system that contains the trapped air.

A low brake pedal caused by air trapped in the hydraulic system should not be confused with a low pedal resulting from drum brakes in need of adjustment. These problems have similar symptoms, however, when the brakes simply need adjusting, the pedal will firm up within one or two strokes, and will not have the spongy feel typical of trapped air.

Bleeding Sequences

Brake bleeding can sometimes be performed on only part of the brake hydraulic system. For example, as explained in the next section, master cylinders are commonly bled off the car. In addition, if a single leaking wheel cylinder or brake caliper is replaced, it is often necessary to bleed only that portion of the hydraulic system. However, when the entire brake system is bled, the operation must be carried out in a specific order called a bleeding sequence.

Using the proper sequence allows the brakes to be bled most efficiently and with the least possible chance of air remaining trapped in the system. The correct bleeding sequence at the wheels varies from car to car depending on the design of the brake system; however, on all cars, the overall bleeding sequence is:

- Master cylinder
- Combination valve
- Wheel cylinders and brake calipers
- Load-sensing proportioning valve.

Combination valves and load sensing proportioning valves are only bled when present in the system and equipped with bleeder screws.

Wheel bleeding sequences are covered in greater detail later in the chapter. Because the master cylinder is always bled first, the procedures used to do this job are described immediately below. Master cylinders can be bled off the car on the workbench, or while in position on the car.

MASTER CYLINDER BENCH BLEEDING

New or rebuilt master cylinders are bled on the workbench before they are installed. Bench bleeding is necessary because many master cylinders mount in the car at an angle. Once they are installed, any air they contain rises to the high points in the cylinder and becomes trapped away from the fluid outlets or bleeder screws. There are three types of bench bleeding, manual bench bleeding, basic bench bleeding, and reverse bench bleeding. The first can be done in any shop, but the latter two methods require special tools available from the aftermarket.

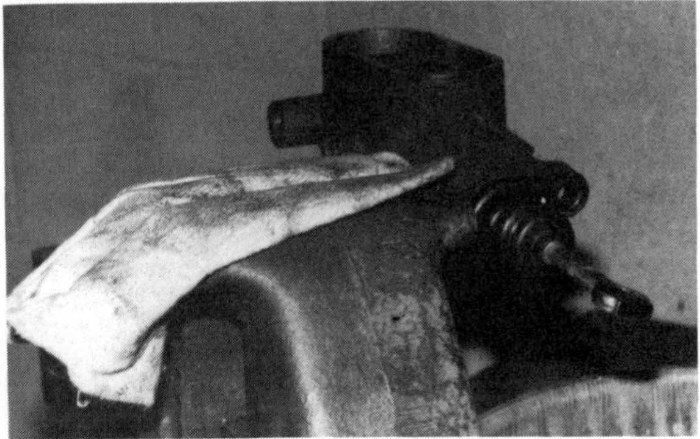
Figure 3-9. A master cylinder properly mounted in a vise for bench bleeding.

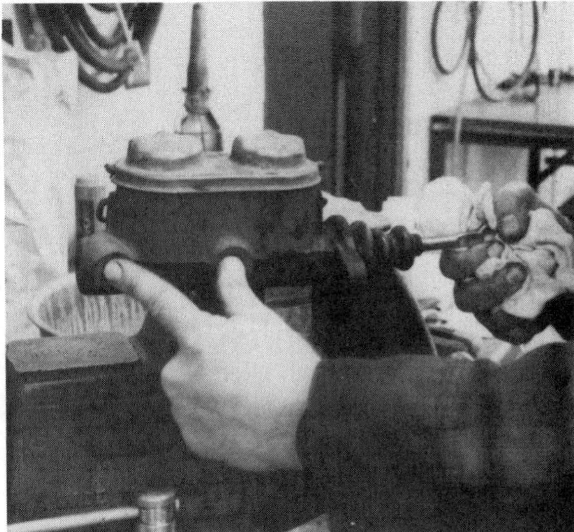
Figure 3-10. Manually bench bleeding a master cylinder.

Manual Bench Bleeding

Manual bench bleeding is just what its name implies — bleeding the master cylinder by hand. To do this, the pistons are bottomed in the cylinder bore and the fluid outlets are plugged. As the pistons are then released, a low pressure area is created in the cylinder bore, which allows atmospheric pressure to force fluid into the cylinder.

This procedure can be messy, and sometimes results in fluid spray, so have a suitable basin available to catch excess brake fluid. Also, wear safety glasses to protect your eyes, and take precautions to avoid getting fluid on painted surfaces. To bench bleed the master cylinder in this manner:

1. Clamp the master cylinder in a vise by its mounting flange so you have access to the fluid outlet ports, figure 3-9. Do not clamp on the master cylinder body or the bore may distort. If the vise has sharp teeth, use smooth-faced jaw protectors to prevent damage to the cylinder mounting flange.
2. Position the basin to catch any fluid leakage from the outlet ports, then fill the master cylinder reservoirs with new brake fluid of the proper type and grade.
3. Use the master cylinder push rod, or other round-ended rod, to slowly stroke the master cylinder pistons inward until they both bottom.
4. Plug the fluid outlets with your fingertips, figure 3-10, then slowly allow both pistons to fully return on the back stroke. Remove your finger tips from the fluid outlets.
5. Repeat Steps 3 and 4 until the fluid coming from the outlets is air free, and bubbles no longer emerge from the compensating and replenishing ports in the reservoir.

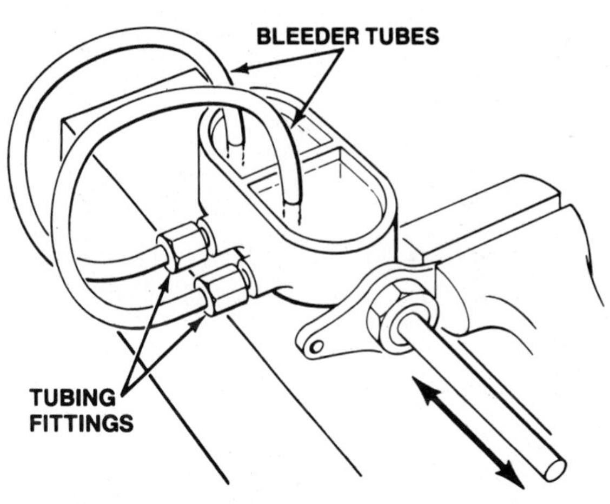

Figure 3-11. Bench bleeding tubes route fluid from the outlet ports back into the reservoir.

Basic Bench Bleeding

Basic bench bleeding is the most common method used to bleed master cylinders. This procedure requires a vise and two short lengths of special tubing. One end of each piece of tubing has a fitting that threads into a master cylinder outlet port; the other end twists around to route fluid from the port back into the reservoir. To bench bleed a master cylinder in this manner:

1. Clamp the master cylinder in a vise by its mounting flange so you have access to the fluid outlet ports, figure 3-9. Do not clamp on the master cylinder body or the bore may distort. If the vise has sharp teeth, use smooth-faced jaw protectors to prevent damage to the cylinder mounting flange.
2. Connect the lengths of special tubing to the outlet ports of the master cylinder, figure 3-11,

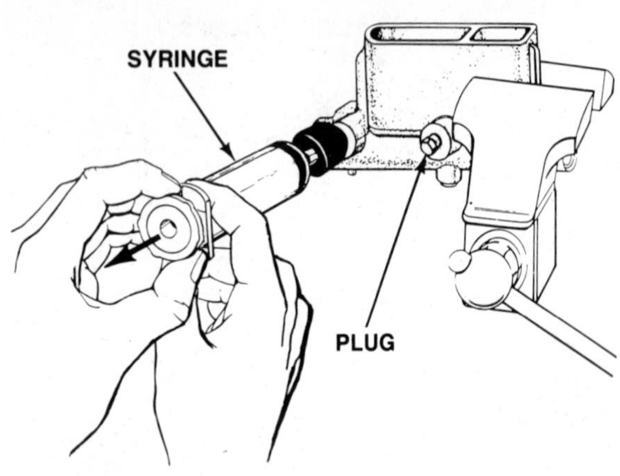

Figure 3-12. Extend the syringe plunger to draw brake fluid and air bubbles out of the master cylinder.

and route the tubing ends back into the fluid reservoir.

3. Fill the reservoir with new brake fluid of the proper type and grade.

4. Use the master cylinder push rod, or other round-ended rod, to slowly stroke the master cylinder pistons in and out. Be sure to bottom both pistons on the forward stroke, and allow both pistons to fully return on the back stroke. Continue until air bubbles no longer emerge from the tubing ends.

5. Stroke the pistons with short sharp movements of approximately ½ inch (13 mm) until bubbles no longer emerge from the tubing ends or the compensating and replenishing ports in the reservoir.

Reverse Bench Bleeding

Reverse bleeding does the same job as manual or basic bench bleeding. However, instead of stroking the pistons to move fluid through the master cylinder and force out the air, a special large plastic syringe is used to create a low-pressure area in the cylinder bore. This allows atmospheric pressure to force fluid through the cylinder and into the syringe. The same fluid (minus any trapped air) is then forced back through the cylinder in the opposite direction. To reverse bleed a master cylinder:

1. Clamp the master cylinder in a vise by its mounting flange so you have access to the fluid outlet ports, figure 3-9. Do not clamp on the master cylinder body or the bore may distort. If the vise has sharp teeth, use smooth-faced jaw protectors to prevent damage to the cylinder mounting flange.

2. Seal the fluid outlet ports using the plugs supplied with the syringe, then fill the reservoir half full of new brake fluid.

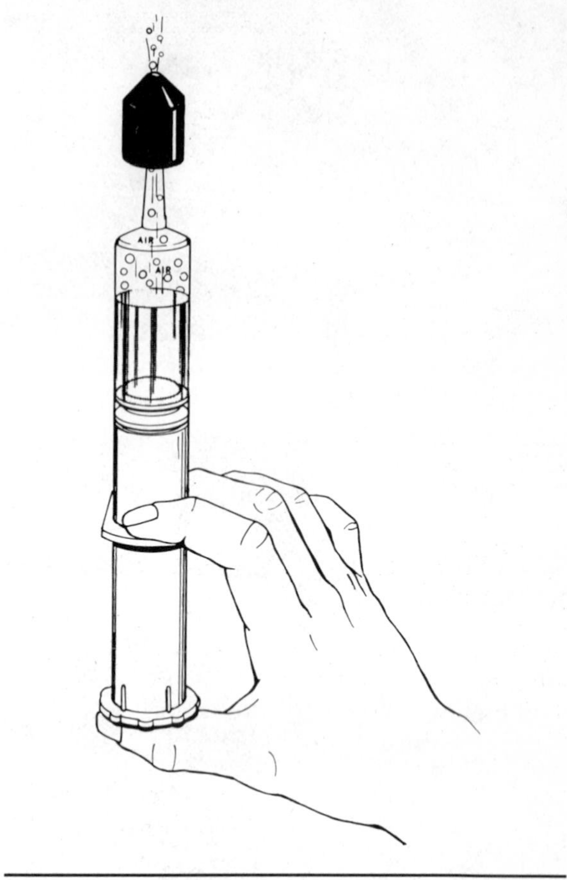

Figure 3-13. With the syringe in an upright position, slowly depress the plunger to expel any air.

3. Fully depress the plunger of the syringe

4. Remove the plug from one of the fluid outlets and press the syringe firmly against the opening. Extend the plunger until the syringe is half full of fluid, figure 3-12.

5. Remove the syringe and temporarily block the fluid outlet. Point the tip of the syringe upward and slowly depress the plunger until all of the air is removed, figure 3-13.

6. Press the syringe to the same outlet and force the fluid back into the master cylinder by depressing the plunger, figure 3-14. Air bubbles will exit the compensating port and rise to the surface of the reservoir.

7. Repeat steps 3 through 6 until the air bubbles stop, then plug the fluid outlet. Repeat these same steps at the other fluid outlet or outlets.

8. Use the master cylinder push rod, or other round-ended rod, to stroke the pistons with short sharp movements of approximately ½ inch (13 mm) until bubbles no longer emerge from the compensating and replenishing ports in the reservoir.

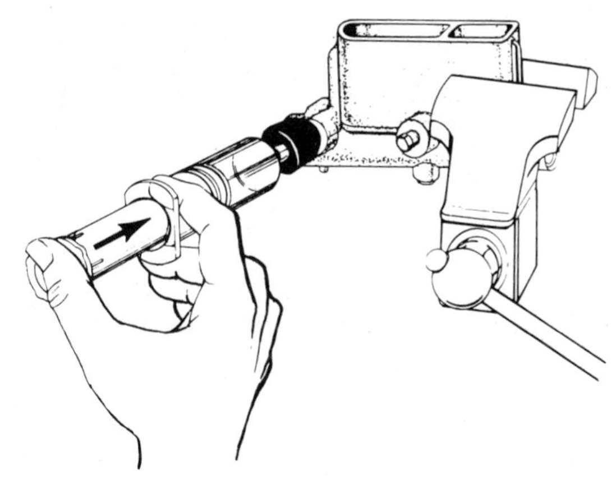

Figure 3-14. Force the air-free brake fluid back into the master cylinder to dislodge trapped air bubbles.

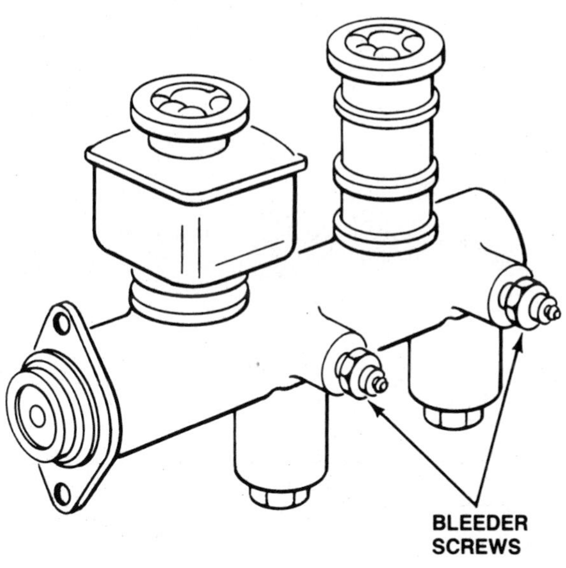

Figure 3-15. A master cylinder with bleeder screws.

MASTER CYLINDER ON-CAR BLEEDING

Bleeding the master cylinder on the car purges any last bit of trapped air in it. On-car bleeding is sometimes necessary to remove small pockets of air that can develop when the plugs are removed from the master cylinder fluid outlets and the hydraulic lines are connected. On-car bleeding is done after the master cylinder is bench bled, and prior to bleeding the rest of the hydraulic system.

The pressure used to move fluid through the master cylinder can come from either a pressure bleeder or force applied to the brake pedal. In both cases, brake fluid under pressure can spray from the fittings where the cylinder is bled. Always wear safety glasses or other eye protection when bleeding a master cylinder on the car. In addition, use a fender cover to prevent brake fluid spray from getting on the vehicle finish.

On-Car Bleeding with Bleeder Screws

Some master cylinders that mount in a level position are designed to be bled on the car and have bleeder screws for this purpose, figure 3-15. Bleeding a master cylinder with bleeder screws requires a bleeder wrench of the proper size, a length of clear plastic hose with an inside diameter small enough to fit snugly over the bleeder screws, and a jar partially filled with clean brake fluid. To bleed such a master cylinder:

1. Discharge the vacuum or hydraulic power booster (if equipped) by pumping the brake pedal with the ignition OFF until the pedal feels hard.
2. Fill the master cylinder reservoir with new brake fluid, and make sure it remains at least half full throughout the bleeding procedure.
3. Connect the plastic hose to the forward bleeder screw and submerge the end of the hose in the jar of brake fluid.
4. Loosen the bleeder screw approximately one-half turn, and have your assistant slowly depress the brake pedal and hold it to the floor. Air bubbles leaving the bleeder screw will be visible in the hose to the jar.
5. Tighten the bleeder screw, then have your assistant slowly release the brake pedal.
6. Repeat steps 4 and 5 until no more air emerges from the bleeder screw. Transfer the plastic hose to the rear bleeder screw and repeat these same steps until the master cylinder is completely bled.

On-Car Bleeding without Bleeder Screws

Most master cylinders do not have bleeder screws. Once they are bench bled and installed on the car, any remaining air is bled from the brake line fittings at the fluid outlets. The majority of master cylinders have a single fitting for each hydraulic circuit, however, some newer designs have dual outlets for each circuit. On these newer designs, both fittings of a circuit must be opened when bleeding the master cylinder on the car. This job requires a flare-nut wrench to loosen the tubing fittings, and a basin or rag to catch the fluid pumped

Figure 3-16. Bleeding a master cylinder on the car.

out of the system. To bleed a master cylinder without bleeder screws:

1. Discharge the vacuum or hydraulic power booster (if equipped) by pumping the brake pedal with the ignition OFF until the pedal feels hard.
2. Fill the master cylinder reservoir with new brake fluid and make sure it remains at least half full throughout the bleeding procedure.
3. Loosen the forward brake tube fitting approximately one-half turn, and have your assistant slowly depress the brake pedal and hold it to the floor. Brake fluid and air bubbles will flow out from between the tubing and tubing fitting, figure 3-16. Hold a basin or rag under the fitting to catch the fluid released.
4. Tighten the tubing fitting, then have your assistant slowly release the brake pedal.
5. Repeat steps 3 and 4 until no more air bubbles emerge from the fitting. Repeat these same steps at the rear tubing fitting until the master cylinder is completely bled.

BLEEDING THE WHEEL BRAKES

When bleeding the wheel brakes, the proper bleeding sequence must be followed. Generally, the wheel cylinder or caliper farthest from the master cylinder is bled first, followed by the next closest cylinder or caliper, and so forth. On cars with front-rear split brake systems, the rear brakes are bled first, then the front brakes. On cars with diagonal-split brake systems, one rear brake is bled first, then the opposite front brake, then the other rear brake, and finally the remaining front brake. Not all manufacturers follow these sequences, however. Chrysler recommends that both rear brakes be bled first regardless of the hydraulic system split. Figure 3-17 shows the proper bleeding sequences for most late-model cars and light trucks. The correct sequence for a specific model can also be found in the factory shop manual.

The many variations in brake hydraulic systems make some easier to bleed than others. In addition, certain brake systems respond better to one bleeding procedure than another. There are actually five methods that can be used to bleed the wheel brakes:

- Manual bleeding
- Gravity bleeding
- Vacuum bleeding
- Pressure bleeding
- Surge bleeding.

Manual Bleeding

Manual bleeding uses hydraulic pressure created by the master cylinder to pump fresh fluid through the brake system. In this procedure, one technician opens the bleeder screws while another applies the brake pedal; air, contamination, and old fluid is pushed out of the bleeder screws ahead of the new fluid.

It is extremely important when manually bleeding a brake system that the pedal be applied and released slowly and gently. Rapid pedal pumping can churn up the fluid and reduce the size of trapped air bubbles, making them more difficult to bleed from the system. There are, however, special situations where rapid pedal pumping is helpful in brake system bleeding. These are covered in the *Surge Bleeding* section later in the chapter.

Manual bleeding requires an assistant to apply and release the brake pedal, a bleeder screw wrench, approximately two feet of clear, plastic hose with an inside diameter small enough to fit snugly over the bleeder screws, and a clear jar partially filled with clean brake fluid. To manually bleed the brake system:

1. Discharge the vacuum or hydraulic power booster (if equipped) by pumping the brake pedal with the ignition OFF until the pedal feels hard.
2. Fill the master cylinder reservoir with new brake fluid and make sure it remains at least half full throughout the bleeding procedure.
3. Slip the plastic hose over the bleeder screw of the first wheel cylinder or caliper in the bleeding sequence, and submerge the end of the tube in the jar of brake fluid, figure 3-18.

Brake Bleeding Sequences

DOMESTIC MODELS

AMC/Jeep/Renault (1)

1978-87 All Jeep	RR-LR-RF-LF
1978-87 passenger cars	RR-LR-RF-LF
1983-87 Alliance, Encore;	
Manual bleed	RR-LF-LR-RF
Power bleed	RR-LR-RF-LF

Chrysler (1, 2)

1978-87 All	RR-LR-RF-LF

Ford (1, 3)

1981-87 All w/FWD	RR-LF-LR-RF
1978-87 All w/RWD	RR-LR-RF-LF

General Motors (1)

1978-87 w/standard split	RR-LR-RF-LF
1980-87 w/diagonal split	RR-LF-LR-RF
1978-87 Corvette only (4)	LR-RR-LF-RF

IMPORT MODELS

Chevrolet

1978-82 LUV Truck (5)	LF-RF-LR-RR
1985-87 Sprint	RR-LF-LR-RF
1985-87 Spectrum	LF-RR-RF-LR

Chrysler

1978-79 Arrow (8)	RR-RF-LF
1978-83 Challenger, Sapporo, w/rear drums	RR-RF-LF
1978-83 Challenger, Sapporo, w/rear discs	RR-LR-RF-LF
1984-87 Conquest	RR-LR-MOD (12)-RF-LF
1978 Colt, Champ	RR-LR-RF-LF
1979-82 Champ, RWD Colt	LR-RF-RR-LF
1979-87 FWD Colt, Vista	LR-RF-RR-LF
1982-87 Ram 50;	
w/2WD	RR-LR-RF-LF
w/4WD	RR-LR-RF-LF-LSVP (9)

Ford

1978-82 Courier (3, 6)	RR-LR-RF-LF
1978-80 Fiesta (7, 10)	RF-LR-LF-RR

Honda

1978-87 All	LF-RR-RF-LR

Mazda

1978-87 All w/RWD	RR-LR-RF-LF
1981-87 All w/FWD	RF-LR-RR-LF

Nissan/Datsun

1978-87 All w/RWD	LR-RR-RF-LF
1978-87 All w/FWD	LR-RF-RR-LF

Subaru

1978-84 All	RR-LR-RF-LF
1985-87	LF-RR-RF-LR

Toyota

1978-87 Cars	RR-LR-RF-LF
1978-82 Trucks	RR-LR-RF-LF
1983-87 Trucks, Vans	RR-LR-RF-LF-LSPV (9)

Volkswagen

1978-87 All	RR-LR-RF-LF

Volvo (11)

1978-87 240, 260 DL, GL	LF-RF-RR-LR
1983-87 740, 760	LF-RF-LR-RR

Notes

1. When pressure bleeding, metering valve must be locked open.
2. On models with a vertically mounted combination valve, loosen a front bleed screw before bleeding. On all models depress brake pedal to re-center warning light switch.
3. Re-center pressure differential valve by depressing brake pedal with ignition switch turned on.
4. When bleeding rear brakes, raise the front of the car to prevent air from being trapped in the calipers.
5. System must be manually bled with the engine running.
6. When bleeding rear brakes, bleed lower wheel cylinder first.
7. Catch bottle must be at least one foot (300 mm) above the bleed screw.
8. The left rear cylinder has no bleeder, both rear cylinders bleed through the right side.
9. Load sensing, proportioning, and bypass valve.
10. Apply hand brake while bleeding rear wheel cylinders.
11. Calipers with multiple bleed screws must be bled from all ports simultaneously.
12. When equipped with anti-lock brakes, bleed the modulator before the front brakes.

Figure 3-17. Manufacturer recommended brake bleeding sequences.

4. Loosen the bleeder screw approximately one-half turn, and have your assistant slowly depress the brake pedal and hold it to the floor. Air bubbles leaving the bleeder screw will be visible in the hose to the jar.
5. Tighten the bleeder screw, then have your assistant slowly release the brake pedal.
6. Repeat steps 4 and 5 until no more air bubbles emerge from the bleeder.
7. Transfer the plastic hose to the bleeder screw of the next wheel cylinder or caliper in the bleeding sequence, and repeat steps 4 through 6. Continue around the car in the specified order until the brakes at all four wheels have been bled.

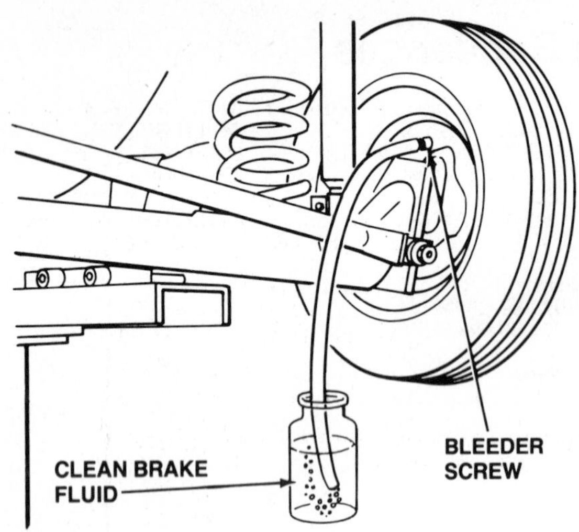

Figure 3-18. When brakes are bled through clear tubing into a jar, air bubbles are easy to spot and fluid spills are prevented.

Vacuum Bleeding

Vacuum bleeding uses a special suction pump that attaches to the bleeder screw. The pump creates a low-pressure area at the bleeder screw which allows atmospheric pressure to force brake fluid through the system when the bleeder screw is opened. Vacuum bleeding requires only one technician, however, it can only be used on wheel cylinders with cup expanders, and brake calipers with O-ring seals. On wheel cylinders without cup expanders, and calipers with stroking seals, the low pressure will pull the lips of the seals away from the cylinder or caliper bore, allowing air to enter. To vacuum bleed a brake system:

1. Fill the master cylinder reservoir with new brake fluid and make sure it remains at least half full throughout the bleeding procedure.
2. Attach the plastic tube from the vacuum bleeder to the bleeder screw of the first wheel cylinder or caliper in the bleeding sequence, figure 3-19. If necessary, use one of the adapters provided with the vacuum bleeding kit.
3. Squeeze the pump handle 10 to 15 times to create a partial vacuum in the catch bottle.
4. Loosen the bleeder screw approximately one-half turn. Brake fluid and air bubbles will flow into the bottle. When the fluid flow stops, tighten the bleeder screw.
5. Repeat steps 3 and 4 until no more air bubbles emerge from the bleeder.
6. Transfer the vacuum bleeder to the bleeder screw of the next wheel cylinder or caliper in the bleeding sequence, and repeat steps 3 and 4. Continue around the car in the specified order until the brakes at all four wheels have been bled.

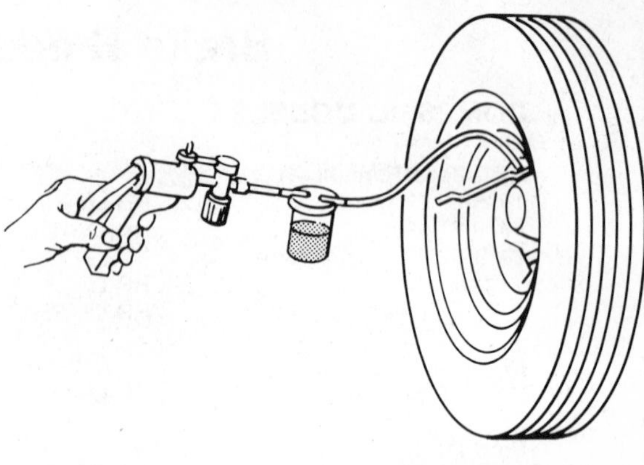

Figure 3-19. A vacuum bleeder attached to a bleeder screw.

Gravity Bleeding

Gravity bleeding uses the force of gravity to pull new brake fluid through the hydraulic system. In this process, the bleeder screws at all four wheels are opened at the same time, and the system is allowed to drain naturally until the fluid coming from the bleeders is free of air.

Gravity bleeding is a slow process that can take an hour or more. In addition, this procedure cannot be used on brake systems with residual pressure check valves because the valves restrict the fluid flow. The advantage of gravity bleeding is that it can be done by a single technician, who is freed to attend to other jobs while the brakes bleed. When other bleeding procedures fail, gravity bleeding can sometimes be effective on brake systems that agitate the fluid and trap small pockets of air.

Gravity bleeding requires a bleeder wrench, four lengths of plastic hose which fit snugly over the bleeder screws, and four jars to catch the dripping fluid. Unless a plastic hose is used to "start a siphon" at each bleeder screw, it is possible that air may enter the system rather than be bled from it. This can occur because the total opening area of the four bleeder screws is somewhat larger than that of the two compensating ports through which the fluid must enter the system. To gravity bleed the brake system:

1. Fill the master cylinder reservoir with new brake fluid. During the bleeding process, check the fluid level periodically to ensure that the reservoir remains at least half full.

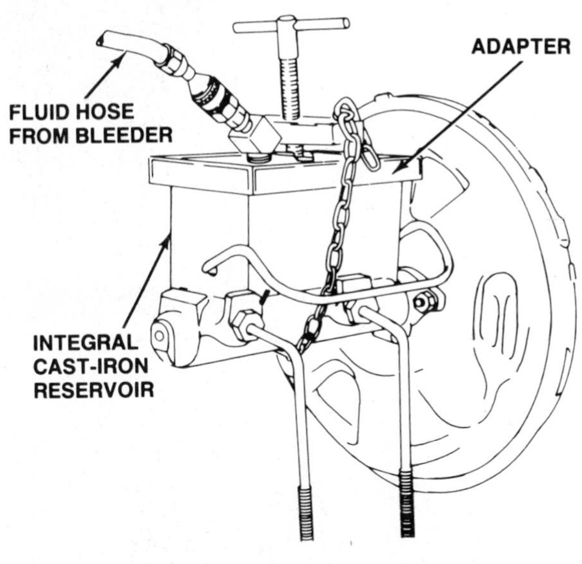

Figure 3-20. A pressure bleeder adapter for an integral master cylinder reservoir.

2. Attach a length of plastic tubing to each bleeder screw, and place the ends of the tubes in jars to catch the drainage.
3. Open each bleeder screw approximately one full turn and make sure that fluid begins to drain. Allow the system drain until the fluid flowing from the bleeder screws is free of air bubbles.
4. Close the bleeder screws and top up the fluid level in the master cylinder reservoir.

Pressure Bleeding

Pressure bleeding, sometimes called power bleeding, is the fastest, most efficient, and most common method used to bleed the brake hydraulic system. In this process, a pressure bleeder attached to the master cylinder forces brake fluid through the system under pressure to purge any trapped air. Pressure bleeding is done by one person; once the hydraulic system is pressurized, the technician simply opens the bleeder screws in the prescribed order and allows fluid to flow until it is free of air bubbles.

The tools required for pressure bleeding include a plastic hose and fluid catch jar as used in manual bleeding, as well as a pressure bleeder, a source of air pressure to charge the bleeder, and an adapter to attach the pressure bleeder to the master cylinder fluid reservoir. Cast-metal cylinders with integral reservoirs commonly use a flat, plate-type adaptor that seals against the same surface as the reservoir cover, figure 3-20. Some plastic master cylinder reservoirs also use plate-type adapters, but

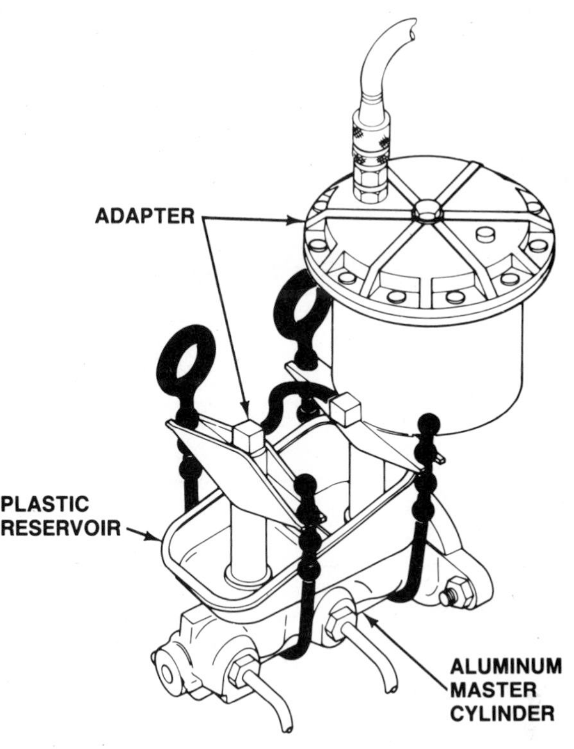

Figure 3-21. A pressure bleeder adapter for a plastic master cylinder reservoir.

others require adapters that seal against the bottom of the reservoir, figure 3-21. Pressure bleeder manufacturers offer many adapters to fit specific applications.

Metering valve override tools

In addition to the tools described above, a metering valve override tool is required when pressure bleeding the front brakes of certain vehicles. The override tool is used to deactivate the metering valve because the operating pressure of power bleeders is within the range where the metering valve blocks fluid flow to the front brakes. Metering valves that require an override tool have a stem or button on one end that is either pushed in or pulled out to hold the valve open. The override tool performs this service.

To install the override tool used on General Motors vehicles, figure 3-22, loosen the combination valve mounting bolt and slip the slot in the tool under the bolt head. Push the end of the tool toward the valve body until it depresses the valve plunger, then tighten the mounting bolt to hold the tool in place.

Some full-size Ford cars manufactured after 1979 have a metering valve with a stem that must be pushed in to bleed the front brakes, much like the General Motors design. Ford

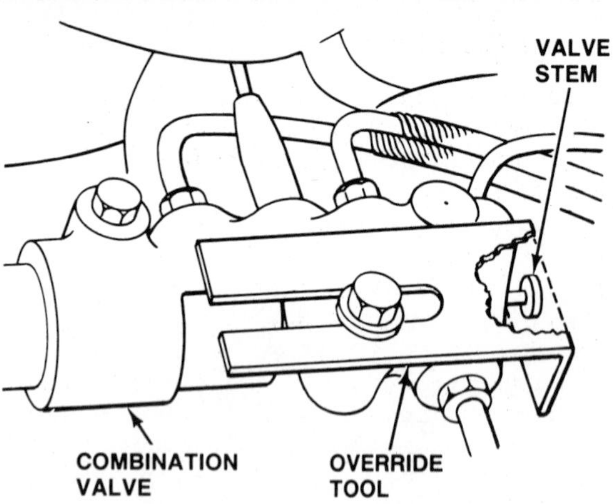

Figure 3-22. Installation of a General Motors metering valve override tool.

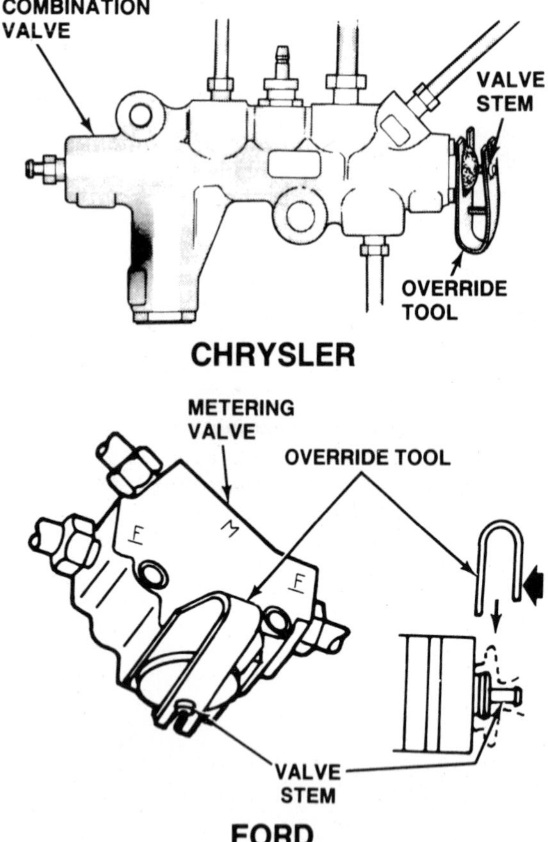

Figure 3-23. Installation of Chrysler and Ford metering valve override tools.

Figure 3-24. Compressed air is used to charge the pressure bleeder.

does not offer a special tool for this purpose, however. Have an assistant manually override the valve when you bleed the front brakes on one of these vehicles.

To install the override tool used on Chrysler and early Ford vehicles, figure 3-23, slip one fork of the tool under the rubber boot, and the other fork under the valve stem head. The spring tension of the tool holds the valve open, but allows the valve stem to move slightly when the system is pressurized. If the valve is held rigidly open, internal damage will result.

Pressure Bleeding Procedure

Just as in manual bleeding, it is important to follow the proper sequence when pressure bleeding a brake system. Some manufacturers recommend one sequence for manual bleeding, and another for pressure bleeding. To pressure bleed a brake system:

1. If it has not already been done, consult the equipment manufacturer's instructions and fill the pressure bleeder with the proper type of brake fluid.
2. Make sure the bleeder is properly sealed and the fluid supply valve is closed, then use compressed air to pressurize the bleeder until approximately 30 psi (207 kPa) is indicated on the bleeder gauge, figure 3-24.
3. If the vehicle is equipped with a metering valve, override it with the appropriate tool.
4. Clean the top of the master cylinder, then remove the master cylinder cover and clean

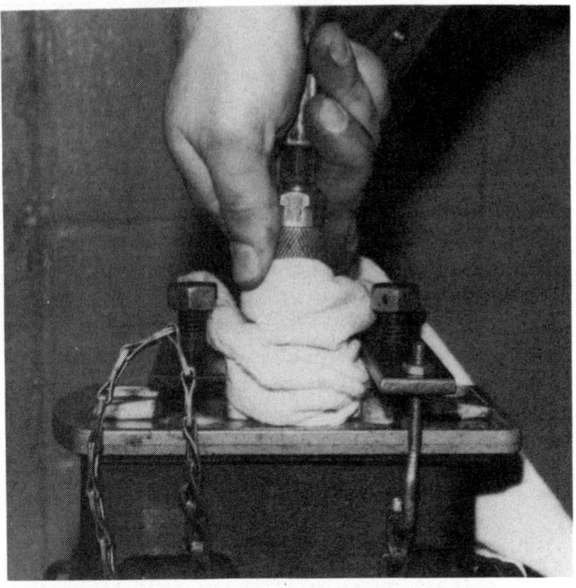

Figure 3-25. When you disconnect the pressure bleeder hose fitting, wrap it in a shop towel to catch any fluid spray.

around the gasket surface. Be careful not to allow any dirt to fall into the reservoir.

5. Fill the reservoir about half full with new brake fluid, then install the proper pressure bleeder adapter on the master cylinder.
6. Connect the pressure bleeder fluid supply hose to the adaptor making sure the hose fitting is securely engaged.
7. Open the fluid supply valve on the pressure bleeder to allow pressurized brake fluid to enter the system. Check carefully for fluid leaks that can damage the vehicle finish.
8. Slip the plastic hose over the bleeder screw of the first wheel cylinder or caliper to be bled, and submerge the end of the tube in the jar of brake fluid.
9. Open the bleeder screw approximately one-half turn, and let the fluid run until air bubbles no longer emerge from the tube. Close the bleeder screw.
10. Transfer the plastic hose to the bleeder screw of the next wheel cylinder or caliper in the bleeding sequence, and repeat steps 9 and 10. Continue around the car in the specified order until the brakes at all four wheels have been bled.
11. Remove the metering valve override tool.
12. Close the fluid supply valve on the pressure bleeder.
13. Wrap the end of the fluid supply hose in a shop towel, and disconnect it from the master cylinder adapter, figure 3-25. Do not spill any brake fluid on the vehicle finish.
14. Remove the master cylinder adapter, adjust the fluid level to the full point, and install the fluid reservoir cover.

Surge Bleeding

Surge bleeding is a supplemental bleeding method used to help remove air bubbles that resist other bleeding processes. In surge bleeding the brake pedal is pumped rapidly to create turbulence in the hydraulic system. This agitation helps dislodge air bubbles that cling to the pores of rough castings, or become trapped at high points or turns in the brake lines. Surge bleeding is *not* recommended for systems filled with silicone DOT 5 brake fluids. These fluids tend to trap tiny air bubbles that are very difficult to bleed from the hydraulic system; the added agitation of surge bleeding only makes the problem worse.

Surge bleeding requires an assistant to pump the brake pedal, a bleeder screw wrench, approximately two feet of clear, plastic hose with an inside diameter small enough to fit snugly over the bleeder screw, and a jar partially filled with clean brake fluid. To surge bleed a brake system:

1. Slip the plastic hose over the bleeder screw of the wheel cylinder or caliper to be bled and submerge the end of the tube in the jar of brake fluid.
2. Open the bleeder screw approximately one-half turn.
3. With the bleeder *open,* have your assistant rapidly pump the brake pedal several times. Air bubbles should come out with the brake fluid.
4. While your assistant holds the brake pedal to the floor, close the bleeder screw.
5. Repeat steps 2 through 4 at each bleeder screw in the recommended order.
6. Re-bleed the system using one of the four other methods described above.

BLEEDING THE TEVES ABS SYSTEM

The Teves anti-lock brake system used on some Ford and General Motors vehicles has a front/rear split hydraulic system. As explained in the *Classroom Manual,* the front brakes are applied by a dual master cylinder, while the rear brakes are applied directly by pressure in the boost chamber of the hydraulic power booster. The front wheels can be bled as in a normal brake system, manually or with a pressure bleeder. The rear brakes, however, may be bled using a pressure bleeder or the accumulator pressure of the booster.

Only Ford actually recommends bleeding the rear brakes with a pressure bleeder. The procedure is essentially the same as described earlier with the following special considerations:

1. Charge the pressure bleeder to 35 psi (240 kPa).

2. When bleeding the brakes, open one rear bleeder screw for 10 seconds, close it, then open the bleeder screw at the opposite wheel for the same length of time. Alternate in this manner until the fluid from both bleeders is free of air bubbles.
3. When bleeding is completed, check the fluid level in the reservoir as described earlier in the chapter.

To bleed the rear brakes of a Teves anti-lock brake system using accumulator pressure requires a bleeder screw wrench, approximately two feet of clear, plastic hose with an inside diameter small enough to fit snugly over the bleeder screws, and a clear jar partially filled with clean brake fluid. To bleed the rear brakes using accumulator pressure:

1. Turn the ignition switch ON. Depress the brake pedal repeatedly until the electro-hydraulic pump motor starts. The pump motor will stop when the accumulator is charged.
2. Slip the plastic hose over the bleeder screw of the right rear caliper and submerge the end of the tube in the jar of brake fluid.
3. With the ignition switch still in the ON position, have an assistant *lightly* depress and hold the brake pedal.
4. Alternately open the rear bleeder screws for 10 seconds at a time until the fluid coming from the bleeders is free of air bubbles. Open the bleeder screws slowly and carefully because the accumulator provides much higher pressure than is available from a pressure bleeder.
5. When bleeding is completed, check the fluid level in the reservoir as described earlier in the chapter.

FLUID CHANGING

In addition to removing air, brake systems need to be bled in order to clean out old and/or contaminated fluid. This process is called fluid changing or flushing. The hygroscopic (water attracting) nature of most brake fluid makes it necessary to flush the brake system periodically because moisture in the fluid drastically lowers its boiling point, water also causes rust and corrosion of system components that can lead to brake failure. To avoid these problems, brake fluid should be changed yearly in humid climates and every other year in dry climates.

Brake fluid changing or flushing is done by bleeding the system until all of the old fluid is purged from the system. Because fluid changing requires the ability to flush out contamination and move a great deal of fluid, the best method to use is pressure bleeding. Other acceptable choices are manual bleeding and vacuum bleeding. Gravity bleeding is *not* recommended for fluid changing because it is slow, and without significant pressure behind the fluid flow, there is no guarantee that contamination will be completely flushed from the system.

To change a vehicle's brake fluid, first use a syringe to remove all of the old brake fluid from the master cylinder reservoir. Fill the reservoir with new fluid, then follow the procedures outlined earlier in the chapter for the chosen method of bleeding. Continue to bleed at each wheel until the fluid that emerges from the bleeder screw is free of any discoloration and contamination.

RECENTERING PRESSURE DIFFERENTIAL SWITCHES

After the brake system has been bled, the pressure differential switch may have to be recentered to turn off the warning light. The reason is that opening a bleeder screw creates a pressure differential between the circuits of the hydraulic system. The switch "sees" this difference as a leak or partial system failure. The piston inside the switch body then moves to one side and completes the warning light circuit.

If the warning light remains on after bleeding, make sure the parking brake is off and the master cylinder reservoir is full of fluid before you assume that the pressure differential switch needs to be recentered. The monitor switches for these parts often share the same warning light as the pressure differential switch, and may be the reason the light is on! There are three types of pressure differential switches; each type requires a different recentering procedure.

Single-Piston Switch without Centering Springs

Some 1967-70 Ford Motor Company cars and light trucks have a single-piston pressure differential switch without centering springs, figure 3-26A. A similar switch is used on some imports. Recentering this type of switch requires an assistant and a bleeder wrench. To recenter the switch:

1. Turn the ignition switch ON. The warning light will come on because the piston has raised the switch plunger to complete the circuit.
2. Determine if the brake hydraulic system of car you are working on is split diagonally or front to rear, then open a bleeder screw in the circuit of the system opposite that which was last bled.

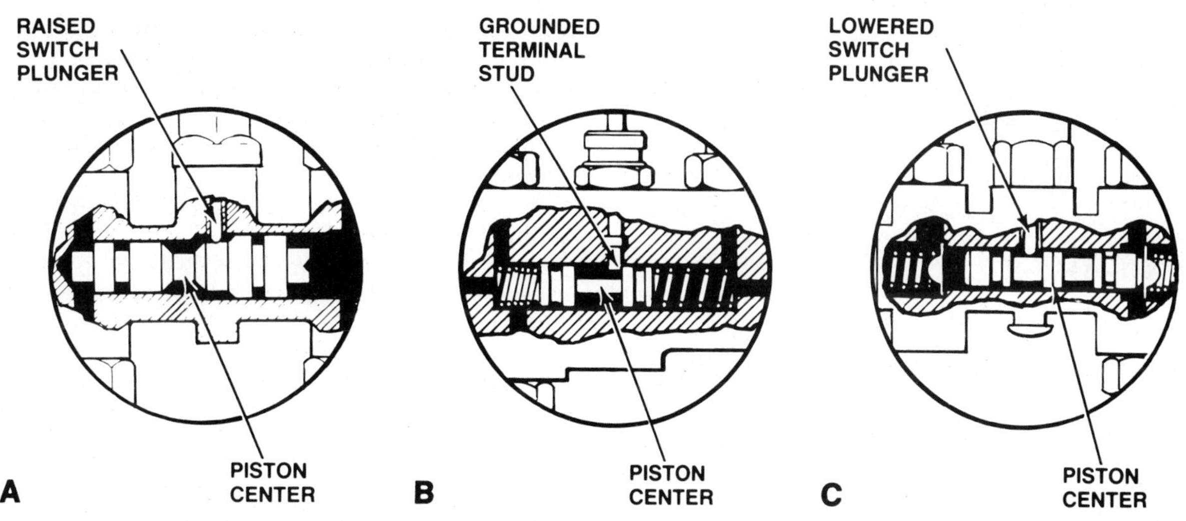

Figure 3-26. Pressure differential switch pistons must be centered to turn off the brake system warning light.

3. Have your assistant slowly push on the brake pedal until the warning light goes out.
4. Close the bleeder screw.

Very little pedal pressure or movement is required for this procedure. If too much is applied, the piston will overcenter and re-illuminate the warning light. Should that happen, open a bleeder screw in the opposite side of the hydraulic system. It is often easier and more accurate to depress the brake pedal by hand when centering the switch in this manner.

Single-Piston Switch with Centering Springs

Most domestic and import cars manufactured after 1970 have a single-piston pressure differential switch equipped with centering springs, figure 3-26B. This type of switch illuminates the warning light only when the brakes are applied and a pressure difference exists between the two circuits of the brake system. The switch recenters itself automatically when the brakes are released, unless the piston sticks in position against the terminal stud. If the warning light remains lit after the brake system has been repaired:

1. Turn the ignition switch ON. The warning light will illuminate.
2. Apply the brake pedal with moderate-to-hard force. Hydraulic pressure will free the stuck piston and the centering springs will position it properly in the bore. The warning light will then go out.

If the warning light remains lit, and the parking brake or fluid level switches are not at fault, you can attempt to free the piston by using the centering procedure given above for switches without centering springs. If this does not free the piston, replace the pressure differential switch.

Two-Piston Switch with Centering Springs

American Motors cars built before 1971, and some imports, have a two-piston pressure differential switch with centering springs, figure 3-26C. This type of switch locks the warning light on until it is recentered. To do this:

1. Turn the ignition switch ON. The warning light will illuminate because piston movement has allowed the switch plunger to extend and complete the circuit.
2. Unscrew the switch plunger assembly from the switch body, and apply the brake pedal with medium to hard force. The centering springs will then recenter the piston.
3. Reinstall the switch plunger assembly in the switch body.

If the warning light remains lit, and the parking brake or fluid level switches are not at fault, you can attempt to free the piston by using the centering procedure given above for switches without centering springs. If this does not free the piston, replace the pressure differential switch.

Chapter

4

Brake Line, Hydraulic Valve, and Electrical Component Service

Brake lines distribute pressurized brake fluid from the master cylinder to the hydraulic components at the wheel friction assemblies. Any brake line failure will cause at least one of the two brake system hydraulic circuits to fail as well. The first portion of this chapter covers the inspection, replacement, and fabrication of brake lines.

Hydraulic control valves in the brake system ensure that brake fluid reaches the wheel friction assemblies at the correct time and/or pressure. A faulty valve can cause premature or excessively hard brake application that can lock the wheels and lead to an accident. The middle portion of this chapter describes the testing and replacement of hydraulic valves.

Brake system electrical components consist primarily of the stoplights and system warning light, along with the switches that control them. The final portion of this chapter details the testing, adjustment, and replacement of these parts.

BRAKE LINE INSPECTION

Brake lines, both reinforced rubber hoses and double-wall steel tubing, are relatively trouble free. However, each can suffer wear and deterioration from a number of factors including ozone in the air, contaminants in the brake fluid, debris thrown up by the tires, abrasion from contact with suspension and steering components, and stress created in an accident. Damaged brake lines can fail at any moment and cause a serious accident.

Because of the wear and deterioration that takes place, brake lines should be checked periodically for damage and leaks. Most manufacturers recommend that brake hoses be inspected twice a year, or any time the brakes are serviced. Steel brake tubing should be inspected yearly, or any time the brakes are serviced. Brake lines that are not in perfect condition must be replaced.

Brake Hose Inspection

Rubber brake hoses suffer damage both from within and without. On the inside of the hose, contaminated brake fluid attacks the rubber lining and causes it to swell, restricting fluid flow. Hoses softened by contaminated fluid also expand under pressure, causing a spongy brake pedal and increased pedal travel. And if a brake caliper is allowed to hang by the hose, the inner hose lining may be torn, creating a lip that blocks the flow of fluid to or from the friction assembly.

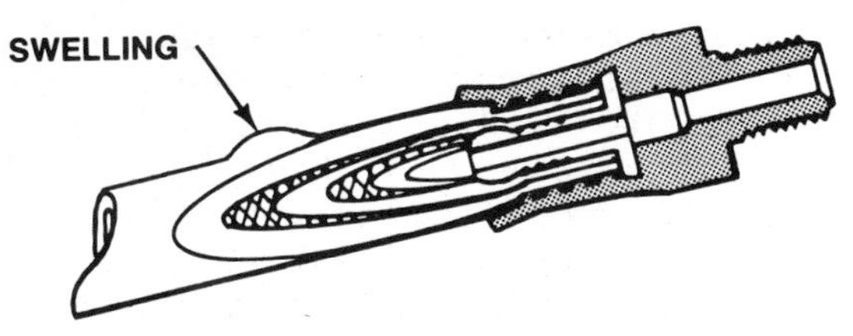

Figure 4-1. A swollen or blistered hose usually has an internal leak.

Figure 4-2. Fluid leaks often appear as a dark stain on the hose.

On the outside of the hose, ozone in the atmosphere attacks the rubber in the hose outer casing, causing it to age and become brittle. Cracks then develop that eventually become leaks. The other main external wear problem is abrasion from contact with suspension and steering parts, as well as debris thrown up by the tires. An improperly routed hose or a hose that is too long and rubs against part of the suspension will quickly be worn through, creating a leak. Non-standard parts such as wide wheels and tires can also rub against the hoses and cause abrasions. A hose that is too short can be stretched or pulled apart during steering or suspension movement.

Most cars have three or four brake hoses. All cars have a hose at each front wheel, and cars with independent rear suspension also have a hose at each rear wheel. Cars with a live rear axle generally have a single rear hose between the axle and the chassis. Visually inspect the brake hoses for swelling, blisters, leaks, stains, cracks, and abrasions. Swelling and blisters, figure 4-1, are signs of internal fluid leakage that has penetrated to the outer hose covering. Obvious leaks, or stains from leaks, figure 4-2,

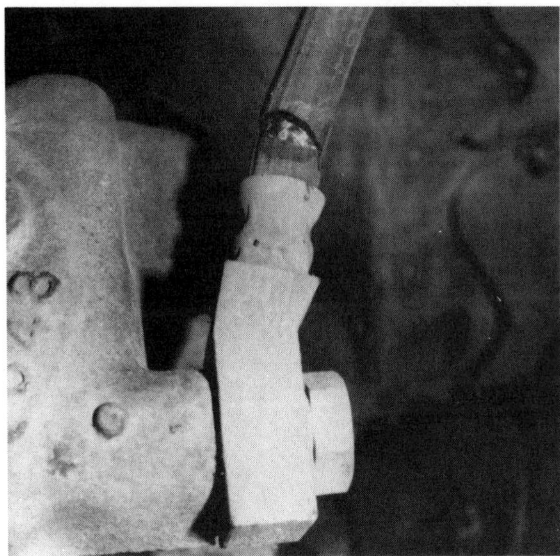

Figure 4-3. Cracked and weathered brake hoses are weakened and must be replaced.

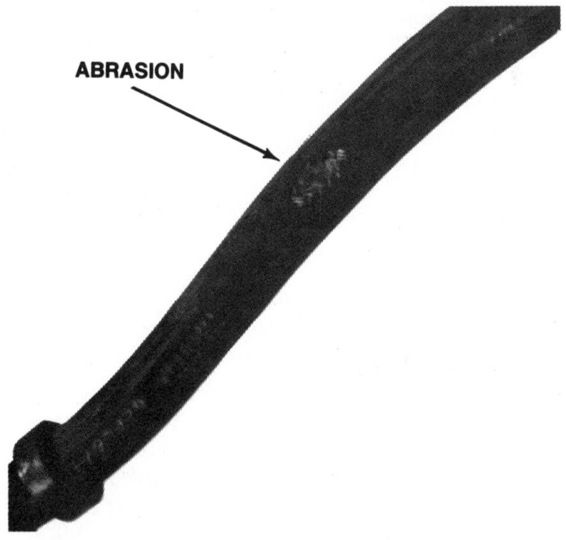

Figure 4-4. Abrasions result when the hose is improperly routed, or the wrong length for the application.

may appear on the surface of the hose and around the fittings on the hose ends. Cracks can appear anywhere on the hose, figure 4-3, as can signs of abrasion, figure 4-4. Finally, check the hose mounting hardware and locating brackets for damage and tightness.

Brake Tubing Inspection

Steel brake tubing is naturally more durable than rubber brake hoses, and usually requires less service. Tubing does, however, suffer from rust and corrosion, cracking, and impact dam-

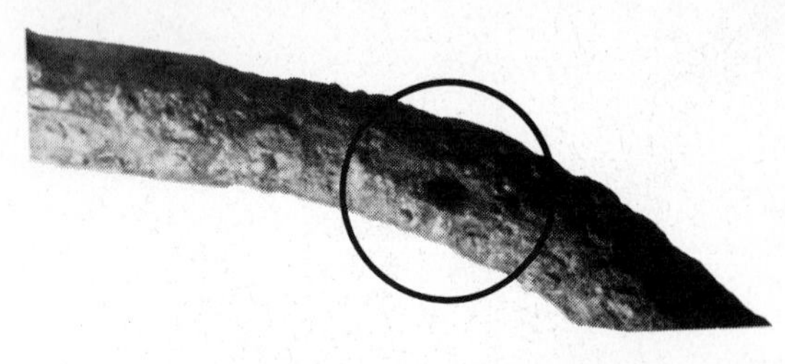

Figure 4-5. Rust and corrosion on brake tubing are most likely along frame rails or around mounting clips.

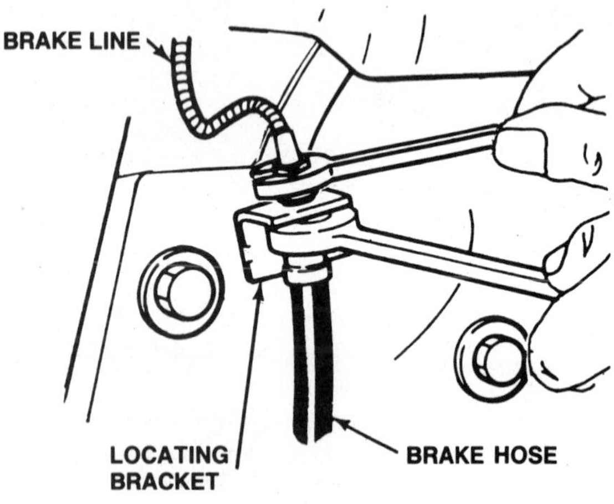

Figure 4-6. Disconnect the female end of a brake hose first.

age. Rust and corrosion, figure 4-5, are caused by water trapped around the brake lines, and the damage is accelerated by salt and other chemicals used to melt ice and snow. Once the problem becomes severe, rusted and corroded tubing will rupture from hydraulic pressure.

Another problem with steel tubing is that it fractures from vibration if not mounted securely. Loose tubing can also snag on objects the vehicle passes over and be ripped away, and unsecured tubing may come into contact with the exhaust system, heating the brake fluid and potentially causing vapor lock.

Impact damage of the brake tubing, mainly dented and restricted tubing, can be caused by debris thrown up by the tires. However, the most common causes of this type of damage are careless attachment of towing hooks, and improper floor jack and hoist positioning.

To inspect the brake lines, begin where they attach to the master cylinder and follow them along their paths to the wheels. Look for rust and corrosion along the frame rails, at the mounting clips, or any place where water, dirt, and road salt accumulate. Make sure the tubing is properly fastened to the chassis; cracks

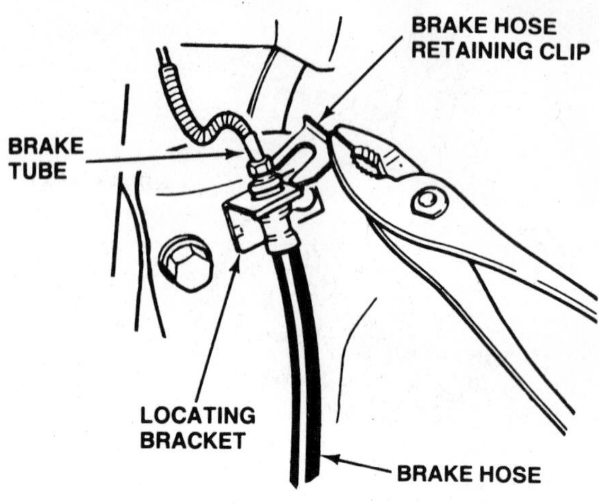

Figure 4-7. A retaining clip secures the brake hose in the locating bracket.

caused by vibration will be indicated by obvious fluid leaks. Physical damage is most likely in the areas directly behind the wheels, or where the tubing crosses below an axle or frame member.

BRAKE HOSE REPLACEMENT

Brake hoses that fail an inspection are replaced with a new hose. Replacement is the only common form of service because the tools necessary to fabricate new brake hoses are not commonly available in the field.

Hose Removal

Many front brake hoses, and rear hoses on cars with independent rear suspensions, have a male fitting on one end and a female fitting on the other; these fittings are swaged or crimped onto the hose and do not turn. With this type of hose, the female end must be disconnected first. Other brake hoses have a banjo fitting in place of the male hose end fitting. With this type of hose it does not matter which end is disconnected first. To remove a brake hose:

1. Clean any dirt from around the fittings at the ends of the hose to prevent it from entering the hydraulic system.
2. Locate the fitting where the steel brake tubing attaches to the female end of the hose. Loosen the tubing nut with a flare-nut wrench, figure 4-6, and unscrew it from the hose. Although the hose fitting is usually prevented from rotating by its locating bracket, it is best to always use a second wrench on the fitting to make sure it does not twist and cause damage to the hose or bracket.
3. Remove the hose retaining clip with a pair

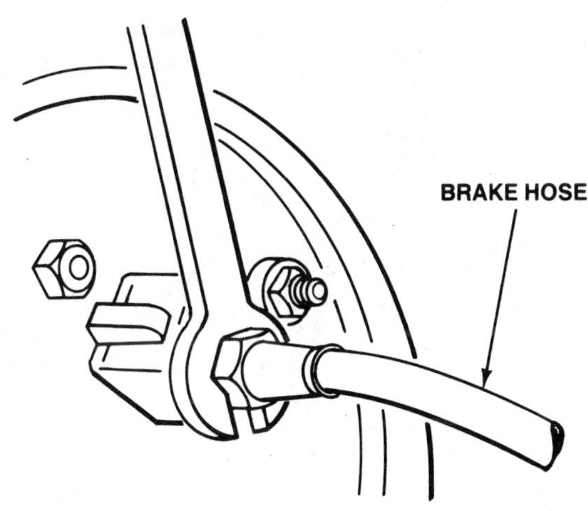

Figure 4-8. Disconnect male hose fittings with a flare-nut wrench.

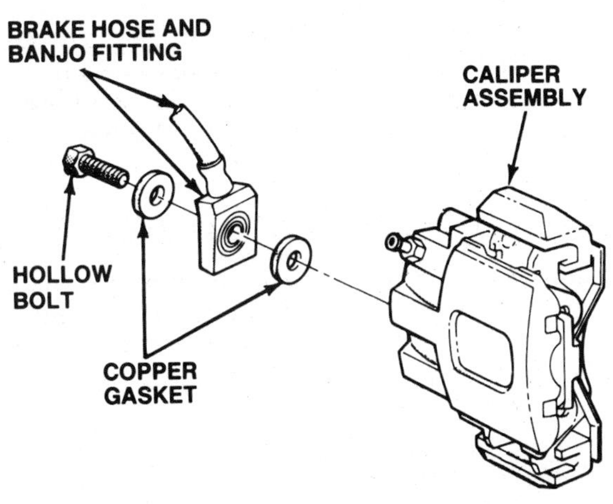

Figure 4-9. Remove the hollow bolt to disconnect a hose with a banjo fitting.

of pliers, figure 4-7, and separate the hose from the locating bracket.
4. Detach the hose from any other locating devices.
5. Use a flare-nut wrench to disconnect the other end of the hose from the wheel cylinder or caliper. On hoses with male fittings, figure 4-8, simply loosen the fitting and unscrew the hose from the caliper. On hoses with banjo fittings, figure 4-9, remove the hollow bolt from the center of the fitting, taking care not to lose the sealing rings on each side of the fitting.

Some rear hoses on cars with live axles have two female ends; in this case, either end can be disconnected first. Use one flare-nut wrench to prevent the hose from twisting, and a second to loosen the tubing nut on the steel brake line. Repeat for the other end of the hose.

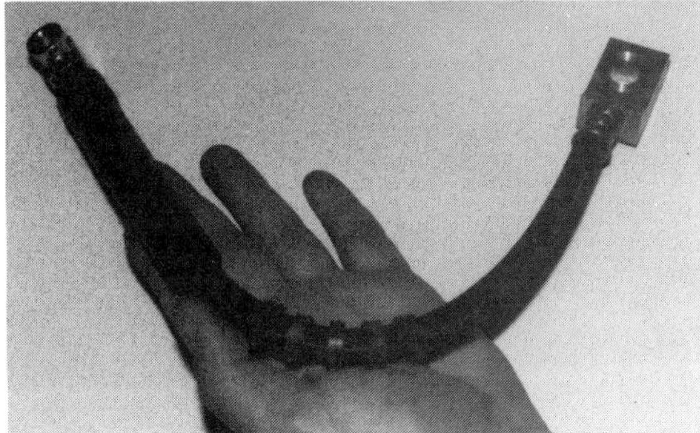

Figure 4-10. Observe the colored stripe or raised rib on the outer casing to prevent twisting a hose when installing it.

Hose Installation

Before you install a new front brake hose, make sure it is the proper part for that side of the car; left- and right-side hoses are not always interchangeable. The new hose must also be the proper length, and when the original equipment hose has special armoring and bracket fittings, the replacement part must have them as well.

During installation, carefully route the new hose in the original location. Make sure the hose maintains a distance of at least ¾ inch (20 mm) from all steering and suspension parts throughout the full range of their movement so there is no danger of the hose being chafed. Never route brake hoses near exhaust systems where heat will harm the rubber casing or increase brake fluid temperatures. If copper sealing gaskets are used at either of the hose fittings, use only new parts when installing the hose. Copper gaskets take a set when they are first used, and reusing an old one may cause a leak. To install a brake hose:

1. If the hose has a male end, thread it into the wheel cylinder or brake caliper; use a new copper gasket where required. Tighten the fitting to the proper torque.
2. If the hose has a banjo fitting, place the hollow bolt through the fitting and thread it into the caliper; use a new copper gasket on each side of the fitting. If the fitting mounts to the caliper in only one position, tighten it to the proper torque. If the position can vary, leave the fitting slightly loose at this time.
3. Route the hose through any locating devices.
4. Place the open end of the hose through the locating bracket, and start the threads of the steel brake line tubing nut into the female fitting.
5. Observe the colored stripe or raised rib on the hose outer casing, figure 4-10; position the hose so that it is not twisted.

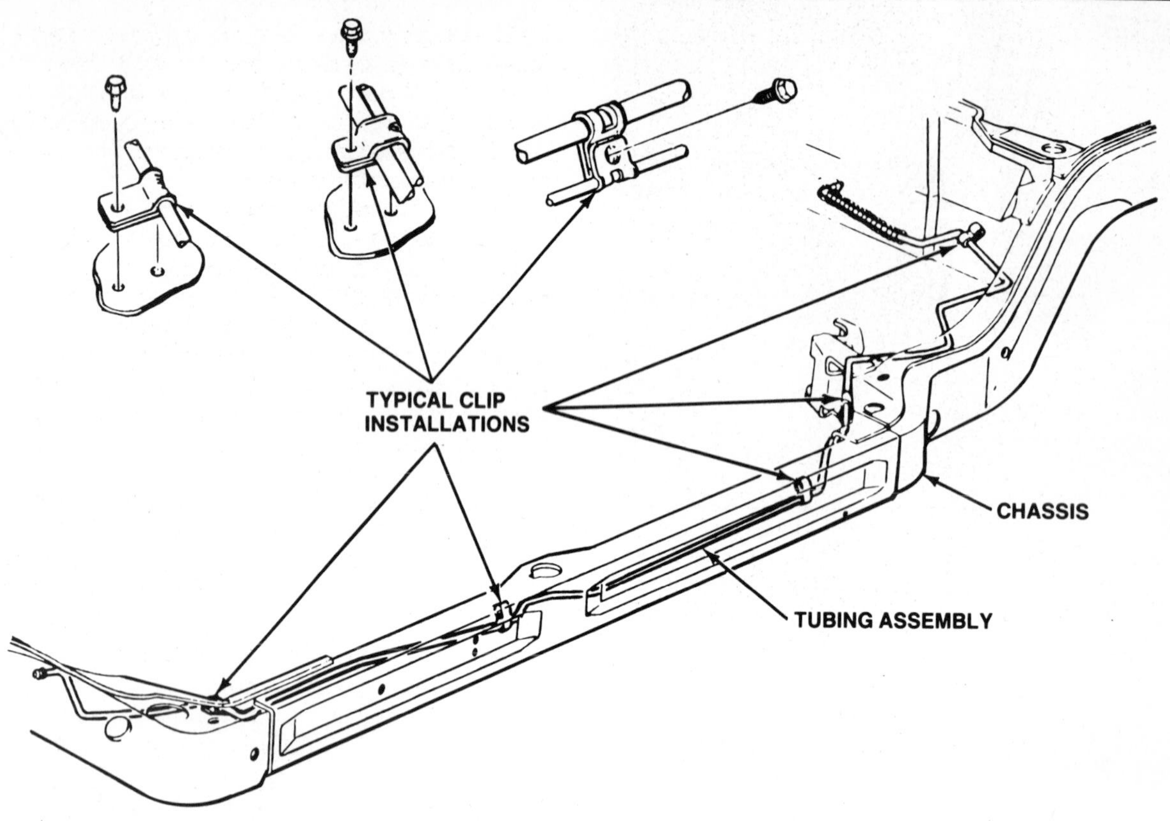

Figure 4-11. Brake tubing is attached to the chassis with special mounting clips.

6. Hold the female hose fitting in position with a flare-nut wrench, and tighten the tubing nut with a second wrench.
7. Again check that the hose is not twisted and that it does not contact any suspension or steering parts. If necessary, loosen the tubing nut, reposition the hose, and retighten the connection.
8. Install the retaining clip to secure the tubing and brake hose to the locating bracket.
9. If the banjo fitting was left loose in Step 2, position it so that the hose clears all obstacles, then tighten the fitting to the proper torque.

When a hose is equipped with two female ends, one end is assembled and tightened, then the other. Start the tubing nut on the steel brake line into one end of the hose, and use a pair of flare-nut wrenches to tighten the connection. Observe the colored stripe or raised rib on the outer casing, and position the hose so that it is not twisted. Start the tubing nut of the other steel brake line into the open end of the hose, then use one flare-nut wrench to prevent the hose fitting from turning, and a second wrench to tighten the tubing nut into the fitting.

BRAKE TUBING REPLACEMENT

Brake tubing service consists of two operations, replacement and fabrication. In some cases, custom replacement tubing that is pre-bent and flared can be purchased for a specific application; this type of tubing is generally available only from the vehicle manufacturer. Most of the time, the technician in the field fabricates replacement tubing from raw tubing stock.

Brake tubing is held in place by clips bolted to the chassis at various points along the line, figure 4-11. To replace a section of tubing, you need flare-nut wrenches to loosen the fittings at both ends of the line, and suitable sockets, wrenches, or screwdrivers to remove the retaining clip bolts or screws. To remove a section of brake tubing:

1. Clean any dirt from around the tubing fittings.
2. Disconnect the fittings at each end of the tubing with a flare-nut wrench. Use a second wrench where two brake lines connect to prevent twisting the tubing or brake hoses.
3. Unbolt or unscrew the mounting clips from the chassis, and detach the brake line.

Figure 4-12. A special tubing cutter is required to accurately cut brake tubing.

4. Remove the mounting clips for use on the new replacement tubing. If the new section of tubing must be fabricated, save the old tubing to use as a bending guide.

To install a section of brake tubing:

1. Attach the mounting clips to the new tubing.
2. Position the tubing in place, and loosely install the retaining clip bolts or screws.
3. Connect the fittings at both ends of the tubing, and tighten them with the appropriate flare-nut wrenches.
4. Tighten the retaining clip bolts or screws.

BRAKE TUBING FABRICATION

As mentioned earlier, most sections of brake tubing replaced in the field are fabricated from straight tubing stock. To minimize the chance of leaks and ensure the strongest possible repair, a single length of tubing should be used to connect two hydraulic components wherever possible. Sometimes, when a brake line is extra long or routed through a hard-to-reach area, the only practical repair is to splice two lengths of tubing together. This should only be done using connectors that seal with flare fittings; *never* use spherical sleeve compression fittings to splice brake tubing.

Brake lines should be fabricated only from special double-walled steel tubing designed for brake system service. *Never* use copper or aluminum tubing for brake lines because these materials do not have sufficient burst strength or resistance to corrosion. The replacement tubing should also be armored if the original equipment tube had this feature.

Sections of straight tubing stock come in a variety of lengths, and the ends are usually fitted with tubing nuts and formed into SAE flares. Occasionally, a piece of replacement tubing will be exactly the right length for the job, and its fittings and flares will match those of the original tubing. In these cases, the only fabrication needed is to bend the new tubing to the appropriate shape. More often, however, you will have to select a length of tubing slightly longer than required, then remove one or both fittings to adjust the length of the tubing, install a different type of tubing nut, or form a different type of flare. As a result, brake tubing fabrication consists of three operations, those used to cut, bend, and flare the tubing.

Tubing Cutting

Brake tubing must be cut when a damaged section is removed from the vehicle, or when a piece of replacement tubing is trimmed to the proper length. When tubing is cut on the car, the cut is usually made in a location that allows good access. This makes the job easier, and also simplifies the flaring operation that will be required later. When cutting a length of new replacement tubing, measure the old piece as accurately as possible, then add approximately one-eight inch (3 mm) for each flare that needs to be formed.

Brake tubing should never be cut with a hacksaw; the uneven pressure the blade exerts will distort the tubing shape and leave a ragged edge that cannot be properly flared. Metal chips and flashing are also likely to be left in the line where they can later come loose and cause problems in the hydraulic system. Always cut tubing with one of the many special tools made for this purpose. To cut a piece brake tubing:

1. Determine the exact length of replacement tubing needed, and mark the straight tubing stock at that point.
2. Position the tubing cutter so that the cutting wheel is directly over the mark, figure 4-12. Tighten the knob on the cutter until the cutting wheel makes firm, but not tight, contact with the tubing.
3. Rotate the cutter around the tube while slowly tightening the knob to maintain cutting wheel pressure against the tubing. Continue until the tube is severed.

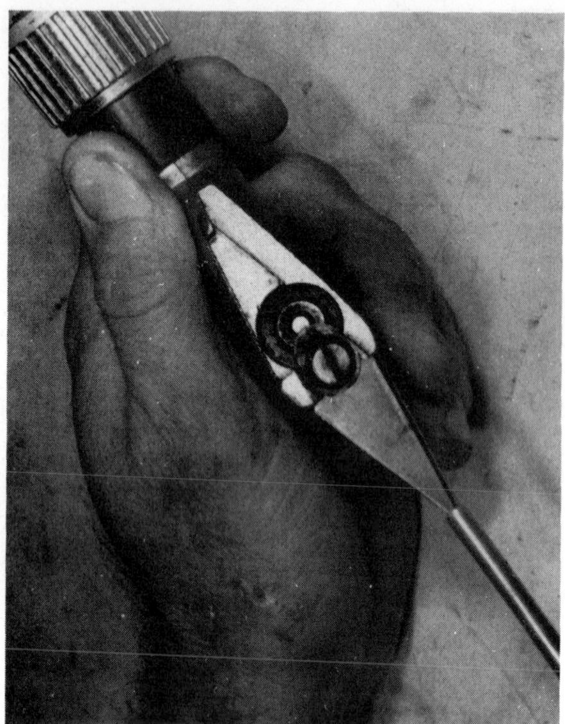

Figure 4-13. Ream the cut end of the tubing with a de-burring tool before flaring.

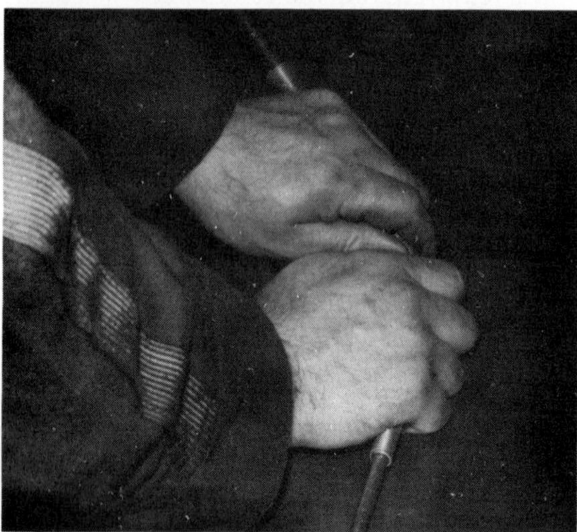

Figure 4-14. Forming brake tubing with a spring-type bending tool.

4. Use the de-burring tool on the cutter, or a similar separate tool, to ream the inner bore of the tube and remove any metal flashing, figure 4-13.
5. Blow through the tubing with compressed air to remove any loose metal particles.

Tubing Bending

Once a section of replacement tubing has been cut to the proper length, it can be bent to the proper shape. Unless it is specifically designed to be bent by hand, brake tubing must be bent with the proper bending tool or it will be distorted and kink. A kink creates a weak spot, and the tubing may eventually crack at that location. Kinks also restrict the flow of brake fluid through the tubing.

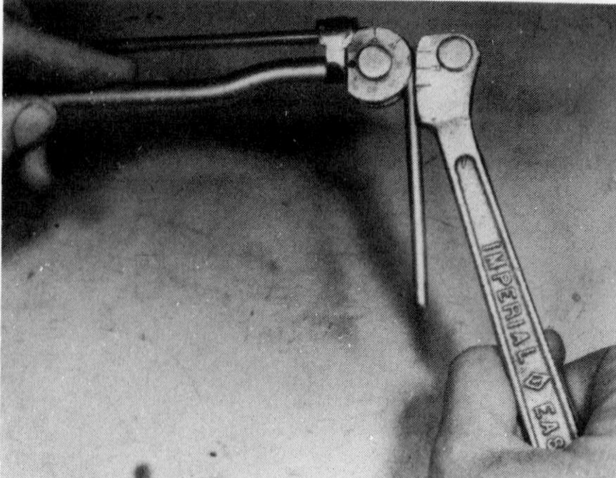

Figure 4-15. Forming brake tubing with a lever-type bending tool.

If a spring-type bender is used, the brake tubing must be bent while at least one end remains unflared. Slide the coiled spring over the unflared end of the tubing to the location where the bend must be made. Using the original section of tubing as a guide, carefully bend the replacement part to the proper shape, figure 4-14.

If a lever-type bender is used, the tubing can be bent either before or after flaring. Slip the tubing into the bender until the location where the bend is to be made is at the fulcrum point. Using the original section of tubing as a guide, carefully exert force on the bender levers to form the replacement tubing into the desired shape, figure 4-15.

Tubing Flaring

Once the cutting and bending operations are completed, flares must be formed on any unfinished tubing ends. Brake tubing is flared with either the SAE double flare or the ISO "bubble" flare, figure 4-16. These flares, and their fittings, are not interchangeable and cannot be mixed. Make sure you form the proper type of flare for the application.

A different special tool is required to form each of the type of flare. The procedures for using these tools are described below. Before forming any type of flare, make sure you install the tubing nut onto the tubing with the threaded portion facing the end of the tubing.

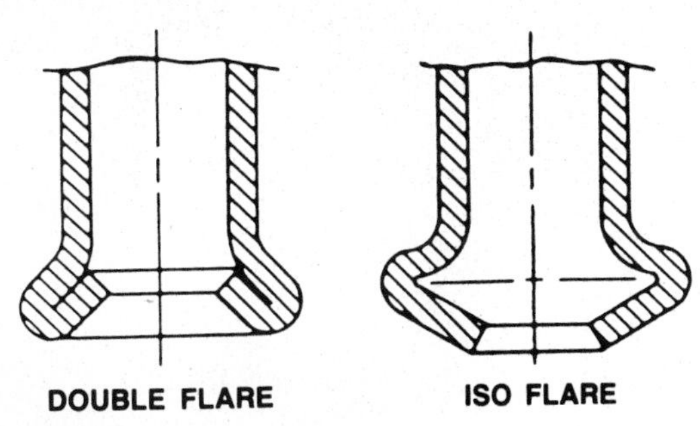

Figure 4-16. SAE and ISO flares are not interchangeable, and require different tools to form.

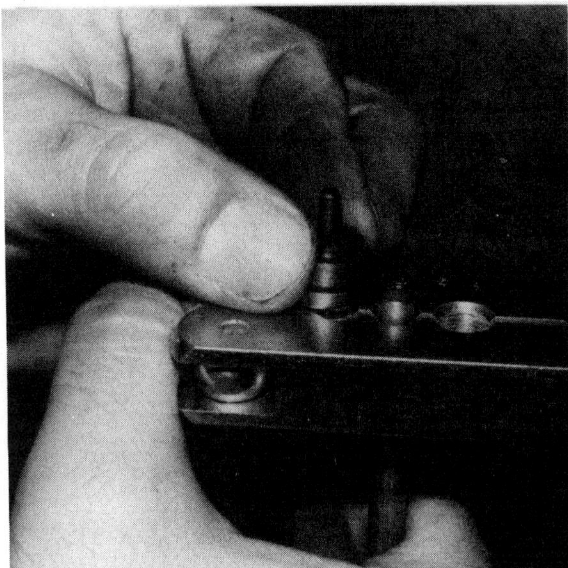

Figure 4-17. Install the tubing to the proper height in the tubing holder.

Figure 4-18. The first step in forming an SAE double flare.

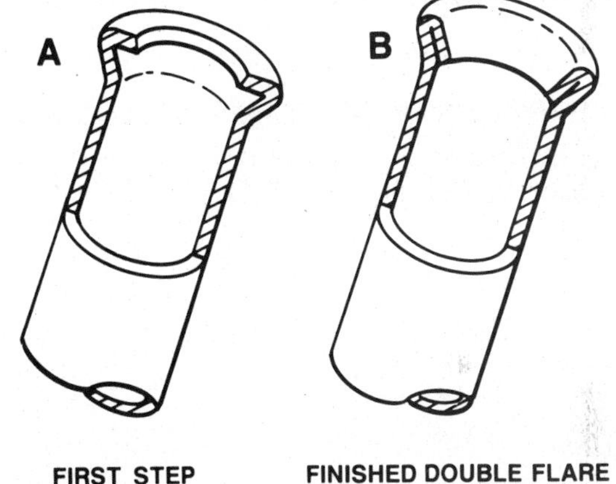

Figure 4-19. The SAE double flare is formed in two steps.

Forming SAE flares

The SAE double flare is formed in a two-step process:

1. Select the forming die from the flaring kit that corresponds to the diameter of the tubing to be flared.
2. Make sure the tubing nut is installed onto the tube, then clamp the tubing in the appropriate hole of the tubing holder. The end of the tube should protrude from the tapered side of the hole, and should extend above the holder surface a distance equal to the height of the first ring on the forming die, figure 4-17.
3. Install the forming die onto the tubing by placing the die pin into the tubing bore.
4. Position the press over the forming die on the tubing holder, and turn the threaded rod of the press until the cone-shaped anvil contacts the die. Continue to tighten the rod until the forming die bottoms on the tubing holder, figure 4-18.
5. Loosen the press and remove the forming die. The end of the tubing should appear as shown in figure 4-19A.
6. Re-tighten the threaded rod of the press to force the cone-shaped anvil into the tube, figure 4-20. Apply light to medium pressure until the lip formed in the first step is in full contact with the inner surface of the tubing.
7. Loosen and remove the press. The end of the tubing should appear as shown in figure 4-19B. If the flare does not have the correct shape, or any cracks are visible, you must cut off the end of the tubing and form a new flare.

Figure 4-20. The second step in forming an SAE double flare.

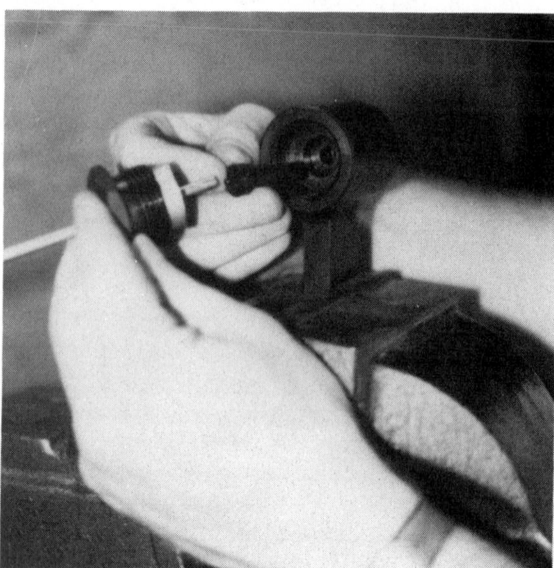

Figure 4-21. Install the tubing into the ISO flaring tool.

Forming ISO flares

The ISO flare is made in a one-step process using a different type of special tool. To form an ISO flare:

1. Clamp the flaring tool in a vise, and select the proper collet and forming mandrel for the size of tubing to be flared, figure 4-21.
2. Insert the forming mandrel into the tool body. While holding the mandrel in place with your finger, thread in the forcing screw until it begins to move the mandrel. Then, unthread the forcing screw one full turn.
3. Insert the tubing into the collet so that approximately ¾ inch (19 mm) extends from the end of the collet, and slide the clamping nut over the collet, figure 4-21.
4. Insert the assembly into the tool body.
5. Tighten the clamping nut. The nut must be very tight or the tubing may be pushed out during the flare forming process.
6. Tighten the forcing screw until it bottoms, figure 4-22.
7. The completed ISO flare should be shaped like that shown in figure 4-23. If the flare does not have the correct shape, or any cracks are visible, you must cut off the end of the tubing and form a new flare.

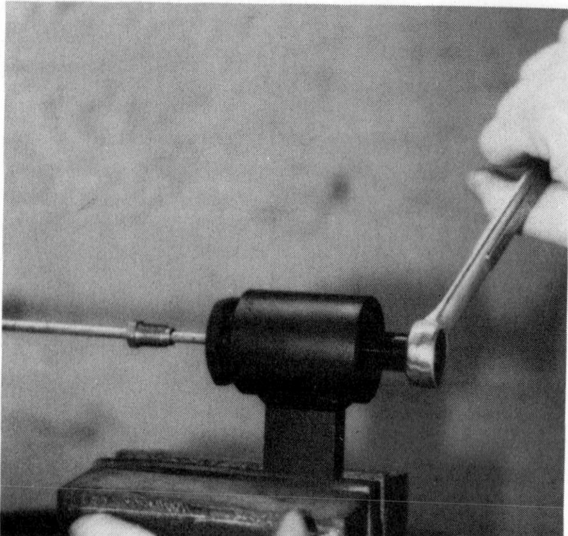

Figure 4-22. Tighten the forcing screw to create the ISO flare.

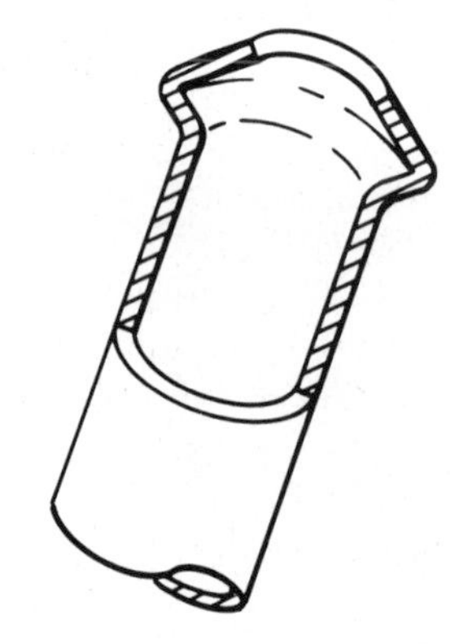

Figure 4-23. A properly formed ISO flare.

HYDRAULIC CONTROL VALVE SERVICE

The hydraulic control valves in the brake system, the metering valve and proportioning valve, are used to regulate the timing and

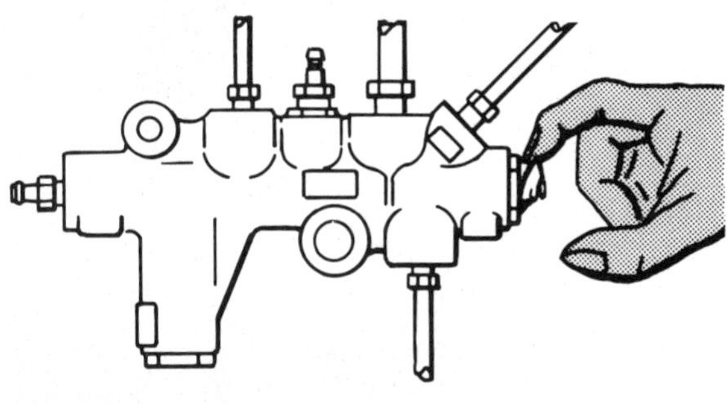

Figure 4-24. Visual inspection of a metering valve for leakage.

amount of hydraulic pressure supplied to the wheel friction assemblies. The control valves are usually located on or near the master cylinder, and can be separate valves or joined together into a combination valve. Height-sensing proportioning valves are located under the car near the rear axle.

Hydraulic control valve service includes procedures for testing and replacement. With the exception of certain height-sensing proportioning valves, neither metering valves nor proportioning valves can be repaired or adjusted; leaking or faulty valves must be replaced. If the defective valve is part of a combination valve, the entire valve assembly must be replaced.

Metering Valve Tests

The metering valve withholds hydraulic pressure from the front disc brakes until the shoes of the rear drum brakes have contacted the drums. A faulty metering valve will usually allow the front brakes to apply early and lock prematurely, especially on slick pavement. A metering valve can also seize shut and prevent any pressure from reaching the front brakes, although this is an uncommon type of failure.

If front brake locking causes you to suspect a faulty metering valve, give the valve a visual inspection. Check around the rubber boot at the valve stem for leakage, figure 4-24; a trace of moisture is normal, but an excessive amount indicates a defective valve. Have an assistant apply the brake pedal while you watch the valve stem. As pressure to the front brakes builds, the valve stem should move. If it does not, replace the valve. More accurate metering valve tests can be performed in two ways, with a pressure bleeder or with pressure gauges, as explained below.

Pressure bleeder metering valve test

A brake system pressure bleeder can be used to test a metering valve because the pressure pro-

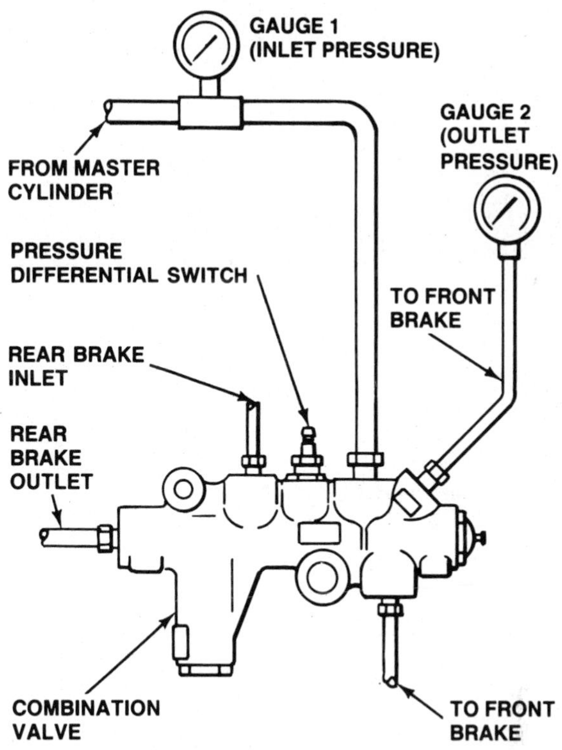

Figure 4-25. Proper gauge connections for testing a metering valve.

vided by the bleeder is sufficient to close the valve, but not high enough to reopen it. To test a metering valve with a pressure bleeder:

1. Connect the pressure bleeder to the master cylinder as described in Chapter 2, and charge the bleeder to 40 psi (275 kPa).
2. *Do not* disarm the metering valve with an override tool.
3. Pressurize the brake system with the power bleeder, and open a front brake bleeder screw. If fluid flows from the bleeder, the metering valve is not closing properly and must be replaced.

Pressure gauge metering valve test

The most precise method of testing metering valve operation is to use a pair of pressure gauges to measure the actual closing and opening points of the valve. This test requires an assistant to apply the brake pedal, two gauges that read from 0 to a minimum of 500 psi (0 to 3450 kPa), and the appropriate fittings to attach the gauges to the hydraulic system. To test a metering valve using gauges:

1. Tee Gauge 1 into the brake line from the master cylinder to the metering valve, figure 4-25, so the gauge does not block the flow of fluid to the metering valve.

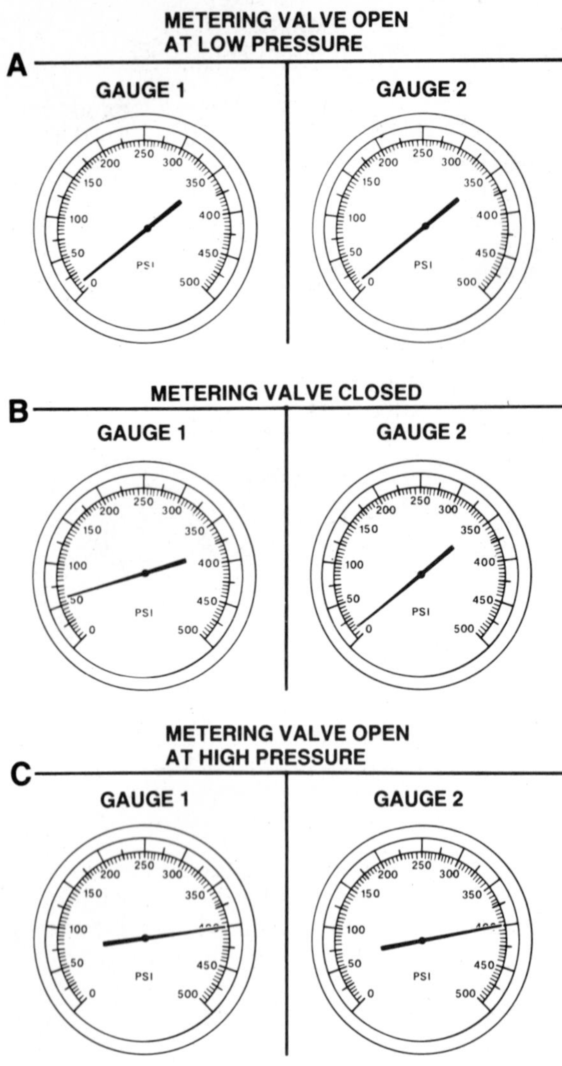

Figure 4-26. Typical gauge readings during a metering valve test.

2. Connect Gauge 2 to one of the metering valve outlets leading to the front brakes.
3. Have an assistant slowly apply the brake pedal while you observe both gauges.
4. If the metering valve is working properly, the gauge readings will rise at the same rate, figure 4-26A, until they reach the valve closing point. Depending on the vehicle this will be somewhere between 3 and 30 psi (20 and 210 kPa).
5. When the metering valve closes, the reading on Gauge 2 will remain constant; however, the pressure reading on Gauge 1 will continue to increase, figure 4-26B.
6. At a reading of approximately 75 to 300 psi (520 to 2070 kPa) on gauge 1, the metering valve will open. The reading on Gauge 2 will then

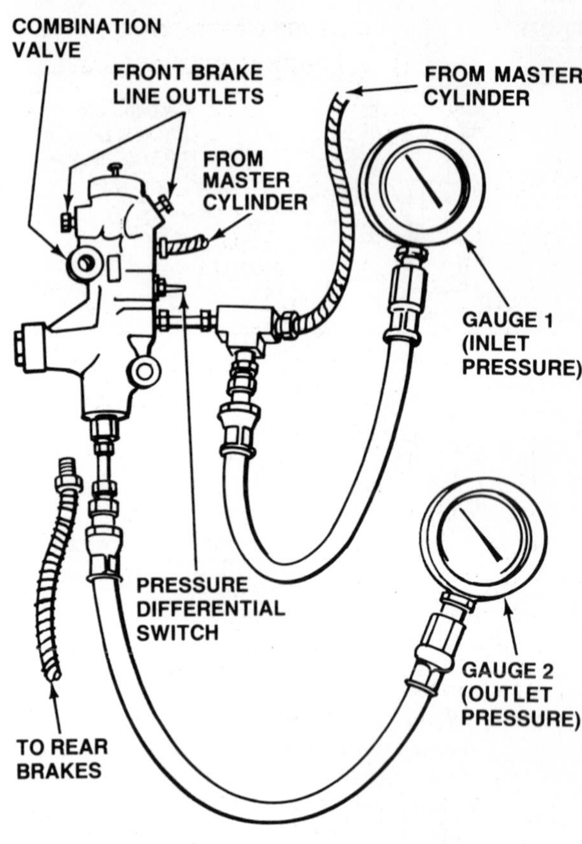

Figure 4-27. Proper gauge connections for testing the proportioning valve on a front/rear-split brake system.

increase to match that on Gauge 1, and from that point on, both gauges will have identical readings, figure 4-26C.

If the pressures indicated on the gauges do not follow the patterns described above, the metering valve is defective and must be replaced.

Proportioning Valve Tests

The proportioning valve slows the rate of pressure increase to the rear brakes once a certain pressure, called the split point, has been reached. A faulty proportioning valve usually allows rear brake pressure to increase too rapidly, causing the rear wheels to lock prematurely during hard stops or on slippery pavement. The proportioning valve can also fail in such a way that no pressure is allowed to the rear brakes, although this is an uncommon type of failure.

Proportioning valve operation can only be tested with pressure gauges. You will need an assistant to apply the brake pedal, two gauges that read from 0 to 1000 psi (0 to 6900 kPa), and

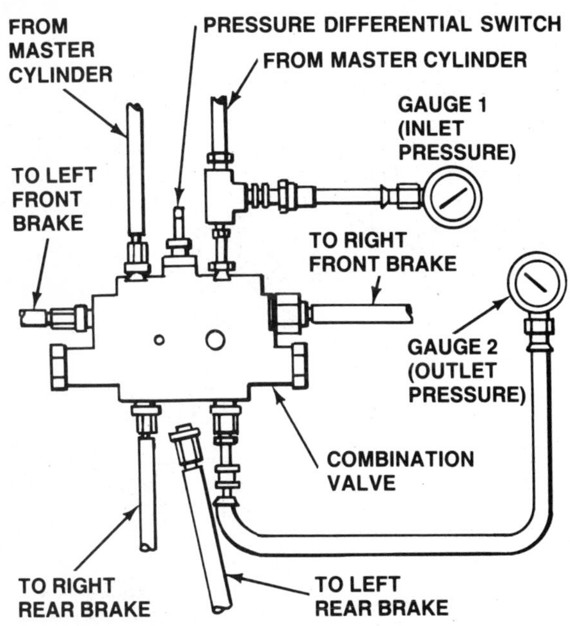

Figure 4-28. Proper gauge connections for testing the proportioning valve to the left rear brake on a diagonal-split brake system.

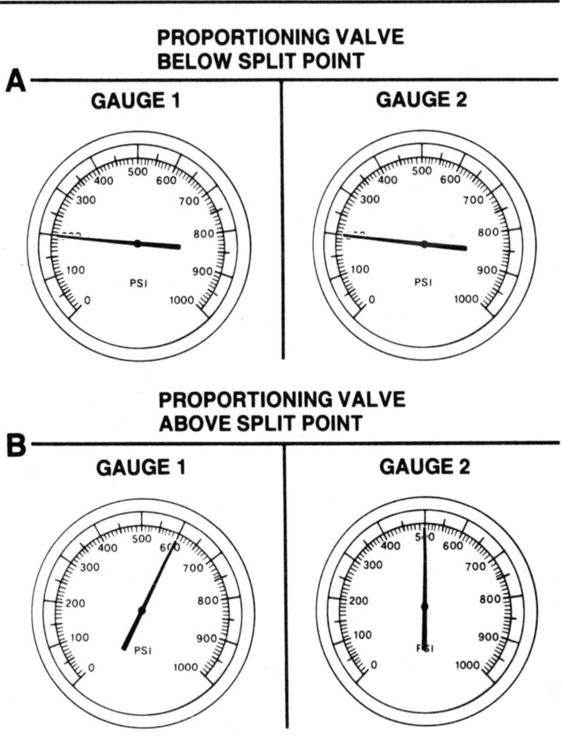

Figure 4-29. Typical gauge readings during a proportioning valve test.

the appropriate fittings to attach the gauges to the hydraulic system. You also need to know the split point of the proportioning valve on the particular make and model of vehicle being tested. Most split points are between 300 and 500 psi (2070 and 3450 kPa), but check the factory shop manual to be sure.

On cars where the dual braking system is split front to rear, only a single gauge hookup and test is required, figure 4-27. On cars with diagonal-split braking systems and dual proportioning valves, the tests will have to be performed twice, once for each half of the hydraulic system, figure 4-28. To test the proportioning valve:

1. Tee Gauge 1 into the brake line from the master cylinder to the proportioning valve so that the flow of brake fluid to the valve is not restricted.
2. Connect Gauge 2 to the rear brake outlet of the proportioning valve.
3. Have an assistant slowly apply the brake pedal and observe both gauges.
4. The readings on both gauges should rise at an identical rate until the split point pressure is reached, figure 4-29A.
5. After the split point is reached, the pressure reading on Gauge 2 will increase at a slower rate than the reading on Gauge 1, figure 4-29B. In other words, less pressure will be allowed to the rear brakes than is being produced by the master cylinder.

If the pressures indicated on the gauges do not follow the patterns described above, the proportioning valve is defective and must be replaced.

Metering and Proportioning Valve Replacement

Metering and proportioning valve replacement can involve changing a separate valve or a combination valve. The special brake tools required for valve replacement are flare-nut wrenches and brake bleeding equipment. To replace a valve:

1. Disconnect the brake lines from the valve with flare-nut wrenches.
2. If the valve contains a pressure differential switch, disconnect the vehicle wiring harness from the switch.
3. Unbolt the valve from its mounting, and remove it taking care not to spill brake fluid on the vehicle finish.
4. Bolt the new valve to the mounting bracket.
5. Connect the brake lines to the valve and tighten them to the manufacturer's specifications.
6. If the valve contains a pressure differential switch, connect the vehicle wiring harness to the switch.

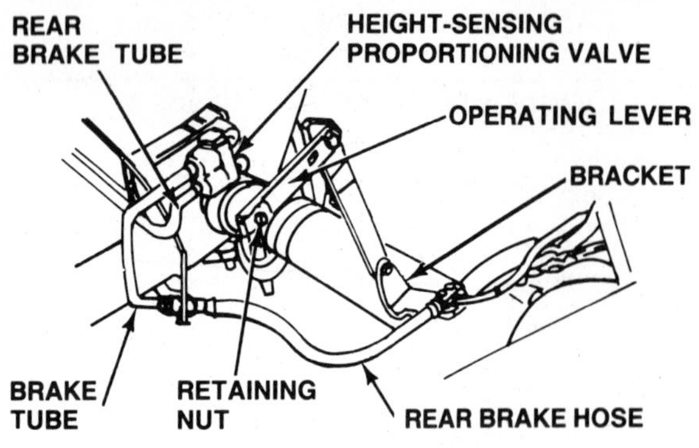

Figure 4-30. The Chevrolet/GMC height-sensing proportioning valve.

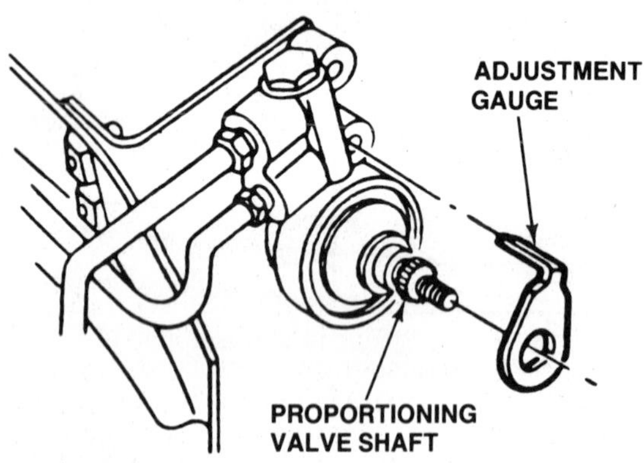

Figure 4-31. Install the plastic adjustment gauge to properly position the proportioning valve shaft.

7. If the valve being replaced is a height-sensing proportioning valve, adjust it as described below.
8. Bleed the brake hydraulic system using one of the methods detailed in Chapter 2.
9. If the valve is equipped with a pressure differential switch that must be manually re-centered, do so as described in Chapter 2.

Proportioning Valve Adjustment

In some cases, a height-sensing proportioning valve must be adjusted when it is replaced. The adjustment ensures that the proportioning action takes effect at the correct hydraulic pressure in relation to vehicle loading. There are nearly as many adjustment procedures as there are variable proportioning valves. The adjustment procedures are given below for two types of valves in use today. Consult the factory shop manual for the exact procedure on other types of valves.

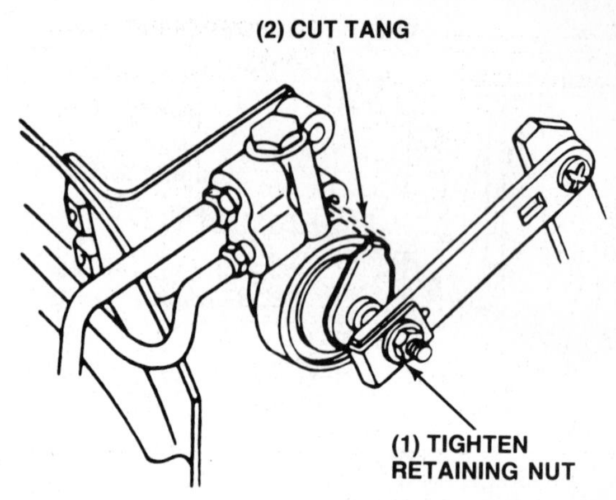

Figure 4-32. Cut the tang from the adjustment gauge after the operating lever is installed.

The height-sensing proportioning valve on Chevrolet/GMC 30 series trucks requires a special plastic adjustment gauge available from the dealer. This gauge is a one-time use item that is installed on the valve to hold it in position during installation. Once the valve operating lever is tightened in place, a tang on the gauge is cut away to allow unrestricted valve operation. To adjust the Chevrolet/GMC height-sensing proportioning valve:

1. Raise the vehicle and support it so that the axle hangs free.
2. Remove the retaining nut and operating lever from the valve shaft, figure 4-30.
3. Rotate the valve shaft as needed to install the adjustment gauge, figure 4-31. The D-shaped hole in the center of the gauge fits over a matching shape on the valve shaft, and the tang of the gauge fits into the lower hole on the valve body.
4. Position the operating lever on the shaft and use a C-clamp or Channel-Lock pliers to press the lever and its plastic bushing over the serrations on the shaft until it is fully seated. Do not use the retaining nut to seat the lever; this can rotate the valve shaft and disturb the adjustment.
5. Install the retaining nut and torque it to 70 to 98 in-lb (8 to 11 Nm).
6. Sever the tang on the adjustment gauge to allow the valve shaft to rotate freely, figure 4-32.

The height-sensing proportioning valve on the Dodge Ram 50 pickup truck adjusts differently:

1. Park the unloaded vehicle on a level surface. Do not raise or support the vehicle.

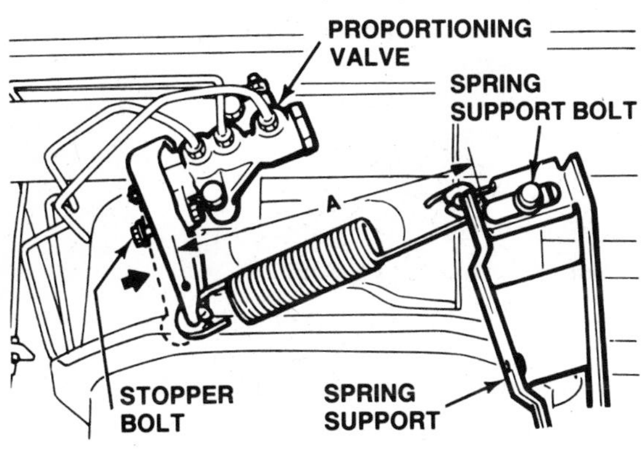

Figure 4-33. This height-sensing proportioning valve is adjusted by setting the spring length, dimension "A."

2. Push the operating lever inward toward the proportioning valve and away from the stopper bolt, figure 4-33.
3. Measure the distance "A" from the lever hole to the spring support hole. The distance should be 163 to 167 mm (6.42 to 6.57 inch).
4. To adjust the distance, loosen the spring support bolt and slide the spring support until the distance is correct.

BRAKE ELECTRICAL COMPONENT TESTS

The electrical components in the brake system can be broken down into two main circuits. The first is the stoplight circuit which is made up of the stoplights and the stoplight switch. The second is the brake system warning light circuit which consists of the warning light, the parking brake switch, and either a pressure differential switch or a brake fluid level switch. Defective parts in either of these circuits must be replaced, they cannot be repaired.

The first step in diagnosing *any* brake system electrical problem is to check the fuses, light bulbs, bulb sockets, wiring, and connectors for obvious problems. Burned out fuses and bulbs, corroded sockets and connections, and broken or grounded wiring are all common problems that cause electrical circuit failure.

The procedures described below are for common types of switches and lights in the brake system; they do not deal with problems involving the vehicle wiring harness or electronic brake system controls. As a general rule, unless you are familiar with the design of a particular system, it is best to consult a factory shop manual for detailed electrical troubleshooting instructions.

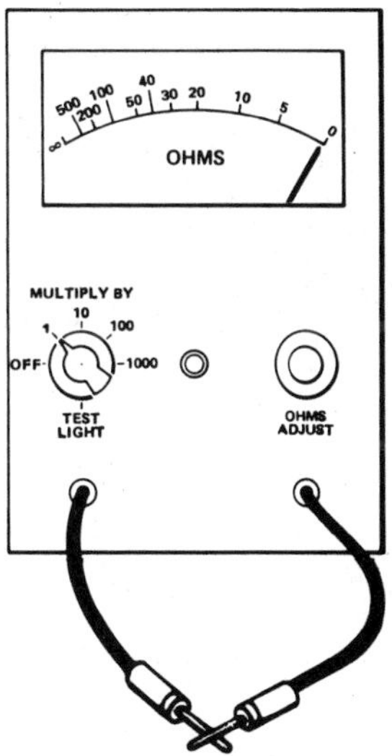

Figure 4-34. An ohmmeter must be calibrated before each use.

Electrical Troubleshooting Tools

Only two basic tools are required for most brake system electrical troubleshooting, a jumper wire and an ohmmeter. A jumper wire is simply a length of insulated wire with an alligator clip or probe on each end. A jumper wire is commonly used to bypass a switch in a circuit to help determine whether the switch or another part of the circuit is faulty.

An ohmmeter, figure 4-34, measures the resistance of an electrical circuit in ohms. In brake system electrical troubleshooting, ohmmeters are commonly used to check switch operation. When a switch is closed it has full continuity, or 0 ohms of resistance. When a switch is open it has no continuity, or an infinite amount of resistance. Ohmmeters can be battery powered or they can plug into commercial alternating current. In either case, any current flow from an outside source will damage the meter. For this reason, the ignition switch should always be OFF when using an ohmmeter to make the tests in this section. If a brake light circuit is wired to operate even when the ignition is OFF, the battery must be disconnected to prevent ohmmeter damage.

Ohmmeters must be adjusted or calibrated before each use. If there are batteries in an ohmmeter they will weaken with age and change voltage with temperature. The voltage applied to a.c.-powered ohmmeters also can vary. To calibrate the meter, figure 4-34, connect its two test leads together. The meter should read zero resistance. If it does not, rotate the

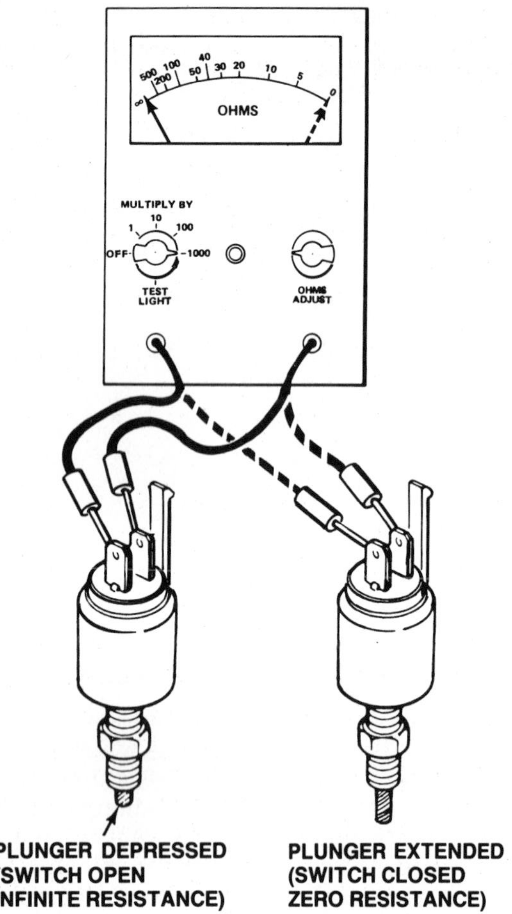

Figure 4-35. Using an ohmmeter to check the operation of a brake light switch.

"ohms adjust" knob until the needle indicates zero ohms. The meter will now give accurate test measurements.

STOPLIGHT CIRCUIT TESTS AND ADJUSTMENTS

There are two potential problems in the stoplight circuit, either the stoplights are always on, or they fail to come on when the brakes are applied. Most of these problems result from burned out fuses or bulbs, a faulty stoplight switch, or a switch that is out of adjustment.

Some early cars have hydraulic stoplight switches mounted on the brake master cylinder or in a nearby junction block. However, virtually all late-model cars have mechanical switches on the brake pedal linkage. Hydraulic switches close once a preset level of pressure is reached in the brake system; hydraulic switches do not require adjustment. Mechanical switches close when the plunger is depressed, or when the plunger extends under internal spring pressure. Most mechanical switches require adjustment when they are installed.

Stoplight Switch Test — Lights Always On

1. Turn the ignition switch ON and confirm that the stoplights are illuminated.
2. Locate the stoplight switch on the pedal linkage or under the hood and detach the wiring harness connector.
   a. If the stoplights stay on, locate and repair the short in the wiring harness.
   b. If the stoplights go out and the car has a hydraulic stoplight switch, go to Step 3.
   c. If the stoplights go out and the car has a mechanical stoplight switch, proceed to Step 4.
3. Check the brake pedal for proper freeplay to make sure the master cylinder is not being held in the applied position.
   a. If there is freeplay, replace the stoplight switch.
   b. If there is no freeplay, adjust the pedal linkage to establish the proper freeplay. Reconnect the stoplight switch and repeat the test starting at Step 1.
4. Loosen the stoplight switch retaining bolts so that the switch plunger no longer contacts the brake pedal linkage.
5. Connect the test leads of an ohmmeter to the stoplight switch, figure 4-35, then move the switch plunger in and out through its full range of travel.
   a. If the meter reads zero resistance at one end of the switch travel and infinite resistance at the other, adjust the switch as described below.
   b. If the meter reads zero resistance at all times, replace the switch and adjust the new switch as described below.

Stoplight Switch Test — Lights Never On

1. Turn the ignition switch ON, have an assistant push on the brake pedal, and confirm that the stoplights are not illuminated.
2. Turn the ignition switch OFF.
3. Locate the stoplight switch on the pedal linkage or under the hood and disconnect the wiring harness connector.

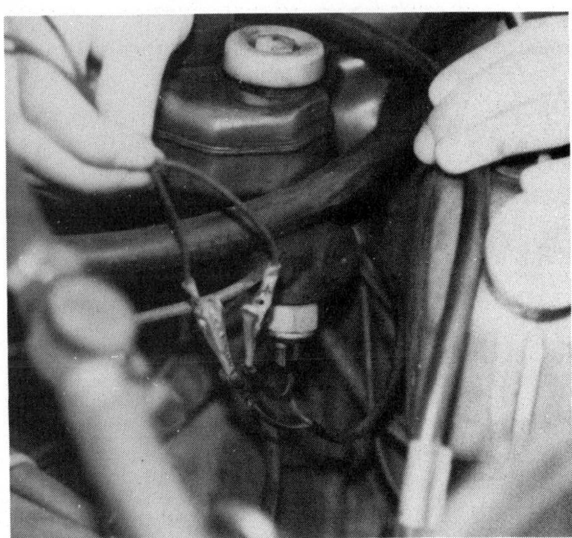

Figure 4-36. Using a jumper wire to bypass a stoplight switch.

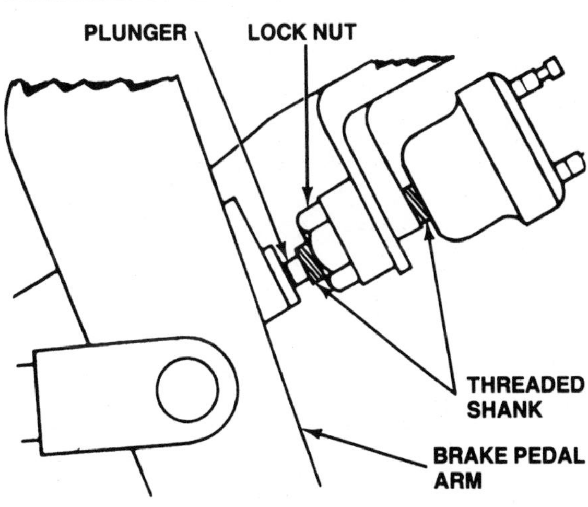

Figure 4-37. To adjust a stoplight switch with a threaded shank, loosen the locknut and thread the switch in or out of its bracket.

4. Connect a jumper wire between the two wires of the harness connector, figure 4-36.
5. Turn the ignition switch ON and check if the stoplights are illuminated.
   a. If the stoplights are on, go to Step 6.
   b. If the stoplights are not on, locate and repair the open circuit in the wiring harness.
6. Turn the ignition switch OFF.
   a. If the car has a hydraulic stoplight switch, go to Step 7.
   b. If the car has a mechanical stoplight switch, go to Step 8.
7. Connect the leads of an ohmmeter to the terminals of the switch. Have an assistant apply and release the brake pedal while you observe the ohmmeter. If the reading is not zero resistance when the brakes are applied, and infinite resistance when the brakes are released, replace the switch.
8. Loosen the stoplight switch retaining bolts so that the switch plunger is no longer in contact with the brake pedal linkage.
9. Connect the test leads of an ohmmeter to the stoplight switch, figure 4-35, then move the switch plunger in and out through its full range of travel.
   a. If the meter reads zero resistance at one end of the switch travel and infinite resistance at the other, adjust the switch as described below.
   b. If the meter reads infinite resistance at all times, replace the switch and adjust the new switch as described below.

Mechanical Stoplight Switch Adjustment

Whenever a mechanical stoplight switch is replaced, the new switch must be adjusted. There are three different ways to adjust these switches. Most imported cars have a stoplight switch with a threaded body that is secured in position by a locknut. Some Chrysler vehicles have a stoplight switch that is adjusted using a spacer gauge. And finally, many cars from Chrysler and General Motors have a semi-automatic adjustment mechanism.

To adjust a stoplight switch with a threaded shank, figure 4-37:

1. Disconnect the wiring harness connector from the switch, and connect an ohmmeter to the switch terminals.
2. Loosen the lock nut on the threaded shank of the stoplight switch body.
3. Screw the switch in or out to adjust the clearance between the brake pedal arm and the switch plunger so that the ohmmeter reads zero resistance (the switch is closed) when the brake pedal is depressed approximately ½ inch (13 mm).
4. Tighten the lock nut and disconnect the ohmmeter.
5. Attach the wiring harness connector to the switch and check the stoplights for proper operation.

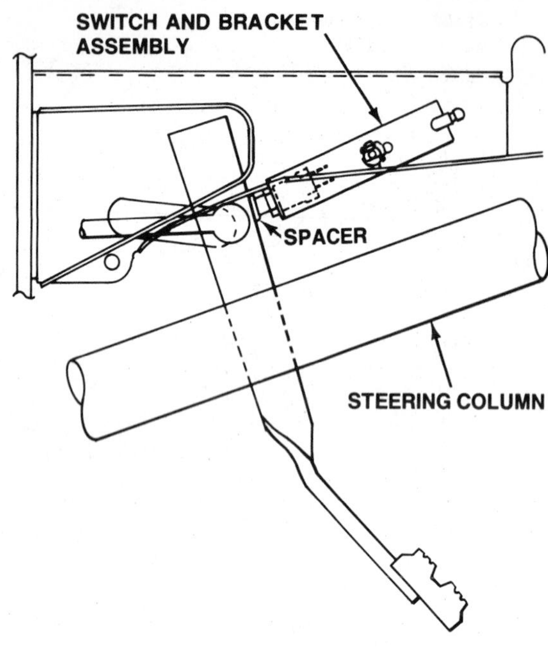

Figure 4-38. Some stoplight switches are adjusted with a spacer between the switch plunger and brake pedal arm.

To adjust a stoplight switch using a spacer gauge figure 4-38:

1. Loosen the screw securing the switch bracket assembly, and slide the assembly away from the brake pedal.
2. Press brake pedal firmly and allow it to return freely. *Do not* pull back on the pedal.
3. Place the spacer gauge between the switch plunger and the brake pedal arm.
4. Slide the switch bracket assembly toward the brake pedal until the switch plunger is bottomed against the spacer gauge.
5. Tighten the screw securing the switch bracket assembly and remove the spacer gauge.
6. Check the stoplights for proper operation.

To adjust a stoplight switch with a semi-automatic adjusting mechanism, figure 4-39:

1. Depress the brake pedal and insert the switch into the tubular clip on the brake pedal mounting bracket; the switch will click when it is fully seated.
2. Pull back on the brake pedal to adjust the switch position. The switch will ratchet in the tubular clip and emit a series of clicks as it does so.
3. Release the pedal and repeat Step 2 until clicks stop.
4. Attach the wiring harness connector to the switch and check the stoplights for proper operation.

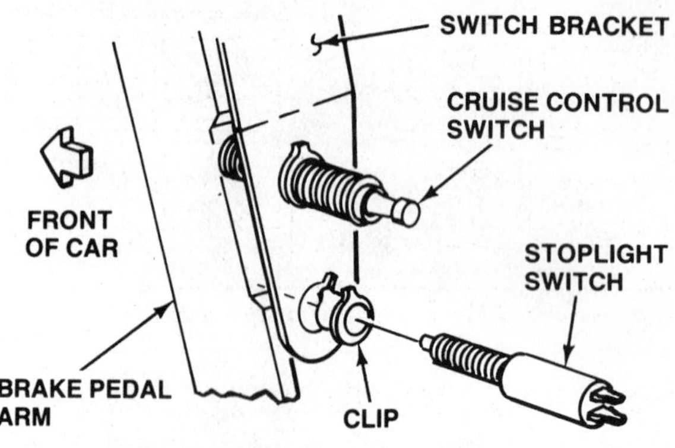

Figure 4-39. This stoplight switch ratchets in tubular clip to adjust semi-automatically.

Stoplight Switch Replacement

Stoplight switch replacement is a fairly basic procedure. Hydraulic switches require only removal and installation. Mechanical switches must be adjusted as described above after they are installed.

To replace a hydraulic stoplight switch:

1. Disconnect the wiring harness connector from the switch.
2. Unscrew the switch from the master cylinder or junction block.
3. Screw the new switch into the master cylinder or junction block.
4. With the switch slightly loose, have an assistant *slowly* apply the brake pedal and hold it to the floor. Brake fluid will be forced out around the threads of the switch along with any air that may have entered the system. Wear safety glasses to prevent eye injuries from fluid spray, and use a rag or drain pan to catch the fluid and prevent damage to the vehicle finish.
5. Tighten the switch and have your assistant slowly release the brake pedal. Repeat Steps 5 and 6 until no air comes from the system.
6. Attach the wiring harness connector to the switch and check the stoplights for proper operation.

To replace a mechanical stoplight switch:

1. Disconnect the wiring harness connector from the switch.
2. Remove any nuts, bolts, or screws securing the switch in place, then remove the switch from its bracket.
3. Place the new switch in the bracket, and install the mounting nuts, bolts, or screws loosely.
4. Adjust switch as described above, then tighten it securely in position.
5. Attach the wiring harness connector to the switch and check the stoplights for proper operation.

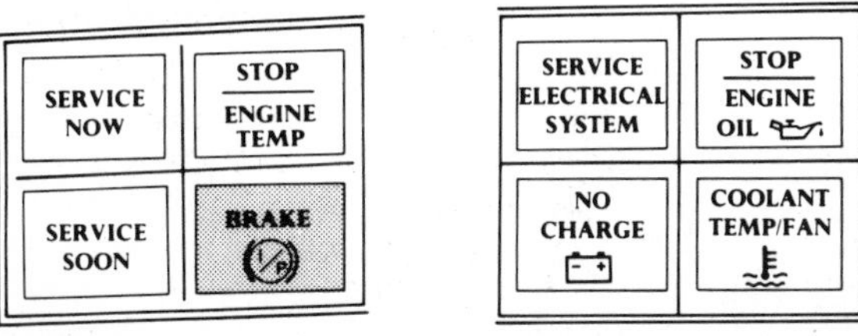

Figure 4-40. A typical brake system warning light.

WARNING LIGHT CIRCUIT TESTS

All cars are equipped with a brake system warning light on the instrument panel, figure 4-40. This light generally serves two purposes, and is activated in two ways. The minor function of the light is to serve as a parking brake warning indicator. A switch on the parking brake linkage turns the light on when the parking brake is applied. The major function of the light is as an indicator of trouble within the brake hydraulic system. For this purpose, the light is activated by either a pressure differential switch or a brake fluid level switch. All cars manufactured since 1967 are equipped with one or the other of these switches.

A pressure differential switch turns on the brake system warning light when there is a significant difference in pressure between the two hydraulic circuits of the dual braking system. The brake fluid level switch turns on the brake system warning light when the fluid level in the master cylinder reservoir falls below a predetermined level; this can be caused by wear of the brake linings or a leak in the hydraulic system.

Problem diagnosis in the brake system warning light circuit involves testing the warning light bulb, the parking brake switch, and either the pressure differential switch or the brake fluid level switch depending on which is fitted to the car being serviced.

Warning Light Bulb Tests

Most cars rely on a single bulb for the brake system warning light, although some imported cars have a back-up bulb that illuminates when the main bulb burns out. The back-up warning light is activated by the change in resistance that results when the main light burns out. Brake system warning lights should be tested twice a year or whenever the brake system is serviced. The exact procedure for testing the warning light bulb varies from car to car, but most tests are performed by turning the ignition switch to a certain position without starting the engine.

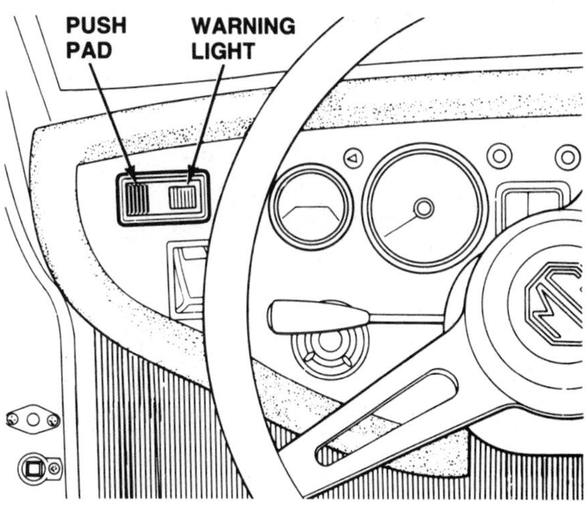

Figure 4-41. A "press-to-test" brake system warning light.

To perform the warning light test on a Ford Motor Company vehicle, turn the ignition switch to either the ON or START position. The warning light will illuminate as long as the engine is not running. On Chrysler products, turn the ignition switch to a point *between* the ON and START positions. Once again, the warning light will illuminate when the engine is not running. On a General Motors car, turn the ignition switch to the START position. The light will illuminate during cranking until the engine starts.

Once the engine is running, the warning light bulb on most cars can be tested simply by applying the parking brake, assuming of course that the parking brake warning light switch and wiring are operating properly. Some older imported cars have a press-to-test warning light, figure 4-41, that illuminates when a test button or the light itself is pressed. Consult the shop or owners manual if you are in doubt about the proper test for a given car.

Parking Brake Switch Test

If the bulb test above indicates that the warning light is operating properly, yet the light does not illuminate when the ignition is ON and the parking brake is applied, there is a problem in the parking brake switch circuit. This is usually caused by a parking brake switch that fails to close and complete the circuit. To test the parking brake switch:

1. Turn the ignition switch OFF.
2. Locate the parking brake switch on the pedal, lever, or handle linkage and disconnect the wiring harness connector from the switch.

Figure 4-42. Ground the pressure differential switch wire to test the wiring harness portion of the circuit.

3. If the harness connector contains a single wire, connect the wire to ground. If the connector has two wires, connect a jumper wire between them.
4. Turn the ignition switch ON, and check if the warning light is illuminated.
   a. If the light is on, replace the parking brake switch.
   b. If the light is not on, locate and repair the open circuit in the wiring harness.

Although it rarely happens, the parking brake switch can also jam in the ON position, or short internally, causing the warning light to stay lit all the time. If you suspect this problem, simply detach the wiring harness connector from the parking brake switch while the ignition is on and the light is illuminated. If the light goes out when the connector is detached, repair or replace the parking brake switch.

FAILURE WARNING SYSTEM SWITCH TESTS

If the brake system warning light flashes when the brakes are applied, or is on constantly when the parking brake is not applied (and the parking brake switch has been ruled out as the problem), the warning system is probably doing its job and indicating that a problem exists in the brake hydraulic system. However, it is also possible that there may be a failure within the warning system itself.

Before you test the failure warning system, perform these preliminary checks to determine if the problem is in the brake hydraulic system. If the car is equipped with a fluid level switch, check the fluid level in the master cylinder reservoir; if it is low, adjust it as necessary. If the car is equipped with a pressure differential switch, detach the wiring harness connector from the switch terminal. If the light no longer comes on when you perform either of the above operations, do a complete brake system inspection. A low fluid level signals a leak in the hydraulic system, or worn brake linings that need to be replaced. A pressure differential switch that is activated indicates there is a leak in the system.

If the brake fluid level is okay and there are no signs of leaks anywhere in the hydraulic system, it is possible that the pressure differential or fluid level switch is defective and keeping the light on. The procedures for testing these switches are given below.

Pressure Differential Switch Test

If the brake system warning light on a car equipped with a pressure differential switch flashes when the brakes are applied, or remains on when there are no external signs of leakage from the hydraulic system, the switch may need to be manually re-centered; this is usually required only on domestic models built before 1971, and certain later imports. However, pressure differential switches with centering springs, as used on all late-model cars, can also require manual centering if the switch piston sticks in its bore. See Chapter 2 for switch centering procedures.

If the warning light comes on again after the switch has been manually centered, the master cylinder is bypassing internally and creating a pressure difference between the two hydraulic circuits. Rebuild or replace the master cylinder to solve the warning light "problem."

It is also possible that a defective pressure differential switch will not turn the warning light on, even though a leak exists in the hydraulic system. If a leak is present and bulb test indicates that the warning light is operating properly, test the pressure differential switch as follows:

1. Turn the ignition switch ON.
2. Detach the wiring harness connector from the pressure differential switch, and use a jumper wire to connect the harness wire to ground, figure 4-42. Observe the warning light:
   a. If the light illuminates, go to Step 3.
   b. If the light does not illuminate, locate and repair the open circuit in the wiring harness.

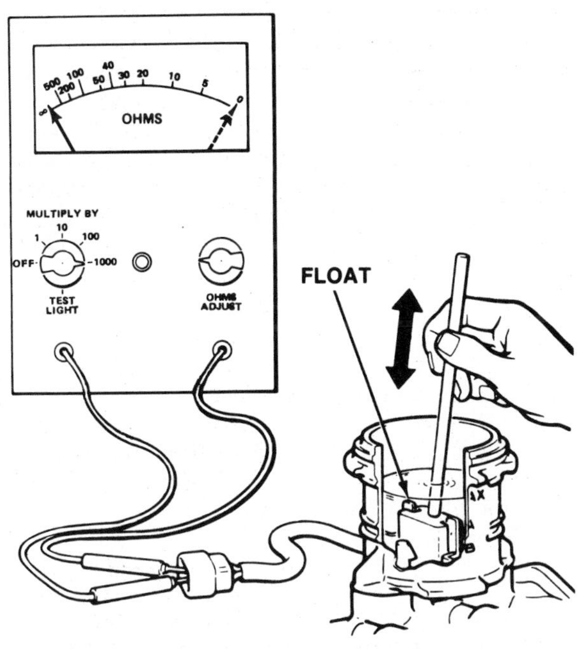

Figure 4-43. Bottom the float manually to close the fluid level switch.

3. Attach the wiring harness connector to the pressure differential switch, and have an assistant apply the brake pedal with light to moderate pressure.
4. Open the left front bleeder screw and have your assistant observe the warning light:
   a. If the warning light illuminates, the switch is good. Close the bleeder screw and recenter the pressure differential switch as described in Chapter 2.
   b. If the warning light does not illuminate, go to Step 4.
5. Close the bleeder screw and have your assistant release the brake pedal, then reapply it with light to moderate pressure.
6. If the car has a front/rear-split braking system, open a rear bleeder screw. If the car has a diagonal-split braking system, open the right front bleeder screw. Have your assistant observe the warning light:
   a. If the warning light illuminates, the switch is good. Close the bleeder screw and recenter the pressure differential switch as described in Chapter 2.
   b. If the warning light does not illuminate, replace the switch.

Fluid Level Switch Tests

Fluid level switches come in two varieties, those where the float remains in the reservoir, and those where the float is attached to the reservoir cap. Both are basic on/off switches, and fail by sticking in either the open or closed position.

Float in reservoir — switch test
To test a fluid level warning switch with a float that remains in the reservoir:

1. Perform a bulb test to confirm that the warning light is operating properly.
2. Adjust the fluid level in the master cylinder reservoir to the full mark.
3. Turn the ignition switch ON and observe the warning light.
   a. If the warning light illuminates, go to Step 4.
   b. If the warning light does not illuminate, go to Step 5.
4. Detach the wiring harness connector from the fluid level switch.
   a. If the warning light goes out, replace the reservoir and switch assembly.
   b. If the warning light remains lit, locate and repair the short in the wiring harness.
5. Remove the master cylinder fluid reservoir cap, and use a clean screwdriver or other probe to gently push the float to the bottom of its travel, figure 4-43, simulating a low fluid level. Observe the warning light:
   a. If the warning light illuminates, the switch is good.
   b. If the warning light does not illuminate, go to Step 6.
6. Detach the wiring harness connector from the switch, and use a jumper wire to bridge the two wires in the connector. Observe the warning light:
   a. If the warning light illuminates, replace the fluid reservoir and switch assembly.
   b. If the warning light does not illuminate, locate and repair the open circuit in the wiring harness.

Float on reservoir cap — switch test
Many imported cars use a fluid level warning switch in which the float is attached to the master cylinder fluid reservoir cap. To test this type of switch:

1. Perform a bulb test to confirm that the warning light is operating properly.

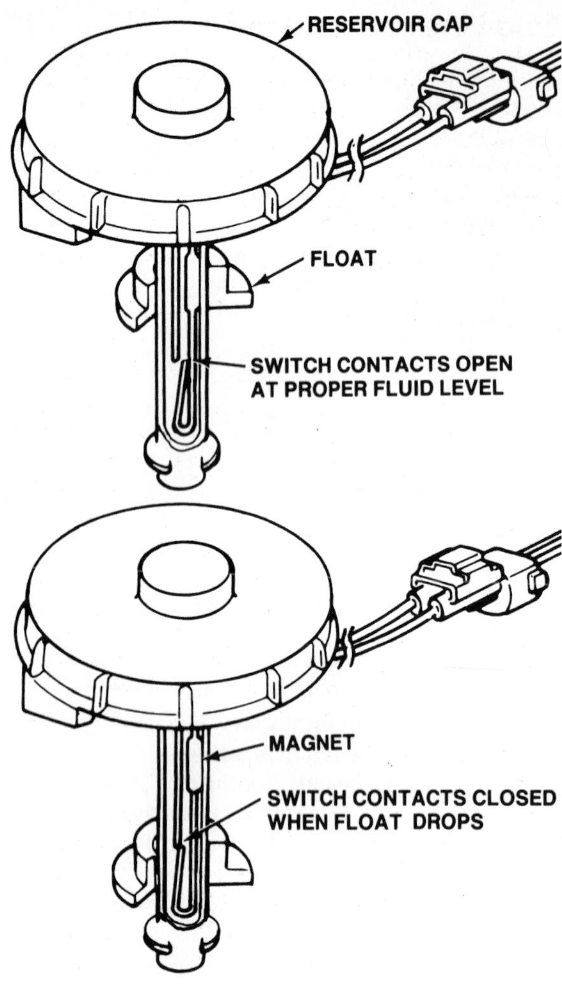

Figure 4-44. Allow the float to drop to the bottom of its travel to manually close this type of fluid level switch.

2. Adjust the fluid level in the master cylinder reservoir to the full mark.
3. Turn the ignition switch ON, and with the reservoir cap in place, observe the warning light.
   a. If the warning light illuminates, go to Step 4.
   b. If the warning light does not illuminate, go to Step 5.
4. Detach the wiring harness connector from the fluid level switch.
   a. If the warning light goes out, replace the cap and switch assembly.
   b. If the warning light remains lit, locate and repair the short in the wiring harness.
5. Remove the master cylinder fluid reservoir cap, and allow float to drop to the bottom of its travel, figure 4-44, simulating a low fluid level. Observe the warning light:
   a. If the warning light illuminates, the switch is good.
   b. If the warning light does not illuminate, go to Step 6.
6. Detach the wiring harness connector from the cap and switch assembly, and use a jumper wire to bridge the two wires in the connector. Observe the warning light:
   a. If the warning light illuminates, replace the cap and switch assembly.
   b. If the warning light does not illuminate, locate and repair the open circuit in the wiring harness.

Chapter

5

Pedal Assembly, Master Cylinder, and Wheel Cylinder Service

The functions of the brake pedal assembly, master cylinder, and wheel cylinders are fundamental to the proper operation of the brake system. The pedal assembly transmits and increases the mechanical force applied by the driver to the brake pedal. The master cylinder converts that force into hydraulic pressure that is routed to the wheel friction assemblies. And the wheel cylinders convert that pressure back into mechanical force that is used to apply the brake shoes against the drum, creating the friction required to stop the car.

An improperly adjusted brake pedal either increases pedal freeplay, or holds the master cylinder in a partially applied position. In the former situation, the brake pedal reserve is reduced; in the latter case, the brakes will initially drag, and can eventually lock completely. A leaking master cylinder or wheel cylinder can greatly reduce stopping power, and in the case of a wheel cylinder, lost brake fluid can contaminate the brake shoe linings as well. If a hydraulic leak is not repaired, the system will eventually lose enough fluid that the brakes will fail altogether, possibly resulting in an accident.

Brake pedal assembly, master cylinder, and wheel cylinder service are common brake repair jobs. This chapter contains service procedures for working on all three of these components. Pedal service is usually limited to adjusting the freeplay. However, master and wheel cylinder service is much more involved and includes procedures for inspection, testing, replacement, and overhaul.

BRAKE PEDAL ASSEMBLY SERVICE

Brake pedal assembly service consists of two operations, pedal and linkage inspection, and pedal freeplay adjustment. The inspection verifies that the pedal assembly is in good condition and works freely. The pedal and linkage must be working properly or pedal freeplay cannot be adjusted accurately.

Brake pedal freeplay is determined by the clearance between the pedal pushrod and the master cylinder primary piston. Proper freeplay is important because too much freeplay causes excessive brake pedal travel, figure 5-1. This reduces the pedal height, and gives the driver less effective travel, or pedal reserve, with which to compensate for vapor lock or brake fade.

If there is no brake pedal freeplay, figure 5-2, the master cylinder pistons cannot return all the way. When this happens, the primary cup seals come to rest slightly in front of, rather than behind, the compensating ports. With the seals in this position, the brakes are held in a

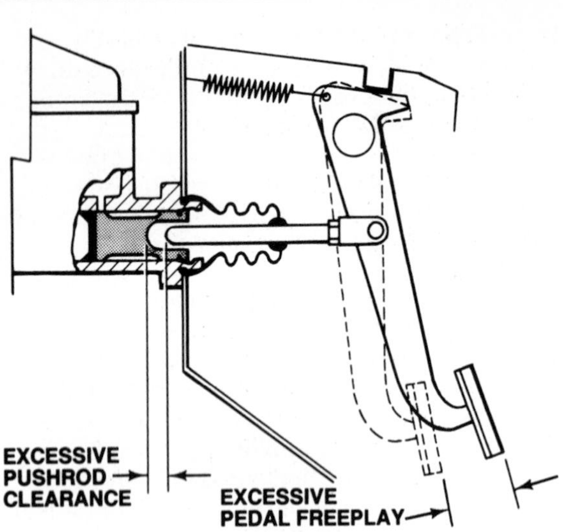

Figure 5-1. Excessive brake pedal freeplay reduces pedal reserve.

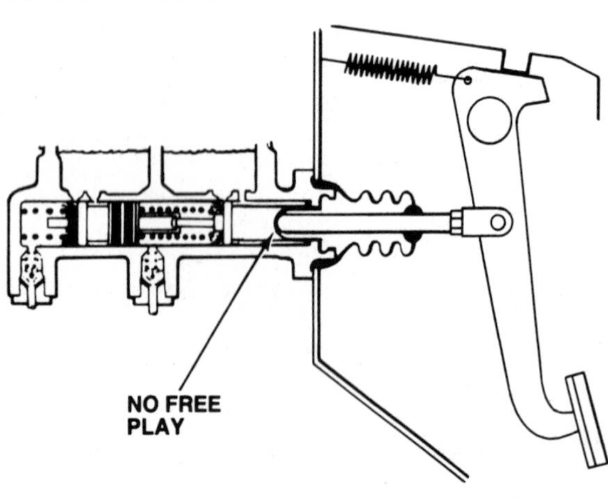

Figure 5-2. If the brake pedal has no freeplay, the brakes may be slightly applied at all times.

partially applied position that causes brake drag, premature wear of the shoes and pads, and possibly heat fade. In addition, fluid cannot pass between the reservoir and the cylinder bore to compensate for temperature changes; as fluid trapped in the lines expands from the extra heat, it eventually causes the brakes to lock.

With the seals held in front of the compensating ports, there is yet another way the brakes can lock. Every time the brake pedal is released, the one-way pumping action of the master cylinder cup seals allows additional fluid into the hydraulic system. After a certain number of stops (the exact number varies with the brake system) the volume of fluid trapped in the system will become so great that the

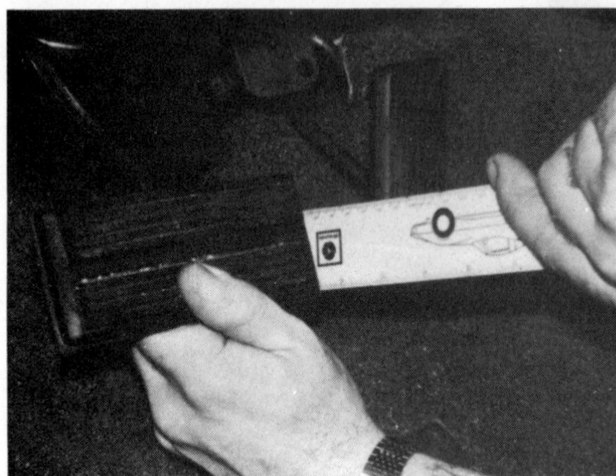

Figure 5-3. Measuring brake pedal freeplay.

shoes and pads will be held in solid contact with the drums and rotors.

Pedal Linkage Inspection

The first step when inspecting the brake pedal linkage is to check the rubber pedal pad. If it is worn, replace it with a new part. Next, grasp the pedal pad and move the pedal arm from side to side to check for wear. A small amount of side play is acceptable; larger amounts indicate that the linkage has loose attaching hardware, or missing or worn bushings. Tighten the hardware and replace worn parts as needed. Finally, depress the pedal by hand and release it. If it does not move in and out smoothly without binding, locate the source of the friction and repair as necessary.

Pedal Freeplay Adjustment

Many cars have no provision for brake pedal freeplay adjustment; tolerances in the design allow an acceptable amount of freeplay with any combination of master cylinder and brake pedal assembly. Brake pedal assemblies that do have provisions for freeplay adjustment are set at the factory, and do not normally need adjustment in the field unless the master cylinder is rebuilt or replaced. To check the pedal freeplay:

1. If the vehicle is equipped with a power booster, pump the brake pedal until the reserve is depleted and the pedal becomes hard.
2. Place a ruler along the axis of brake pedal travel, figure 5-3, and slowly apply the pedal by hand until all of the slack in the linkage is eliminated and a slight resistance is felt.
3. Observe the distance the pedal moved from the relaxed position to the position where resistance is first met; this is the freeplay. Most automakers require that pedal freeplay be between ⅛ and ½ inch (3 and 13 mm). The exact

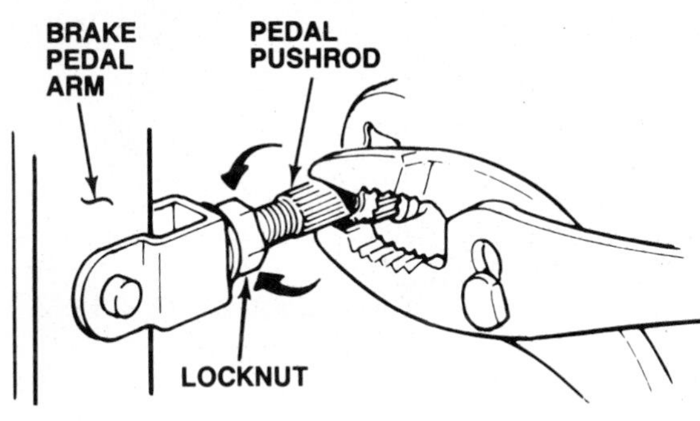

Figure 5-4. Adjusting pedal freeplay at the pedal pushrod.

amount varies from one vehicle to another, however, so always check the manufacturer's specification for the car you are servicing.

Freeplay is adjusted by shortening or lengthening the brake pedal pushrod. To adjust the pedal freeplay on most vehicles, loosen the locknut on the pushrod, rotate the pushrod in one direction or the other, figure 5-4, until you obtain the correct freeplay, then tighten the locknut. Some imported vehicles have an eccentric hinge bolt at the end of the pushrod that connects to the brake pedal. To adjust this type of linkage: loosen the hinge bolt locknut, turn the hinge bolt to obtain the proper freeplay, then tighten the locknut.

Many cars have a mechanical stoplight switch on the brake pedal linkage. If adjusting the pedal freeplay affects the relationship between the switch and the linkage, adjust the stoplight switch as described in Chapter 4.

HYDRAULIC CYLINDER INSPECTION AND TESTING

If a problem is suspected in a master cylinder or wheel cylinder, a detailed external inspection should be made to determine if the cylinder requires additional service. If you determine that additional service is required, an internal inspection should be done to determine whether the cylinder can be rebuilt, or if it must be replaced with a new part. External inspections and tests are described below. Internal inspections are covered in the overhaul section later in the chapter.

The two most common reasons hydraulic cylinders have to be rebuilt or replaced is that they are leaking fluid externally or, in the case of master cylinders, fluid is internally bypassing the piston seals. Both of these problems occur as a result of wear and fluid contamination. Rubber seals naturally wear as they slide against the cylinder bore, and the normal buildup of rubber dust and other contaminants in the brake fluid accelerates this process. The metal cylinder bore walls also wear, although at a much slower rate than the seals. Another problem with piston seals is that the rubber they are made from ages with time and becomes inflexible, which reduces its sealing ability.

Fluid contamination occurs when a foreign liquid enters the hydraulic system. If the contaminant is a petroleum-based product, it quickly attacks the rubber seals and causes them to soften and swell to the point where they lose their sealing ability. Cylinders that suffer this type of damage can usually be rebuilt if the sealing surfaces of the bore are in good condition. If the contaminant is moisture, rust and corrosion form on the cylinder bore and destroy the sealing surface. Although this process takes longer than the damage caused by improper fluids, it is much more likely to require cylinder replacement.

MASTER CYLINDER INSPECTION

To inspect a master cylinder, wipe the fluid reservoir cover clean, and remove it from the master cylinder. Hold the reservoir cover up to the light and make sure the vent holes are open. The holes must be open to maintain atmospheric pressure above the fluid in the reservoir.

Check the condition of the reservoir cover diaphragm seal, figure 5-5; a seal in good condition has no cuts or tears, and the rubber is firm and not distorted in any way. A cut or torn diaphragm allows the brake fluid to absorb moisture from out of the atmosphere and must be replaced. A swollen or misshapen diaphragm is a sign that the brake fluid is contaminated, probably with some type of petroleum-base liquid. In this case, you will have to disassemble the entire brake system, flush it clean of contaminated fluid, and rebuild the hydraulic components using all-new rubber parts.

Next, check the brake fluid level as covered in Chapter 2; a low fluid level may be caused simply by brake lining wear, but it can also be a sign of leaks in the hydraulic system. Inspect the outside of the master cylinder for leaks, paying particular attention to the fluid outlet fittings. If the cylinder has a plastic fluid reservoir, check the seals where the reservoir joins to the cylinder body.

On cars without a power booster, reach under the dashboard and squeeze the wiper boot at the back of the master cylinder to check

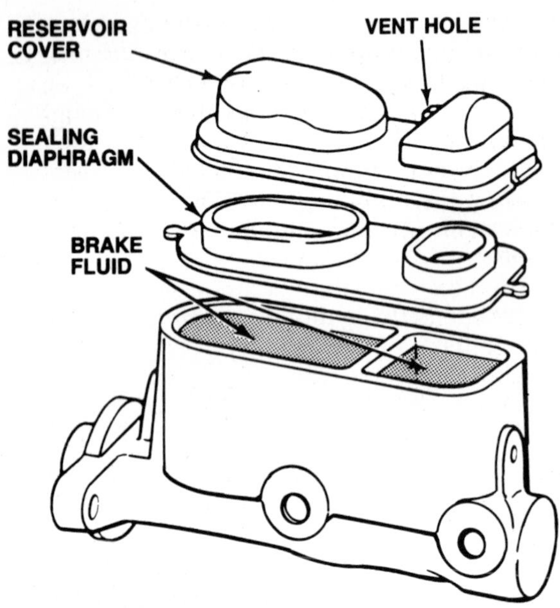

Figure 5-5. The vent holes, diaphragm seal, and fluid level can all be inspected when the reservoir cover is removed.

for fluid leakage from the rear seal. If the car is equipped with a vacuum-assisted power booster, look for fluid stains on the front of the booster below the master cylinder. If you suspect a leak but there are no external signs of one, you will have to loosen the master cylinder mounting bolts and pull the cylinder away from the booster to check for leaks. If a fluid leak is fairly large, or the vacuum supply hose attaches near the bottom of the booster power chamber, you can sometimes remove the hose and inspect inside it for signs of brake fluid that has leaked from the rear master cylinder seal.

MASTER CYLINDER TESTING

A visual inspection checks the external aspects of the master cylinder; however, there are also special tests than can pinpoint whether a particular problem is within the master cylinder, or elsewhere in brake system. The tests below apply to most of the master cylinders now on the market. However, certain cylinders have special features require unique test procedures, or special caution when performing otherwise routine tests. If you are unfamiliar with a particular master cylinder, consult the factory shop manual for any special instructions that may apply to that design.

Quick-Take-Up Valve Test

The Quick-Take-Up (QTU) valve is a special fluid control valve installed in the rear chamber of the fluid reservoir on some General Motors and Ford master cylinders. The QTU valve regulates the flow of fluid between the reservoir and master cylinder bore, and enables the cylinder to move the large initial volume of fluid needed to take up the clearances in the low-drag brake calipers used with this type of master cylinder.

There is no direct method of testing the QTU valve, however, a problem with the valve may be indicated by a couple of common brake system problems. For example, excessive pedal travel will result if fluid is bypassing the valve when the brakes are first applied; check for a damaged, missing, or unseated QTU valve. If the brake pedal returns slowly when the brakes are released, the QTU valve may be clogged so that fluid flow out of the cylinder bore is restricted. If no other cause for these problems can be found, remove the valve, and clean or replace it as needed.

Compensating Port Test

The compensating port test is directly related to brake pedal freeplay as discussed earlier in the chapter. If there is no pedal freeplay, the compensating ports in the master cylinder may be closed. When this happens, fluid cannot flow from the cylinder bore back into the reservoir, and the brakes are always partially applied. The results are brake drag, premature wear of the shoes and pads, brake fade, and possibly brake lockup.

The compensating port test can be used to determine if the ports are open; however,you must use special caution when performing this test. If the brake pedal is not applied *slowly,* fluid can be pumped from the ports with enough velocity to spray onto the underside of the hood. This can damage the vehicle finish, or cause personal injury. Wear proper eye protection while performing this test, and never put your face directly over the master cylinder reservoir.

To check if the compensating ports are open, remove the reservoir cover and have an assistant *slowly* apply the brake pedal while you observe the fluid in the reservoir. As the pedal is applied, a small amount of fluid should be forced out of the cylinder bore through the compensating port in each chamber of the reservoir; this causes a small jet or spurt of fluid to appear on the surface of the brake fluid, figure 5-6.

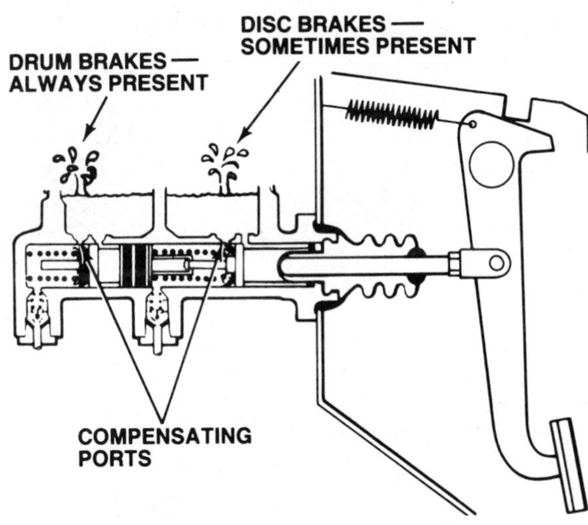

Figure 5-6. Spurts of fluid in the reservoir as the brakes are applied indicate that the compensating ports are open.

The fluid spurt in a reservoir that serves only disc brakes will generally be somewhat smaller than the spurt in a reservoir that serves only drum brakes. In some cases, a spurt may not be present at all. This occurs because the caliper pistons, which are held in place by the relatively light tension of their O-ring seals, will start to move under much less pressure than the wheel cylinder pistons which must work against the much higher tension of the brake shoe return springs.

When performing the compensating port test on a QTU master cylinder, there are some special points and precautions to keep in mind. In these cylinders, the QTU valve initially restricts fluid flow through the rear compensating port, so a jet of fluid will not appear in the reservoir. However, once the clearance in the brake system is taken up and pressure reaches 70–100 psi, the QTU valve check ball unseats and a large quantity of fluid is pumped into the reservoir very rapidly. This can create a safety hazard, so have your assistant apply the brake pedal very lightly (perhaps by hand) so the opening pressure of the QTU valve is not exceeded.

If fluid spurts do not appear when the brakes are applied, the compensating ports are clogged or the pedal freeplay is too tight. Check and adjust the pedal freeplay as described earlier. If spurts still do not appear, and the car has manual brakes, the master cylinder will have to be removed and disassembled to determine why the compensating ports are obstructed. If spurts still do not appear, and the car has a vacuum power booster, back off the master cylinder mounting bolts approximately ⅛ inch (3 mm) and repeat the compensating port test. If the spurts now appear, the power booster output pushrod must be adjusted because it is holding the master cylinder in the applied position. To make this adjustment, you must remove the master cylinder and use a special gauge. See Chapter 10 for details.

External Leak Test

A master cylinder reservoir that is low on fluid, or has run dry and allowed air to enter the hydraulic system, can be an indication of two things: normal brake lining wear, or a fluid leak in the brake hydraulic system. Major external leaks in the hydraulic system are usually obvious. If the leak is at the back of the master cylinder and the car has manual brakes, there will be fluid on the floor inside the car, and an odor of brake fluid in the passenger compartment. If the leak is from a brake line, wheel cylinder, or caliper, there will usually be fluid stains on the backing plate, inside the wheel, or elsewhere under the car.

Minor fluid leaks are not always easy to spot, however. Most cars today have power boosters that can conceal an external leak from the back of the master cylinder. And, because brake fluid is water soluble, small amounts of brake fluid from minor leaks under the car can easily be washed away by car washes or travel on wet roads. Minor wheel cylinder leaks often do not pass enough fluid for the fluid to become visible outside the drum. Finally, a few cars route portions of the brake lines through concealed areas of the car interior, such as under the rear seat, where a leak may not be immediately apparent.

The external leak test is used to check for fluid leakage out of the hydraulic system. To perform the external leak test:

1. Fill the master cylinder reservoir at least half full of brake fluid, and note the exact level.
2. If the reservoir has run dry and air has entered the hydraulic system, bleed the brakes before continuing with the test.
3. Apply the brake pedal several times. If there is a leak in the system, the pedal will go at least half way to the floor under firm pressure.
4. Check the fluid level in the master cylinder reservoir. If the level has dropped, there is an external leak from the system.

Internal Leak Test

In addition to external leaks, the brake system can also suffer internal leakage in the master cylinder. In normal operation, brake fluid in

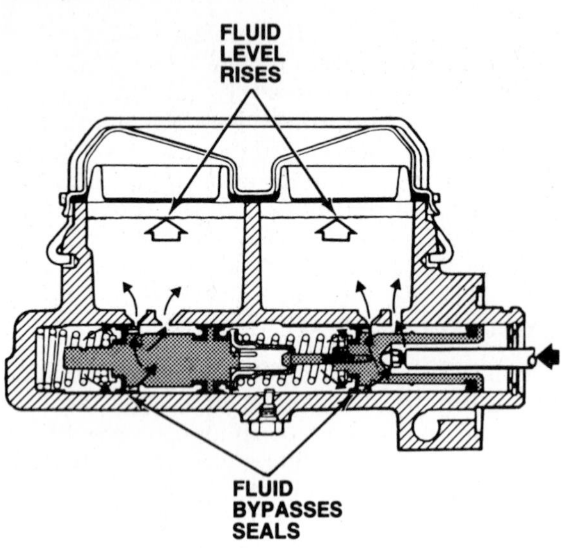

Figure 5-7. Internal leakage past the piston seals causes the fluid level in the reservoir to rise as the brakes are applied.

Figure 5-8. Inspecting a wheel cylinder for leakage.

front of the master cylinder piston primary cup seals is forced into the hydraulic system. However, if the seals cannot contain the pressure created, fluid bypasses around the outside of the seals, figure 5-7, creating an internal leak. Because the volume of the brake hydraulic system is reduced as the master cylinder pistons move forward, fluid displaced past the seals escapes through the compensating and replenishing ports, causing the fluid level in the master cylinder reservoir to rise.

The internal-leak, or bypass, test checks the sealing ability of the master cylinder piston primary seals. This test is usually performed if there is no indication of an external leak in the brake hydraulic system, but the brake warning light is illuminated. To perform the internal leak test:

1. Fill the master cylinder reservoir at least half full of brake fluid.
2. Have an assistant slowly apply and release the brake pedal as you observe the fluid level in the reservoir.
3. If the fluid level rises as the pedal is applied and falls when the pedal is released, fluid is bypassing the seals in the master cylinder.

Another way to test for an internal leak is to hold a constant, firm pressure on the brake pedal for approximately one minute. If the pedal sinks slowly toward the floor, and there are no signs of external leaks, fluid is bypassing the master cylinder seals. This test is effective for diagnosing minor internal leakage because the extended time that pressure is held on the brake pedal allows enough fluid to bypass the seals so that the leak can be detected at the brake pedal.

WHEEL CYLINDER INSPECTION

A leaking wheel cylinder presents two main problems. First, it eventually drains half the brake hydraulic system and allows air to enter; this results in a major loss of braking power. Second, as it leaks, the wheel cylinder contaminates the brake linings with fluid; this reduces stopping power, causes the brakes to slip or grab, and makes the car pull to one side during braking. A leaking wheel cylinder must be rebuilt or replaced. If the brake linings have been contaminated with fluid, the entire wheel friction assembly will require an overhaul.

To inspect a wheel cylinder, remove the brake drum as described in Chapter 6, then grasp the wheel cylinder and attempt to move it. If any movement is detected, make sure all of the mounting hardware is in place and properly tightened. Some imported cars have wheel cylinders that slide in a slot on the backing plate. These cylinders are designed to move, so simply make sure the mounting clips are present and properly installed, and the dust boot that seals the slot is in good condition.

Inspect the outside of the wheel cylinder for signs of leakage; minor stains caused by fluid seepage are considered normal. Fold back the cylinder dust boots and look for signs of liquid, figure 5-8; if you find more than a slight amount of dampness, the cylinder must be rebuilt or replaced.

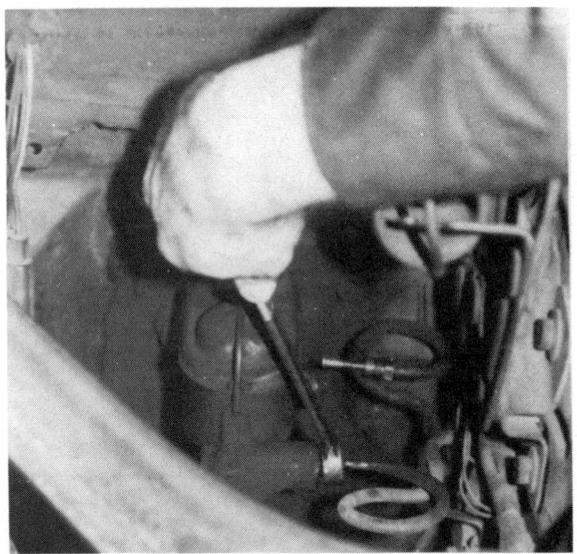

Figure 5-9. The first step when removing a master cylinder is to disconnect the fluid lines.

Next, check for free movement of the wheel cylinder pistons. With the brake drum from only a single wheel removed, have an assistant gently apply and release the brake pedal while you verify that both brake shoes move outward and return smoothly to their stops. On brakes without piston stops, use two large screwdrivers to make sure the pistons are not pushed out of the cylinder bore. Insert the tips of the screwdrivers under the lip at the edge of the backing plate, then lever the screwdriver shafts against the brake shoes to prevent them from moving outward too far.

Depending on the brake design, a frozen wheel cylinder piston can prevent one or both of the brake shoes from applying. This increases the amount of force that must be applied to the brake pedal for a given stop, and causes a pull toward the side of the car that does not have the frozen piston. A sticking piston that slows or prevents full return of the brake shoes will cause the brakes to drag, resulting in rapid lining wear, and possibly heat fade.

MASTER CYLINDER REPLACEMENT

When an external inspection reveals a problem with the master cylinder, the cylinder will have to be removed from the car for further service. To remove a master cylinder:

1. Disconnect the brake lines from the master cylinder fluid outlets with a flare nut wrench, figure 5-9. To prevent fluid spillage, plug the outlets with plastic or rubber plugs available from the aftermarket.

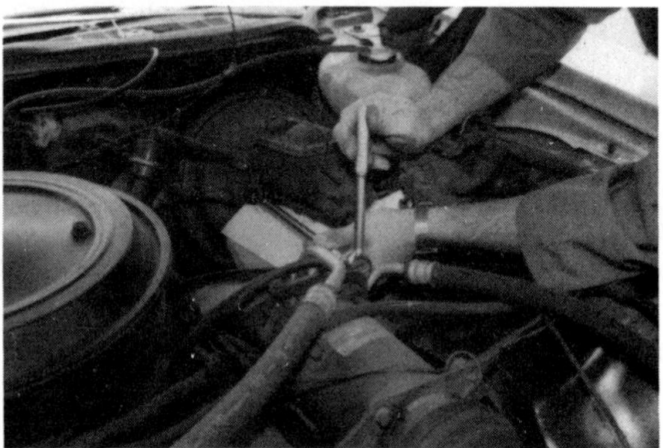

Figure 5-10. Once all connections to the master cylinder are released, it can be unbolted and removed from the car.

2. If the cylinder is so equipped, detach the wiring harness connector from the fluid level warning switch on the reservoir, or the pressure differential switch on the master cylinder body.
3. If the car has manual brakes, determine if the brake pedal pushrod is mechanically connected to the master cylinder. If so, disconnect the pushrod from the brake pedal linkage.
4. Unbolt the master cylinder from the firewall or the power booster, figure 5-10, and remove it from the car. Keep the cylinder upright to prevent fluid spillage.

With the cylinder off the car, make an internal inspection immediately to determine whether the cylinder can be rebuilt. This procedure is covered in the overhaul section later in the chapter. When the rebuilt, or new replacement, master cylinder is ready, install it on the car as follows:

1. Bench bleed the master cylinder as described in Chapter 2.
2. If the vehicle has a vacuum power booster, and the master cylinder was not returning all the way, check the adjustment of the booster output pushrod as described in Chapter 10.
3. Install the master cylinder on the firewall or power booster, and tighten the mounting bolts to the proper torque.
4. Connect the brake lines to the master cylinder fluid outlets and tighten them with a flare-nut wrench.
5. If the cylinder is so equipped, attach the wiring harness connector to the fluid level warning switch on the reservoir, or the pressure differential switch on the master cylinder body.
6. On cars with manual brakes, connect the brake pedal pushrod to the pedal linkage if it was detached when the cylinder was removed.

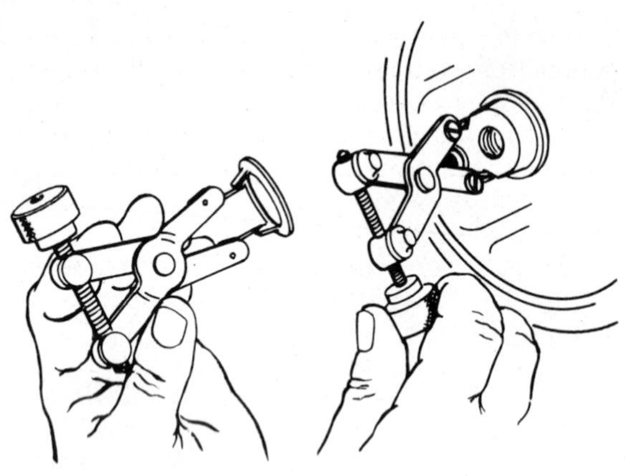
Figure 5-11. A GM wheel cylinder retaining clip being removed with a special tool.

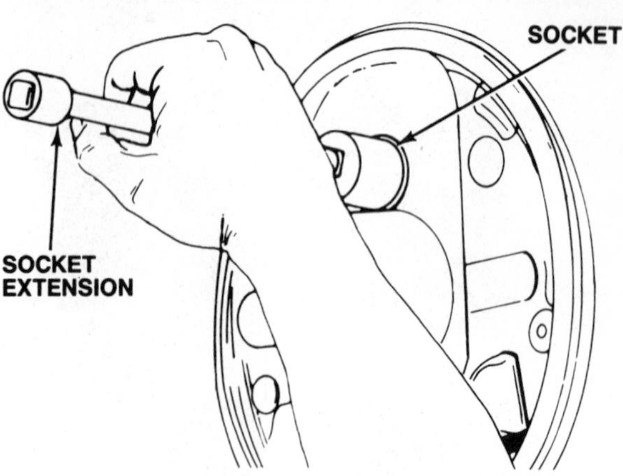

Figure 5-12. Installing the GM wheel cylinder retaining clip.

Check and adjust the pedal freeplay and mechanical stoplight switch as described above and in Chapter 4.

7. Check the factory shop manual for the vehicle manufacturer's recommendations on brake bleeding. If the brake pedal feels high and firm, the brake system may not require additional bleeding. If the brake pedal feels spongy, bleed the master cylinder on the car. If the pedal remains spongy, or the master cylinder was replaced as part of a complete brake system overhaul, bleed the entire system using one of the methods discussed in Chapter 3.

WHEEL CYLINDER REPLACEMENT

When an external inspection reveals that a wheel cylinder needs further service, it does not necessarily mean the cylinder must be removed from the car. Many wheel cylinders can be taken apart, internally inspected, and rebuilt while they are still mounted on the backing plate. Wherever possible, leave the cylinder in place while you perform an internal inspection (described in the overhaul section later in the chapter) to determine whether the cylinder can be rebuilt or if it must be replaced. However, if the wheel cylinder is recessed into the backing plate, or the backing plate incorporates piston stops, the cylinder will have to be removed from the backing plate for any further work.

To remove a wheel cylinder:

1. Remove brake drum and disassemble the friction assembly as described in Chapter 7.
2. Disconnect the steel brake line or rubber hose from the wheel cylinder as described in Chapter 4.
3. Remove the retaining hardware that holds the cylinder to the backing plate. Most cars use a pair of bolts for this purpose, but some GM cars have a retaining clip that requires a special tool, figure 5-11, to bend back the retainer tabs and release the wheel cylinder. A pair of awls can also be used if the special tool is unavailable. On imported cars that use a sliding wheel cylinder, release the retaining clips taking care to note the relative positions of the various parts.

If the cylinder had to be removed from the car to perform an internal inspection, do so at this time. Once the cylinder is rebuilt, or a new replacement part is obtained, install the wheel cylinder as follows:

1. Position the wheel cylinder on the backing plate, and loosely thread the brake hose or tubing fitting into the cylinder.
2. Install the wheel cylinder mounting hardware. If applicable, torque the mounting bolts to the proper tightness. On GM cars with a retaining clip, hold the wheel cylinder in place with a block of wood between the cylinder and axle flange, then tap the retainer into place with one of the special tools available for this purpose, figure 5-12, or a 1⅛-inch 12-point socket; make sure the retainer tabs fully engage the slot on the cylinder. If the car was equipped with a bracket to keep the wheel cylinder from rotating, the bracket *must* be reinstalled. On imported cars with sliding cylinders, install the retaining clips in the correct order and make sure they are properly locked together. Install the dust boot to prevent dirt from entering the friction assembly through the slot.
3. Tighten the brake hose or tubing fitting where it enters the wheel cylinder, and finish installing the brake line as described in Chapter 4.

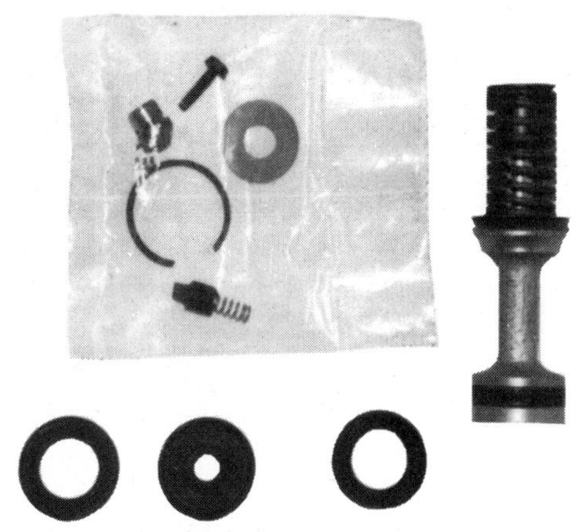

Figure 5-13. A typical master cylinder rebuild kit.

4. Assemble the friction assembly and install the brake drum as described in Chapter 7.
5. Bleed the brakes as described in Chapter 3.

HYDRAULIC CYLINDER OVERHAUL

Once an external inspection has revealed that a master cylinder or wheel cylinder will require further service, the next step is to disassemble the cylinder and determine whether it can be rebuilt or if it must be replaced. In the past, most cylinders were rebuilt and returned to service. Today, the trend is to replace defective cylinders.

There are several reasons hydraulic cylinders are rebuilt less often today. First many newer cylinders have aluminum bodies whose bore *cannot* be honed; this makes overhaul impossible in many cases. Second, hydraulic cylinder overhaul, particularly on dual master cylinders, requires a fair amount of time and skill on the part of the technician. At modern shop labor rates, it is often less expensive for the customer in the long run if the cylinder is replaced. Finally, considering the brake system's critical part in vehicle safety, and the large number of lawsuits filed in recent years, cylinder replacement can be more appealing because it places a portion of the liability for the repair job on the manufacturer of the replacement part.

Overhaul Kits

The job of overhauling a master cylinder or wheel cylinder involves replacing all of the cylinder components that are subject to wear. These parts are purchased in an overhaul or rebuild kit, figure 5-13. A rebuild kit will usually not include every part in the cylinder; the exact contents vary based on the supplier and the cost of the kit. For example, all wheel cylinder rebuild kits contain new rubber cup seals and dust boots, but better quality kits may also include new pistons, cup expanders, and the cup expander spring. Some kits for imported cars even include a rubber dust cover for the bleeder screw.

The situation is more complex with master cylinder rebuild kits. Once again, all kits contain the rubber parts most likely to wear, but better kits contain added items such as a wiper boot with special grease to seal and lubricate it, tubing seats for the fluid outlets, assembly lubricant, and primary and secondary pistons with return springs. The best kits come with the seals already installed on the pistons; this is especially helpful because installing these seals can be difficult, and the seals are easily damaged. Certain special parts, such as QTU valves and reservoir cover diaphragms, are never included in a rebuild kit, and must be obtained separately from a brake parts supplier.

When rebuilding a hydraulic cylinder, it is helpful to know beforehand exactly what parts you need, and what parts are included in the rebuild kit. This ensures that you will have everything available you need to complete the job. It also saves you having to clean old parts that will be replaced anyway.

Cylinder Internal Inspection

As stated in the previous section, overhauling a hydraulic cylinder involves replacing the parts most likely to wear. Essentially, this means replacing everything but the cylinder body, and in some cases, the metal pistons. This means that to determine whether a master cylinder or wheel cylinder can be rebuilt, you must inspect the condition of the cylinder body, and the cylinder bore in particular.

If a master cylinder or wheel cylinder is equipped with bleeder screws, the first inspection step is always to attempt to loosen the screws with a bleeder wrench. It is not unusual for bleeder screws to become rusted or corroded in place so they cannot be removed. These "frozen" bleeder screws are particularly common in aluminum cylinders where the dissimilar metal of the steel bleeder screw causes electrolysis. In most cases, cylinders with frozen bleeder screws should be replaced. The exceptions are when a replacement cylinder is unavailable, or its cost is so high that the time and effort spent on bleeder screw removal can be justified.

Once you are sure the bleeder screws are free, disassemble the cylinder and wipe the bore clean. Shine a light into the cylinder and

Figure 5-14. The condition of the bore is a primary factor in determining whether a hydraulic cylinder can be rebuilt.

inspect the bore, figure 5-14. If the cylinder is made of cast iron and the bore is in good condition, or only lightly scratched, pitted, scored, or rusted, the cylinder can probably be honed and rebuilt. If the bore is deeply scored, the cylinder must be replaced. If deep scores are honed out, so much metal is removed that the cylinder bore becomes oversize and the piston cup seals will not seal properly because they lose tension against the cylinder wall. Also, if a cast-iron cylinder shows any sign of having been rebuilt before, such as honing marks in the bore or non-stock rubber parts, it is best to replace the cylinder rather than risk another rebuild.

If an aluminum hydraulic cylinder is scratched, pitted, scored, or corroded in any way, the cylinder must be replaced. Aluminum cylinders cannot be honed because they have a wear-resistant, anodized finish. Honing would cut through the anodizing and the resulting bare aluminum finish would corrode very rapidly and be too rough for good sealing.

Bleeder screw removal

If it is necessary to remove a frozen bleeder screw, spray the screw threads with penetrating oil and allow it to soak in for at least 10 minutes. Then, place a deep socket over the bleeder screw so it contacts the body of the cylinder around the base of the screw. Wearing safety glasses to prevent injuries, strike the socket several medium-firm blows; this should slightly deform the metal of the cylinder and break the surface tension between the bleeder screw threads and those in the cylinder opening. Attempt to remove the bleeder screw with the appropriate wrench, but do not apply too much turning force or you will twist the end of the screw off. Once this happens, it is virtually impossible to remove the portion of the screw that remains in the cylinder.

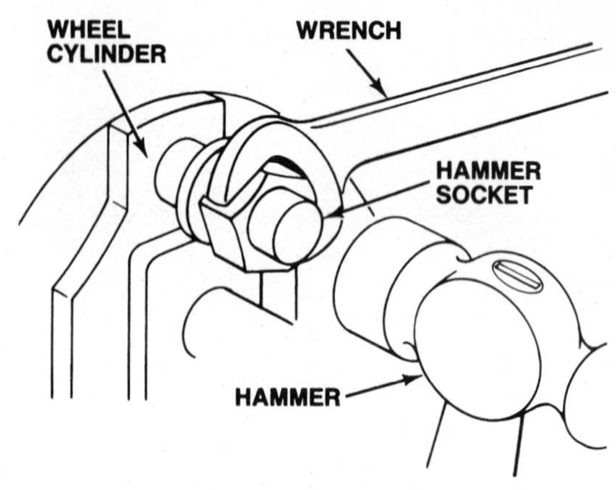

Figure 5-15. A hammer socket is a special tool used to remove frozen bleeder screws.

Repeat the procedures with the penetrating oil, socket and hammer, and bleeder wrench as needed until the screw is free. To make this process easier, special hammer sockets are available from the aftermarket that allow you to apply both hammer blows and turning force at the same time, figure 5-15. Once you remove a bleeder screw, buff its threads clean on a wire wheel, and run a drill bit through its passageways to make sure they are clear.

Cylinder Honing

Honing is a procedure in which abrasive stones are rotated inside the cylinder bore to remove a small amount of metal, along with any scratches, pitting, scuff marks, and rust. Honing restores the sealing surface for the piston seals, however, the new finish is never as smooth as the "bearingized" surface of a new cylinder. As a result, the cup seals in a rebuilt cylinder that has been honed wear faster than those in a replacement cylinder. Many shops today will only rebuild a master cylinder or wheel cylinder if the bore does not need to be honed. In addition, some manufacturers recommend against honing the cylinders on their cars. Check the factory shop manual for the exact recommendations on the car you are servicing.

If the bore of a cast-iron master cylinder or wheel cylinder can be honed, follow this procedure:

1. If a wheel cylinder is to be honed on the car, make sure it is tightly attached to the backing

Figure 5-16. Insert the hone all the way into the cylinder before starting the drill motor.

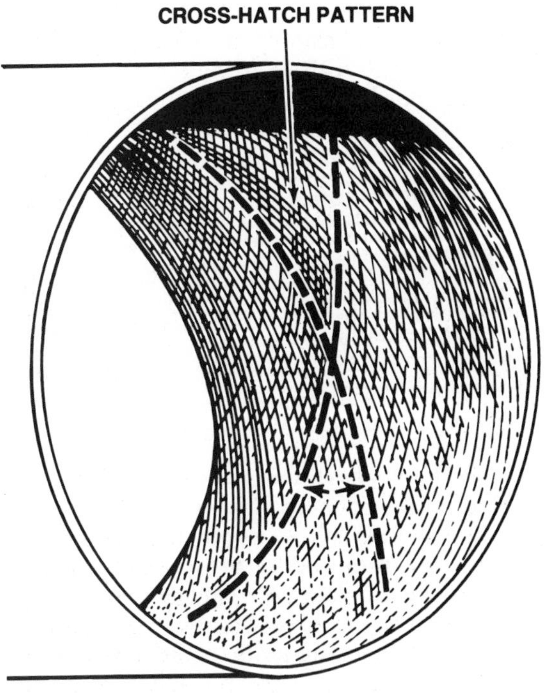

Figure 5-17. Proper honing will leave an even crosshatch pattern in the cylinder bore.

plate. Otherwise, clamp the cylinder in a vise by its mounting flange. Do not clamp on the cylinder body or the bore may be distorted.
2. Select a suitably sized cylinder hone and chuck it in a drill motor.
3. Lubricate the cylinder bore with brake fluid, and insert the hone into the bore, figure 5-16.
4. Operate the drill motor at approximately 500 rpm, and move the hone back and forth with smooth, even strokes. Allow the hone to extend just beyond the open ends of the cylinder to ensure that the entire length of the bore is honed. Use extra caution when honing with a stone hone; if the stones come out of the bore, they can strike the vise or backing plate and be broken. Stones may also be torn loose and thrown free, possibly causing injury. Always wear safety glasses when honing hydraulic cylinders.
5. Keep the bore lubricated with brake fluid, and hone for approximately 10 seconds. Let the hone come to a full stop, then remove it from the cylinder bore.
6. Wipe the bore clean with a rag, and check the surface finish; it should be clean and free of rust, corrosion, and scratches. The hone should also have created an even crosshatch pattern as shown in figure 5-17.
7. If necessary, repeat the honing process for another 10 seconds. If the bore does not clean up after several repetitions of this procedure, replace the cylinder.

After the bore is honed, thoroughly clean the cylinder with a non-petroleum-base brake cleaning solvent to remove all residue and grit.

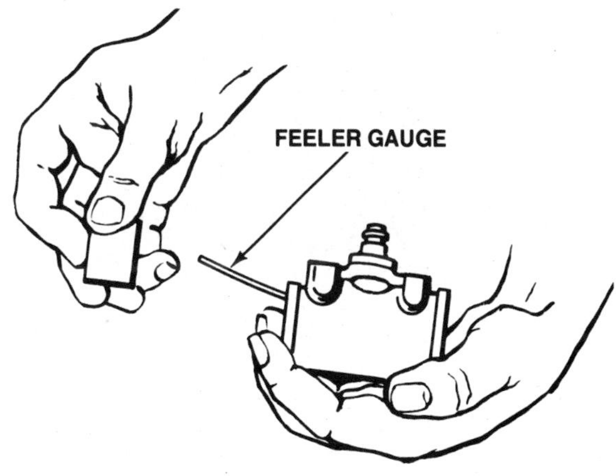

Figure 5-18. Using a feeler gauge to check cylinder bore size.

Cylinder Bore Measurement

After a hydraulic cylinder has been honed, you must measure the size of the cylinder bore to make sure that too much metal has not been removed. There are two methods that can be used to do this, one uses a feeler gauge strip, and the other uses a go/no-go gauge.

To check the bore using a feeler gauge, place a narrow (¼ inch or 6 mm wide) strip of .006 inch (.15 mm) feeler gauge inside the bore. Attempt to insert one of the cylinder pistons into the bore with the feeler gauge in place, figure 5-18; if the piston will fit, the bore is oversize

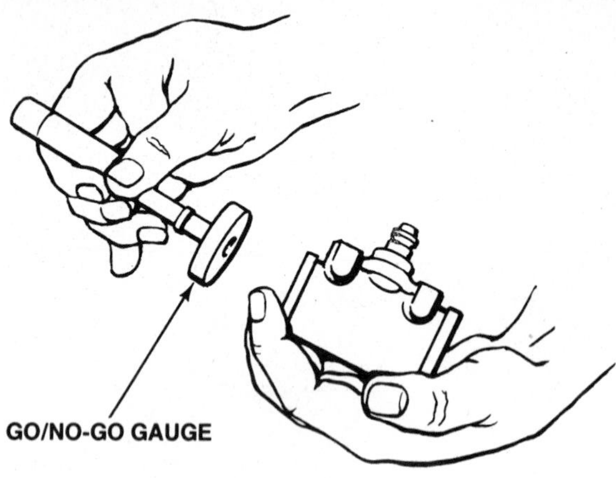

Figure 5-19. Using a go/no-go gauge to check cylinder bore size.

and the cylinder must be replaced. Traditionally, most manufacturers have allowed up to .006" (.15 mm) of piston clearance, however, many newer cylinders with smaller diameters require tighter clearances. If you are unsure of the proper specification, consult the shop manual for the vehicle you are servicing.

To check the bore with a go/no-go gauge, select the proper size of plug from the gauge kit and attach it to the handle. The correct plug is the same size as the nominal cylinder bore plus .006 inch (.15 mm). Attempt to insert the plug into the bore, figure 5-19; if it will fit, the bore is too far oversize and the cylinder must be replaced. Although go/no-go gauges were once quite common, they are seldom used today because of the wider variations in clearance specifications mentioned above.

MASTER AND WHEEL CYLINDER OVERHAUL PROCEDURES

The following pages provide step-by-step overhaul procedures for the Ford cast-iron dual master cylinder, the General Motors aluminum QTU dual master cylinder, and a typical two-piston wheel cylinder. While these are not the only cylinders in use today, they are representative of those found on most vehicles. Before you begin any overhaul, read completely through the procedure to obtain a better idea of the entire job.

The first procedure covers the Ford cast-iron dual master cylinder. This part is a good example of the basic master cylinder design used by manufacturers on virtually all RWD cars and trucks since 1967. Although the Ford dual master cylinder has been made in both cast iron and aluminum versions, the cast iron model is shown in the photo sequence.

The second procedure covers the General Motors aluminum QTU master cylinder. This part has been used on most of GM's FWD cars, and some RWD models beginning in 1980. This modern master cylinder design has a built-in pressure differential switch, and dual proportioning valves that thread directly into the fluid outlets to the rear brakes.

The final procedure covers a typical two-piston wheel cylinder. It is common practice to overhaul wheel cylinders while they are mounted on the backing plate, if possible. This holds the cylinder solidly in position, and reduces the time required for the job. However, as discussed earlier, some wheel cylinders must be removed for rebuilding because they are recessed into the backing plate, or the backing plate is fitted with piston stops. When a wheel cylinder is rebuilt off the car, it is mounted in a vise.

FORD CAST-IRON DUAL MASTER CYLINDER OVERHAUL

1. Remove the master cylinder reservoir cover and pour the brake fluid into a drain pan.

2. Clamp the cylinder in a vise by its mounting flange, and position a drain pan under it to catch any fluid released during disassembly.

3. Depress the primary piston, and remove the retaining snap ring from its groove at the open end of the cylinder.

4. Release the primary piston and allow return spring pressure to force it from the bore.

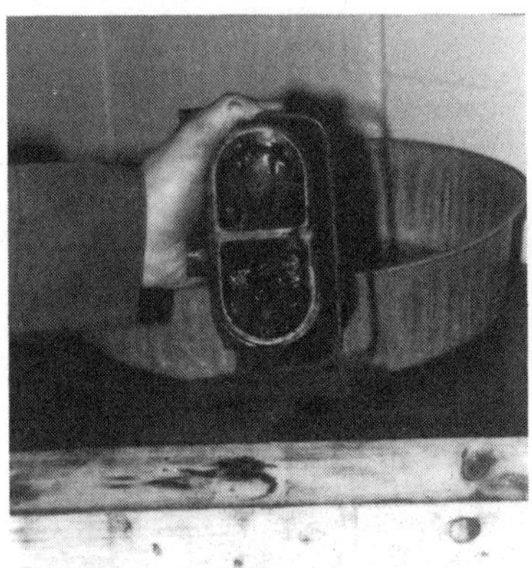

5. Remove the cylinder from the vise, and strike its open end against a wood block to remove the secondary piston.

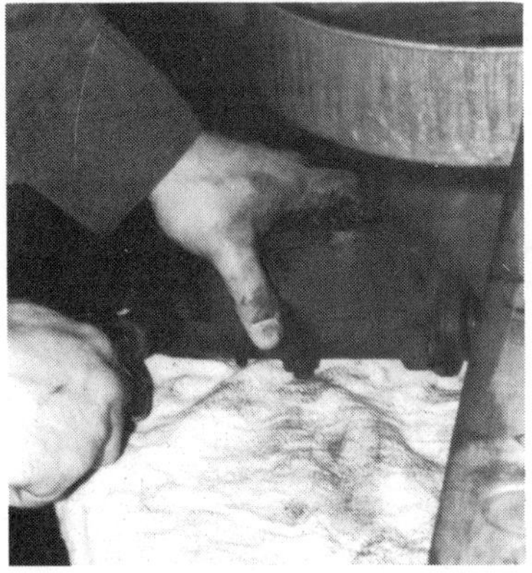

6. To remove frozen pistons, place the open end of the cylinder against a wood block, then apply compressed air to the forward fluid outlet port while blocking the rear port.

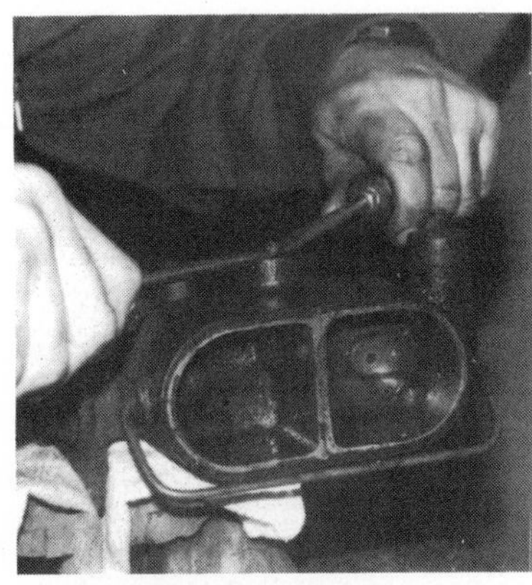

7. Thread self-tapping screws into the tubing seats, and pry out the seats using two screwdrivers. Remove the residual check valves (where used) from under the seats.

8. Inspect the cylinder bore and hone it if necessary.

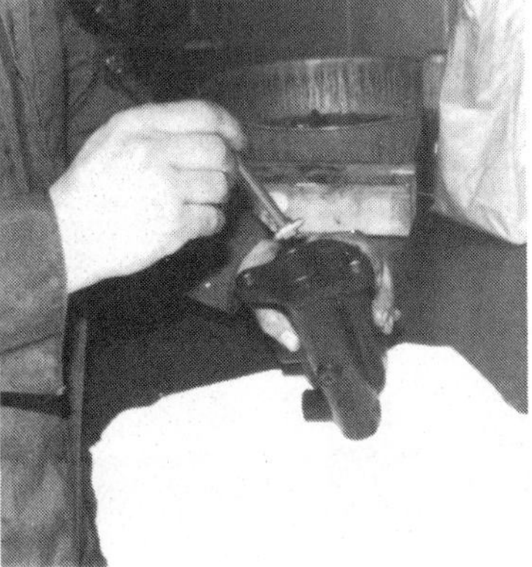

9. Measure the bore size, and replace the master cylinder if the bore is worn, or has been honed, too far oversize.

FORD CAST-IRON DUAL MASTER CYLINDER OVERHAUL

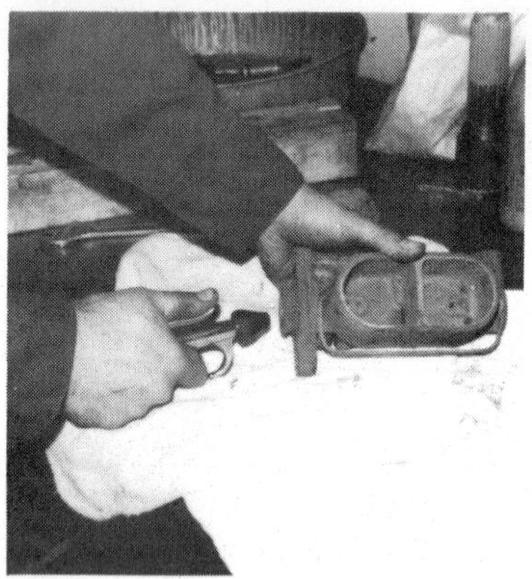

10. Thoroughly clean the master cylinder body, bore, and fluid passages, then blow them dry with compressed air.

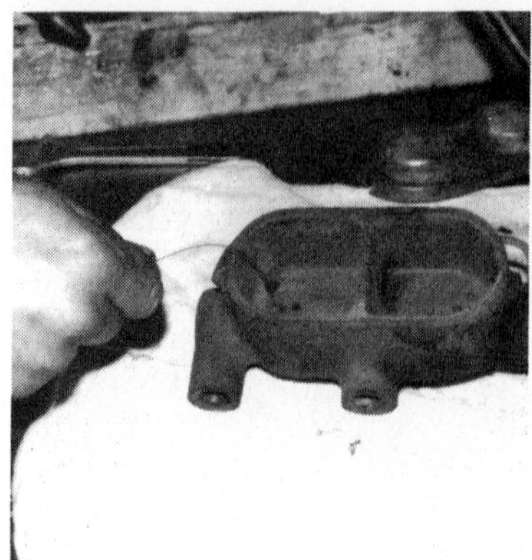

11. Push a thin wire through the compensating and replenishing ports to ensure they are open.

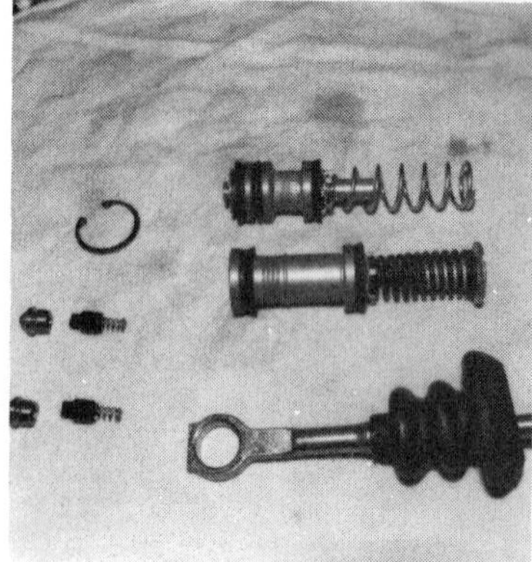

12. Thoroughly clean the cylinder internal parts, including those from the rebuild kit, and lay them out on a clean shop towel.

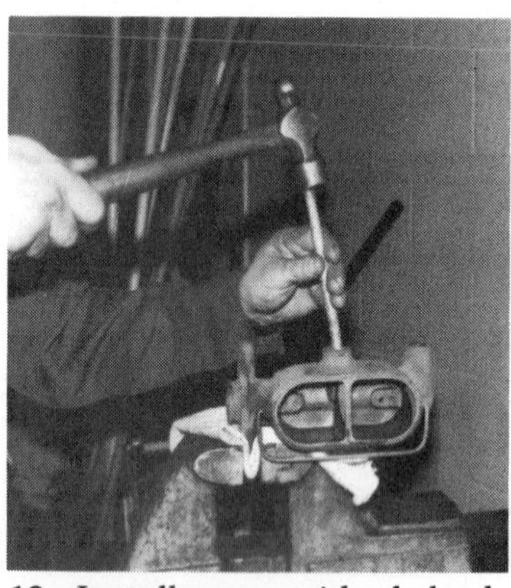

13. Install new residual check valves (where used) into the fluid outlet ports, then drive new tubing seats in over them with a brass or wooden dowel.

14. Lubricate the secondary piston assembly with brake fluid or assembly lube, then install it into the bore, spring end first.

15. Lubricate the primary piston assembly with brake fluid or assembly lube, then install it into the bore, spring end first.

16. Depress the primary piston, and install the retaining snap ring in its groove at the open end of the cylinder.

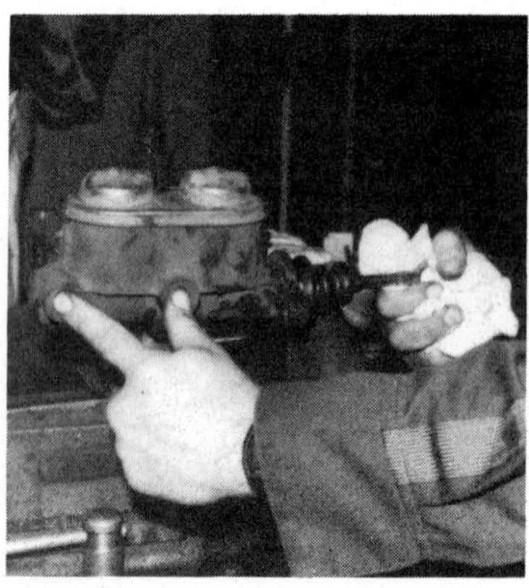

17. Bench bleed the master cylinder to remove trapped air.

18. Install the reservoir cover and diaphragm on the cylinder, and secure them in place with the bail wire.

GENERAL MOTORS ALUMINUM QTU MASTER CYLINDER OVERHAUL

1. Remove the master cylinder reservoir cover and pour the brake fluid into a drain pan.

2. Clamp the cylinder in a vise by its mounting flange, and position a drain pan under it to catch any fluid released during disassembly.

3. Remove the fluid reservoir with a pry bar. Do not remove the Quick-Take-Up valve from the cylinder body unless it is being replaced.

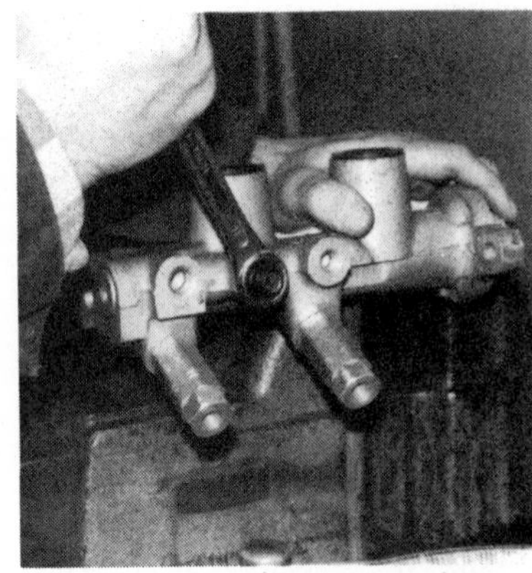

4. Unscrew and remove the pressure differential switch.

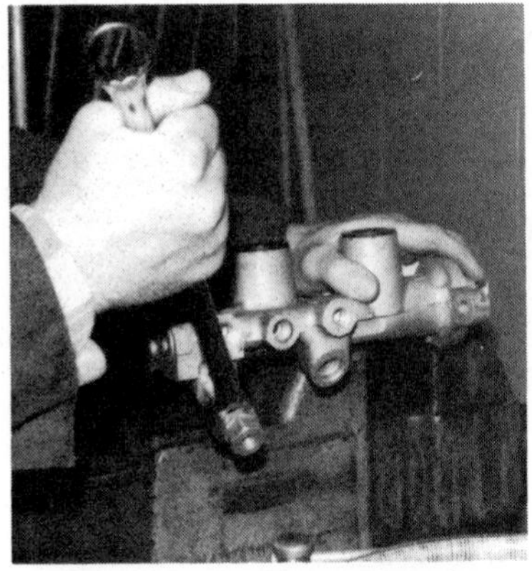

5. Unscrew and remove the two proportioning valves.

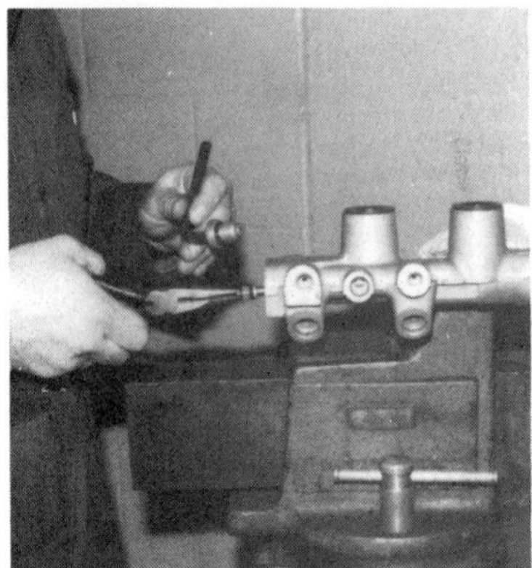

6. Unscrew the Allen-head plug and remove the pressure differential switch piston.

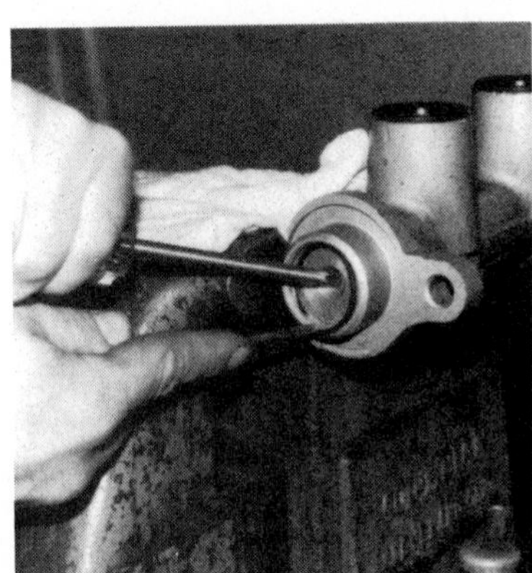

7. Depress the primary piston and remove the retaining lock ring from its groove at the open end of the cylinder.

8. Release the primary piston and allow return spring pressure to force it from the bore.

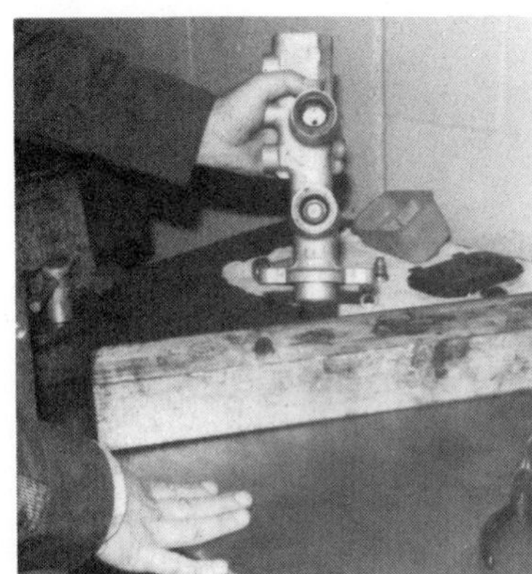

9. Remove the cylinder from the vise and strike its open end against a wood block to remove the secondary piston, spring retainer, and spring.

GENERAL MOTORS ALUMINUM QTU MASTER CYLINDER OVERHAUL

10. To remove frozen pistons, place the open end of the cylinder against a wood block, then apply compressed air to one forward fluid outlet while blocking the other.

11. Inspect the cylinder bore, and replace the cylinder if any damage is visible.

12. Thoroughly clean the master cylinder body, bore, and passages with a non-petroleum-base brake cleaner.

13. Thoroughly clean and dry the cylinder internal parts, including those from the rebuild kit, and lay them out on a clean shop towel.

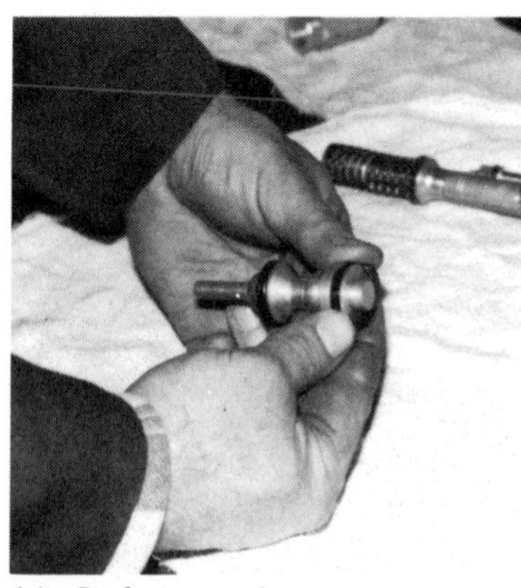

14. Lubricate the new seals with brake fluid or assembly lube and install them on the secondary piston.

15. Lubricate the secondary piston assembly with brake fluid or assembly lube and install the spring, spring retainer, and piston into the cylinder bore.

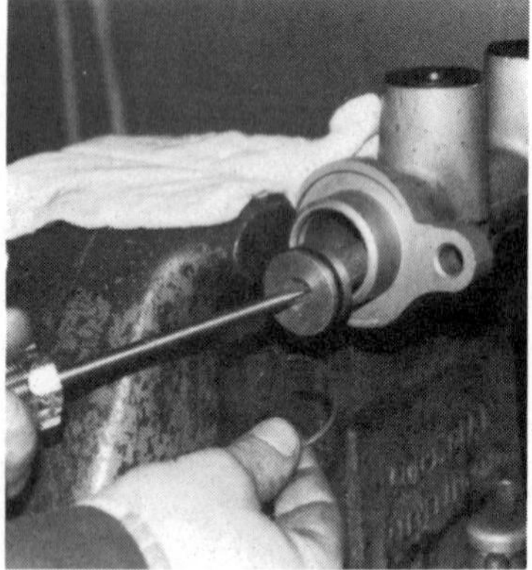

16. Lubricate the primary piston assembly with brake fluid or assembly lube and install the piston into the cylinder bore.

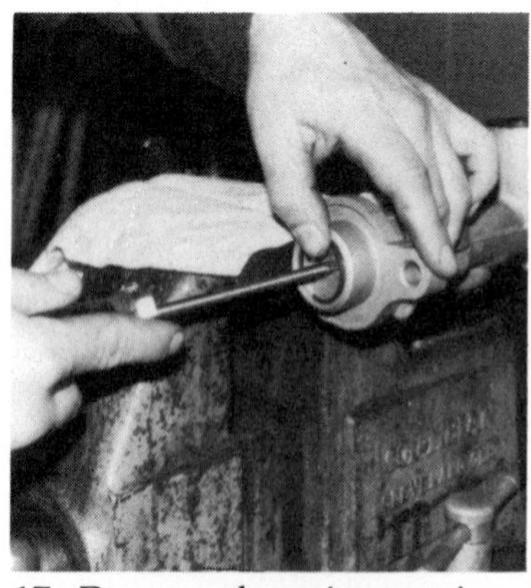

17. Depress the primary piston and install the retaining lock ring into its groove at the open end of the cylinder.

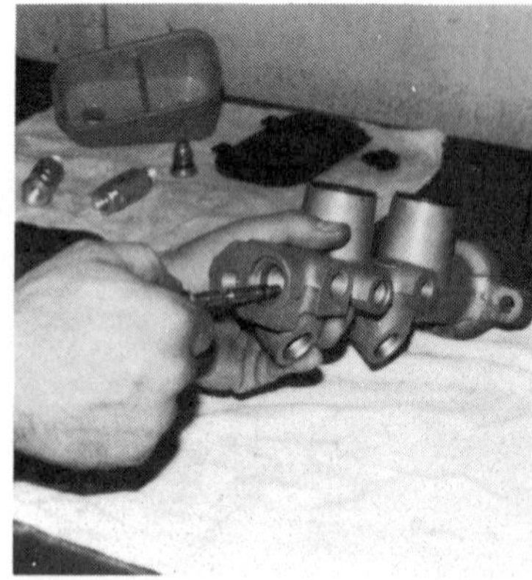

18. Lubricate the pressure differential switch piston with brake fluid or assembly lube and install it into the master cylinder body.

GENERAL MOTORS ALUMINUM QTU MASTER CYLINDER OVERHAUL

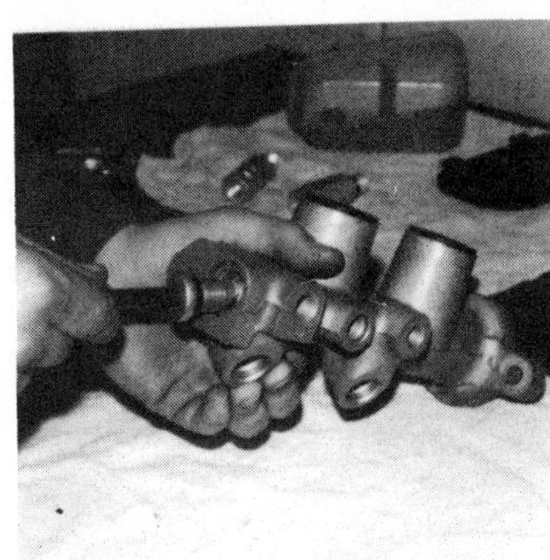

19. Install a new O-ring on the Allen-head plug, and screw the plug into the cylinder body. Torque the plug to 40-140 inch pounds (4.5-16 Nm).

20. Install a new O-ring on the pressure differential switch and screw the switch into the cylinder body. Torque the switch to 15-50 inch pounds (1.7-5.6 Nm).

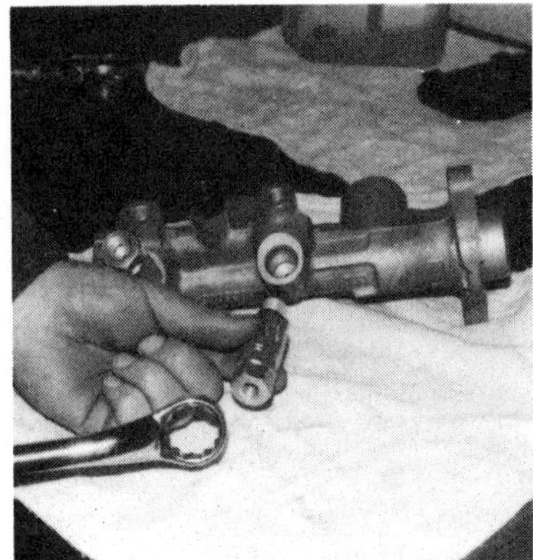

21. Install the proportioning valves onto the master cylinder body. The valves may be aluminum or steel. Never mix the two types; use either two aluminum or two steel valves.

22. Lubricate the new fluid reservoir grommets and install them into the master cylinder body.

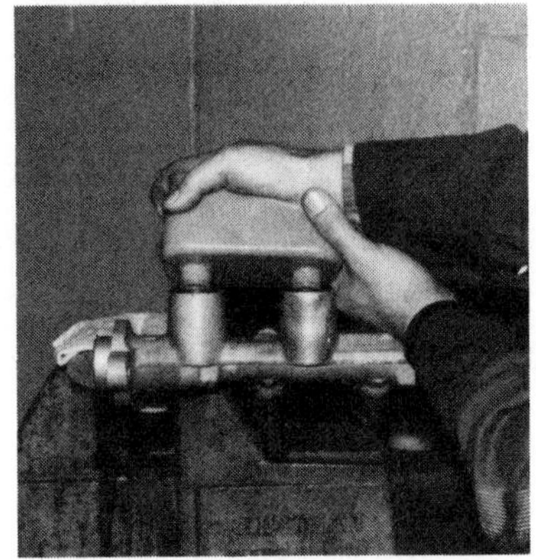

23. Install the fluid reservoir into grommets at an angle, then rock it back and forth under pressure until it is seated.

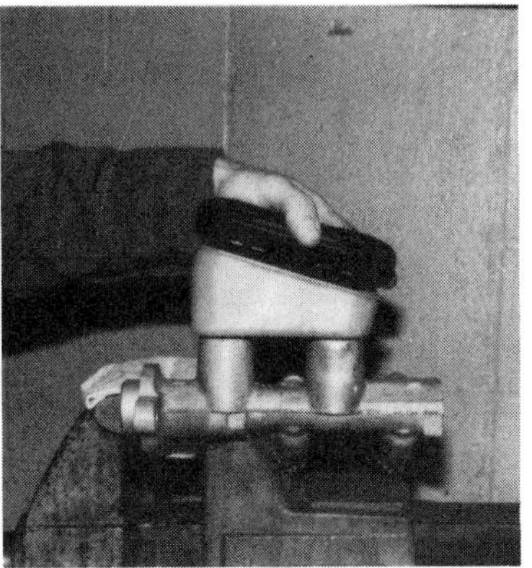

24. Install the reservoir cover and diaphragm on the master cylinder.

TWO-PISTON WHEEL CYLINDER OVERHAUL

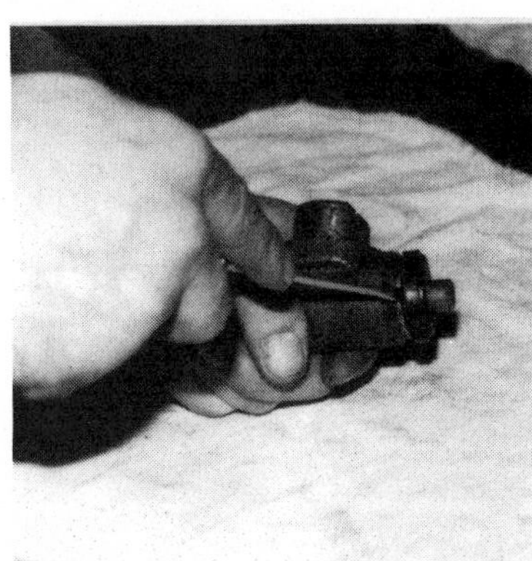

1. Remove the shoe links and rubber boots from the ends of the cylinder.

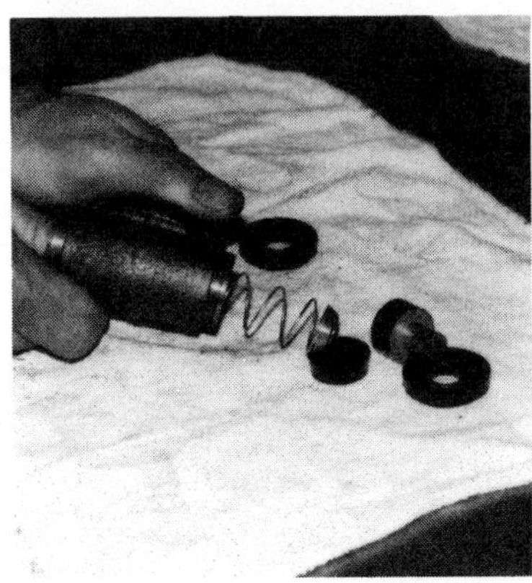

2. Press out the pistons, cups, and cup expander assembly by hand or with a wooden punch.

3. Inspect the cylinder bore. If there is any damage, replace an aluminum cylinder, or hone a cast-iron part.

TWO-PISTON WHEEL CYLINDER OVERHAUL

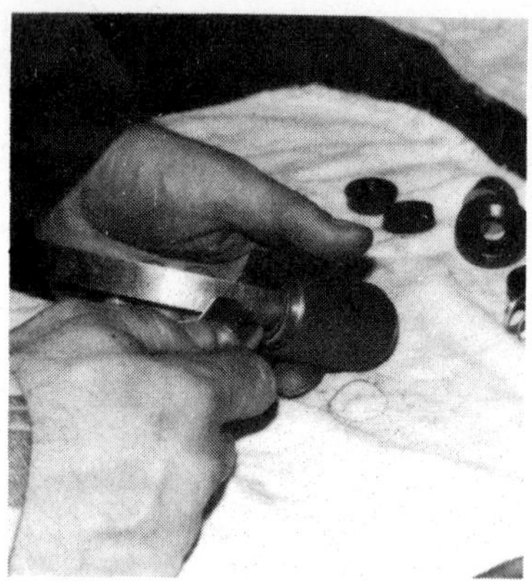

4. Measure the bore and replace the wheel cylinder if the bore is too far oversize.

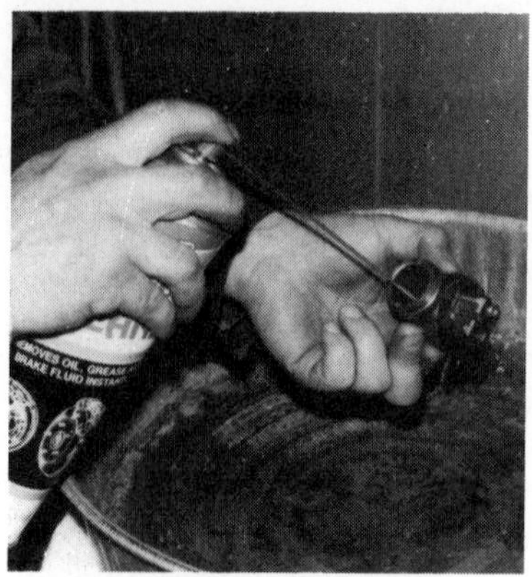

5. Thoroughly clean the wheel cylinder body with a non-petroleum-base brake cleaner.

6. Thoroughly clean and dry the cylinder internal parts, including those from the rebuild kit, and lay them out on a clean shop towel.

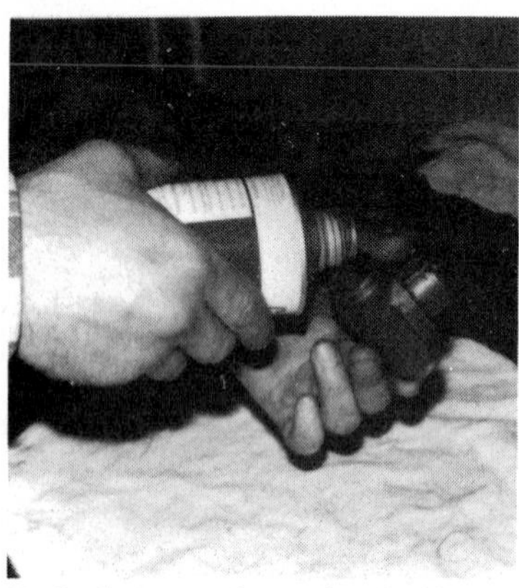

7. Lubricate the cylinder bore with brake fluid or assembly lube.

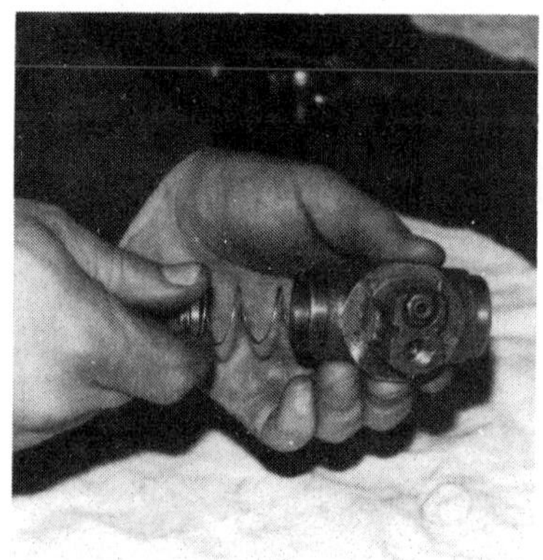

8. Check that the cup expanders are secure on the spring ends, and install the assembly into the cylinder bore.

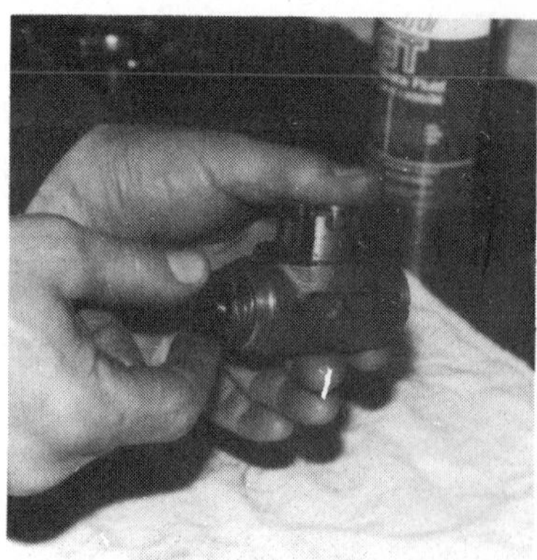

9. Lubricate the cup seals and install them into the cylinder bore, one at each end, with the lip facing inward. Make sure the seals do not cock in the bore.

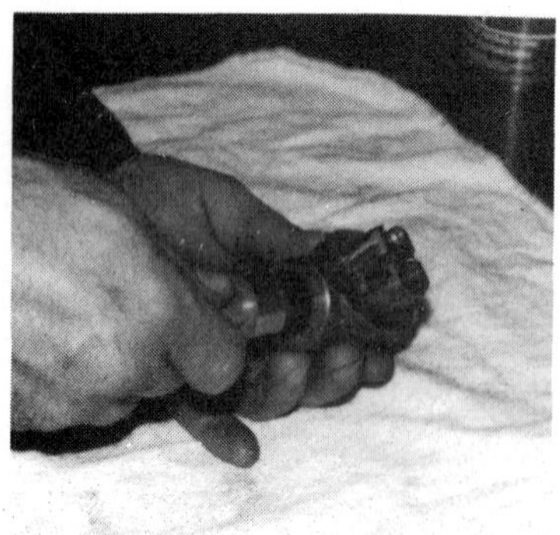

10. Lubricate the pistons and install them into the cylinder body, one at each end, with the flat side contacting the backs of the cup seals.

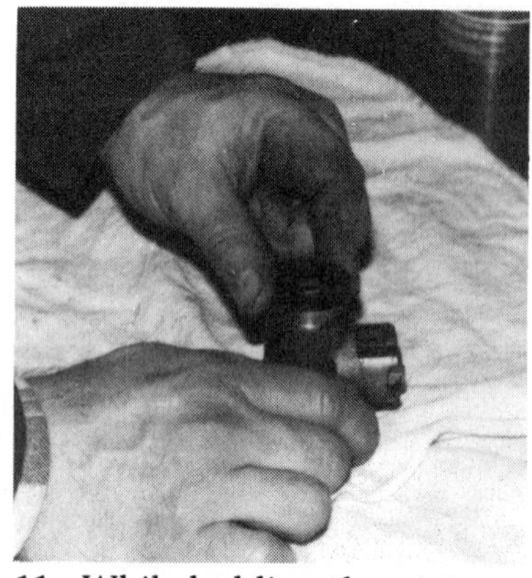

11. While holding the pistons in position, install the dust boots on each end of the cylinder.

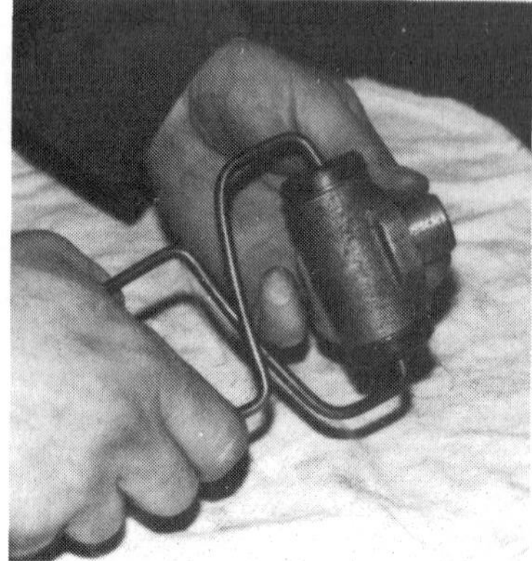

12. If the cup expander spring has very strong tension, use a wheel cylinder clamp to hold the assembly together.

PART THREE

Wheel Friction Assembly Service

Chapter

6

Drum Brake Service

Drum brake service consists of four main operations: drum removal, brake inspection, brake adjustment, and shoe replacement. The operations are listed in this order because you must remove the brake drum before you can inspect the friction assembly or replace the brake shoes. And, while it is possible to adjust the brakes without first inspecting them, this is not recommended. Although wheel cylinder service is also a part of drum brake repair, it is more closely related to the hydraulic system. See Chapter 5 for detailed wheel cylinder service information.

Manually adjusted drum brakes were once used on all four wheels of every car, and brake adjustment and shoe replacement were common service procedures. Today, most brakes are self adjusting, and drum brakes are used exclusively on the lightly loaded rear wheels where the brake linings may last the life of the vehicle. This means that except for routine inspection, rear drum brake service is usually not required until the car has high mileage, a hydraulic problem with the wheel cylinder, or a mechanical failure of some part in the friction assembly. When this occurs, it is best to do a complete brake overhaul rather than a single repair.

This chapter explains the procedures used to remove brake drums, inspect drum brake friction assemblies, manually adjust drum brakes, and replace brake shoes. It also describes detailed shoe replacement procedures for several specific drum brakes. Although this chapter focuses primarily on rear drum brakes, most of the information applies to front drum brakes as well.

BRAKE DRUM REMOVAL

Before you can inspect a drum brake or replace its shoes, you must first remove the drum to gain access to the friction assembly. To do this, it may be necessary to: loosen the parking brake cable and service brake adjustment, remove a drum retaining device, and break the drum loose from accumulated rust and corrosion. Whenever you remove a drum, mark it with an L (left) or an R (right) so you can replace it in the same position.

The removal procedures differ for fixed and floating drums, however, either type can be difficult or impossible to remove unless the parking and service brake adjustments are loosened first. This occurs because wear at the open edge of the drum, or scoring of the brake linings and drum friction surface, creates a ridge or a series of interlocking grooves that hold the drum in

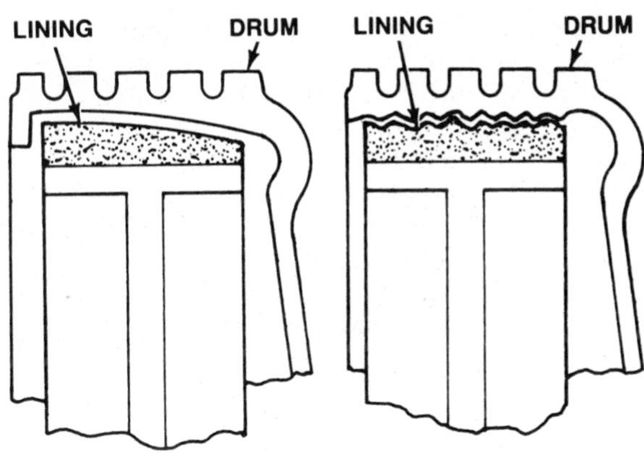

Figure 6-1. Brake lining and friction surface wear can make a drum difficult to remove.

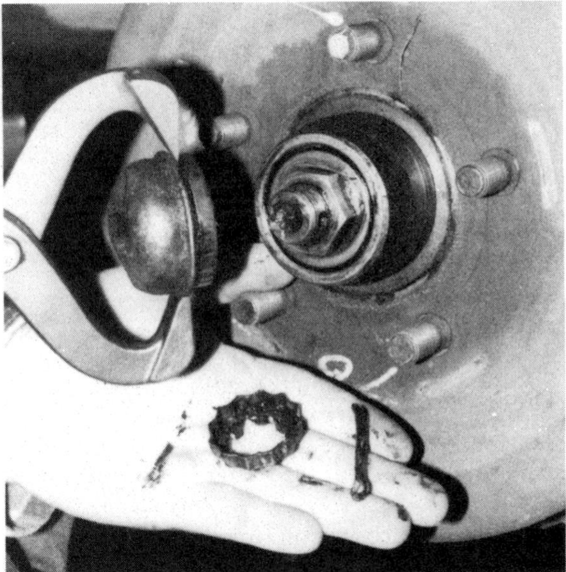
Figure 6-2. This nut adjusts the wheel bearings and retains the drum on the spindle.

place, figure 6-1. If you cannot remove a drum once the retaining devices are removed and the drum is loose on the hub or axle, loosen the parking and service brake adjustments as described later in the chapter so the high points on the linings and drum clear one another.

Fixed Drum Removal

Fixed brake drums that are cast as a unit with the hub assembly are common on the rear axles of FWD cars, and the front axles of older RWD cars with front drum brakes. There are two kinds of fixed drums: those whose hubs contain wheel bearings, and those whose hubs have a tapered opening that fits directly onto the axle. Both types are held in place by a single large retaining nut.

Figure 6-3. A puller may be required to remove a fixed drum if the inner wheel bearing sticks on the spindle.

In most applications, the nut that retains a fixed brake drum also secures the wheel bearings in the hub. When tapered-roller bearings are fitted, the retaining nut is used to adjust bearing end play, and is typically finger tight or torqued to only a few inch-pounds. If ball bearings are used, a spacer and shim pack between the inner and outer bearings sets the bearing clearance, and the retaining nut is tightened to a moderately high torque.

To remove a fixed drum that contains wheel bearings:

1. Use a pair of dust cap pliers to remove the cap from the center of the hub and expose the retaining nut, figure 6-2.
2. Remove the retaining nut. Most cars have a nut secured by a castellated nut lock and a cotter pin; remove the pin along with any other locking devices, then unscrew the nut. Some imported cars have a split nut with a pinch bolt; loosen the bolt, then unscrew the nut.
3. Pull outward on the drum and slide it off the spindle. Take care not to let the thrust washer and outer wheel bearing fall on the ground as they clear the end of the spindle. Also, do not drag the inner wheel bearing across the retaining nut threads on the spindle.

Sometimes, the inner races of the inner wheel bearing will stick or have a light press fit on the spindle. In these cases, the drum will be difficult to remove unless you use a puller, figure 6-3. When doing this, the inner bearing may remain on the spindle, forcing the grease seal out of the hub. Once the drum is off the car, use a puller or a pair of pry bars to carefully remove the inner bearing and grease seal.

Once the drum is off the car, inspect the grease in the hub and on the wheel bearings. If the grease appears in good condition, set the drum on the bench, open side down, and cover

Figure 6-4. Nuts that retain drums with tapered hubs are tightened to high torques.

Figure 6-5. Taper-hub drum pullers are tightened against the axle with a hammer and lever.

the outer bearing and hub opening so the bearings will not become contaminated. If the grease appears old and dirty, repack the wheel bearings before you reinstall the drum. As a general rule, always repack the wheel bearings when performing a complete brake job. See Chapter 11 for information on wheel bearing service.

Taper-hub fixed drum removal

Some cars and light trucks have fixed drums that install directly on the rear driven axles. The hubs of these drums have a tapered opening that fits over a matching taper on the axle; a groove in the hub fits over a locating key in the axle to prevent drum slippage. Once the drum retaining nut is removed, a special puller is required to remove the drum because it becomes tightly wedged onto the axle.

The retaining nuts of fixed drums with tapered hubs are tightened to a high torque and require a great deal of force to remove or install. Unless you take certain precautions, this force may damage the axle and drivetrain, or cause personal injury. Always loosen the retaining nuts, or do final tightening, with the car on the ground and the weight of the vehicle on the wheels. Firmly apply the parking brake, or have an assistant apply the service brakes, to reduce shock loading on the axle and drivetrain as you remove or install the retaining nut.

Figure 6-6. Strike the puller shaft to dislodge the drum from the axle.

To remove a fixed drum that is a taper fit on the axle:

1. Remove the cotter pin from the drum retaining nut.
2. Use an impact wrench or large breaker bar to loosen the retaining nut, figure 6-4.
3. Raise and properly support the vehicle, then remove the wheels from the brake drums.
4. Unscrew the retaining nut from the axle, then loosen the parking and service brake adjustments to ensure that the drum will clear the brake shoes.
5. Attach the drum puller as shown in figure 6-5, then use a hammer and the special lever to tighten the shaft against the axle until the puller is under heavy pressure.
6. Strike the end of the shaft two or three heavy blows with a three- to five-pound hammer, figure 6-6.
7. Check the pressure on the puller shaft; if it is reduced, the drum is moving. Retighten the shaft and strike it several more blows with the hammer. Continue to tighten and strike the shaft until the drum pops free.

If pressure on the puller does not decrease when the shaft is struck with the hammer, and

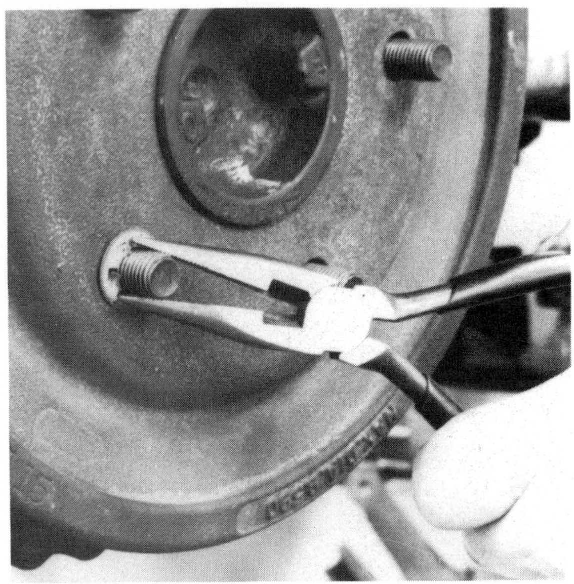

Figure 6-7. Remove speed nuts with pliers.

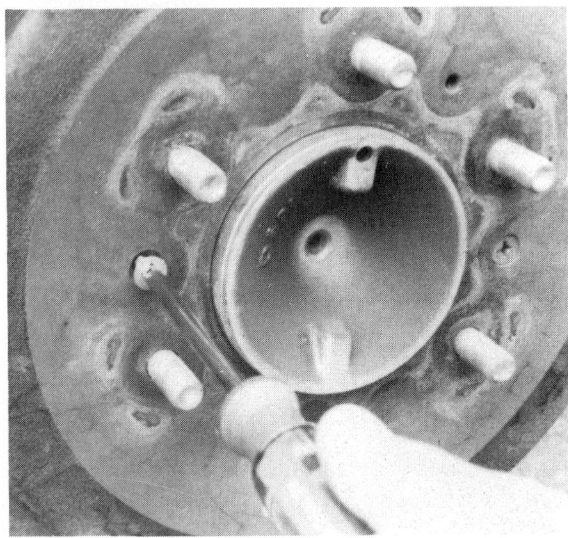

Figure 6-8. Use the appropriate tool to remove screws or bolts that retain drums.

you are trained in the proper use of an oxy-acetylene torch, heat the hub where the drum fits on the axle. Do not apply excessive heat or drum and axle damage may occur. Again strike the shaft and check for drum movement. Continue to heat the hub, and tighten and strike the puller shaft, until the drum pops free.

Floating Drum Removal

Floating brake drums are held in place by the wheel and lug nuts during normal operation. However, they may also be retained by speed nuts over the wheel studs, or bolts or screws threaded through the drum into the hub or axle

Figure 6-9. A few well-placed blows can free a frozen drum from the hub or axle.

flange. To remove speed nuts, grasp them with a pair of pliers and unscrew them from the studs, figure 6-7. On most cars, you can discard the speed nuts because they are unnecessary once the car has left the factory. However, certain Ford cars have high shoulders on the wheel studs that can catch the drum and hold it in a cocked position when the wheel is installed. To prevent the possibility of drum damage in these applications, replace the speed nut after the brake work is completed. To remove bolts or screws that retain a drum, use the appropriate tool, figure 6-8. Set aside the bolts or screws so you can reinstall them when the drum is replaced.

Once the retaining devices are removed, the drum should move freely on the hub or axle, and slip off over the brake shoes. If necessary, loosen the parking and service brake adjustments until the drum will clear the shoes. If the drum does not move on the hub or axle once the brake adjustments are loosened, the drum is rusted or corroded in place. Where the frozen drum mounts on the hub assembly of a non-driven axle, it is often faster and easier to remove the drum and hub as a unit, like a fixed drum, rather than attempt to separate the drum from the hub. However, when the drum is on a driven axle, or must be removed from the hub for replacement, use this procedure:

1. Use a pointed probe to scribe the joint where the drum mounts on the hub or axle. This breaks the surface tension, and like a glass cutter, helps crack the joint free.
2. Strike the outer edge of the drum several solid blows with a three- to five-pound hammer at approximately a 45-degree angle, figure 6-9. A "dead-blow" urethane hammer loaded with lead shot works very well for this purpose.
3. If the drum does not come free, spray penetrating oil around the stud holes and the joint

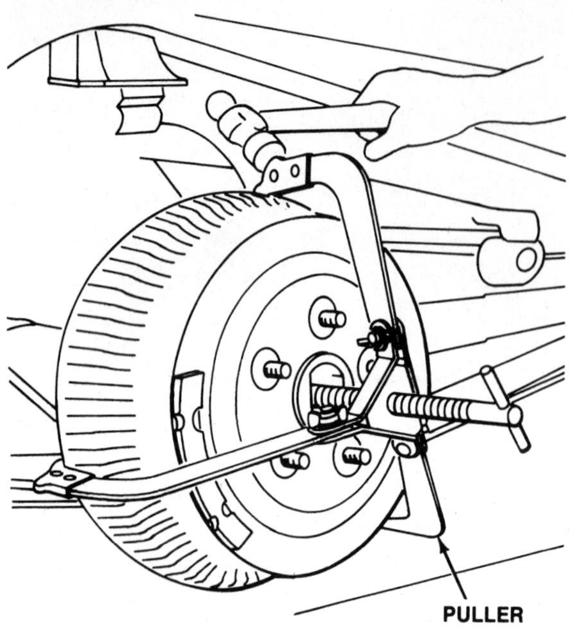

Figure 6-10. A puller is a last resort when removing floating drums.

where the drum mounts on the hub or axle. Allow the oil to soak in for a few minutes, then strike the drum several more blows with the hammer.

4. If the drum still does not come free, install a three-jaw puller on the drum, figure 6-10. Tighten the puller shaft against the end of the spindle or axle, and strike the puller several solid blows with a metal hammer. The special puller shown in the illustration is designed with hammer pads on the puller arms that allow force to be applied at the edges of the drum. On a standard three-jaw puller, strike the puller shaft to free the drum.

5. If the drum still does not come free, and you are trained in the proper use of an oxy-acetylene torch, heat the drum in the area where it mounts on the hub or axle. Continue to heat the drum, and tighten and strike the puller, until the drum comes free.

6. Once the drum is off the car, clean all traces of penetrating oil from the drum and hub or axle flange to prevent contamination of the brake linings.

The methods of drum removal described above should be exercised with caution. Excessive hammer blows, puller force, or heat can warp a drum in the process of removing it. Whenever a drum is removed using one of these methods, it should be checked for damage and distortion as described in Chapter 8 before it is returned to service. In addition, high heat on a hub that contains wheel bearings can melt the grease and lead to seal and bearing failure. If a hub that has been heated contains wheel bearings, repack the bearings and replace the grease seal before you return the car to service.

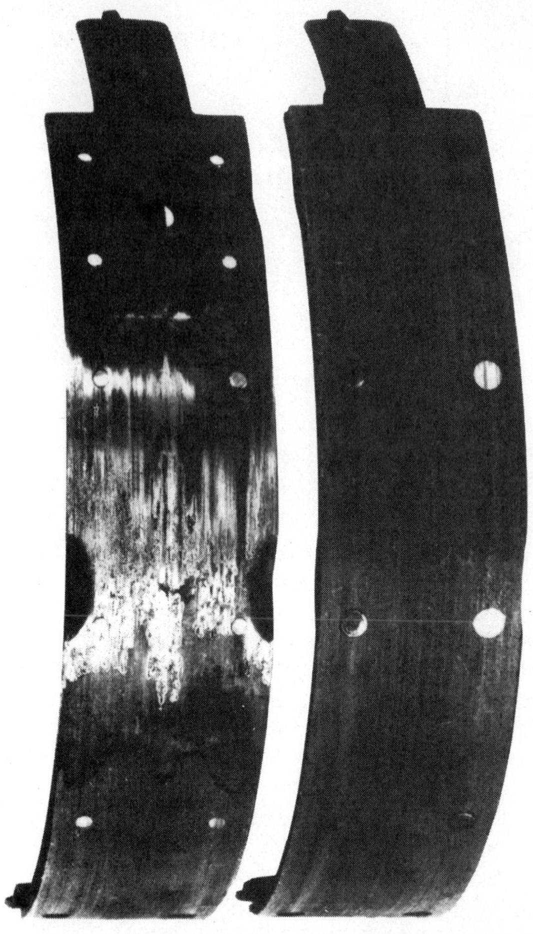

Figure 6-11. Metal-to-metal contact is the destructive result of brake neglect.

DRUM BRAKE INSPECTION

Rear drum brake shoes wear more slowly and require less frequent service than either drum brake shoes or disc brake pads on the front-wheel brakes. This occurs because rear drum brakes provide as little as 20 percent of the total vehicle braking. Also, when compared to disc brake pads, drum brake shoes have greater friction material surface area to absorb heat and distribute wear. Because the rear brake linings wear so slowly, rear brake inspection is an often overlooked part of vehicle maintenance.

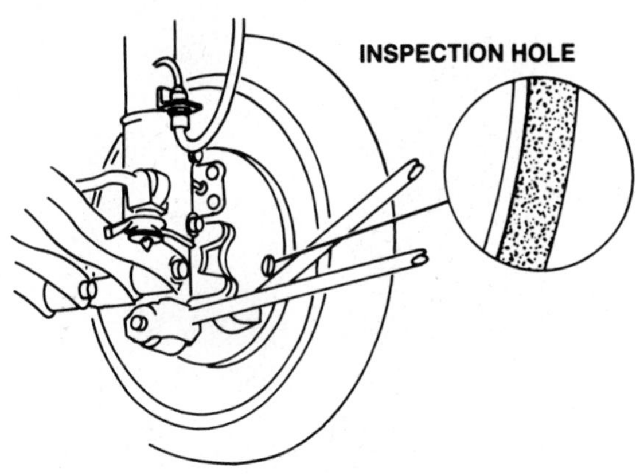

Figure 6-12. A lining inspection hole in the brake backing plate.

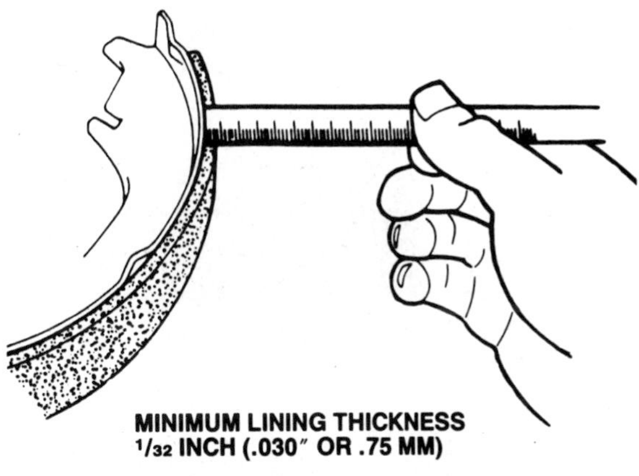

Figure 6-13. Replace the shoes when the linings wear below the minimum safe thickness.

Most vehicle manufacturers recommend that drum brake friction assemblies be inspected every 7,500 miles (12 000 km). You should also inspect the drum brakes before you perform a manual adjustment, or whenever you do any work on the front disc brakes. Unfortunately, many car owners wait to have the brakes inspected until they hear the grinding noise that means a lining is worn completely away, and a shoe lining table is making metal-to-metal contact with a brake drum, figure 6-11. Once this happens, the drum must be machined or replaced as described in Chapter 8. It is much better to replace a set of brake shoes sooner than absolutely necessary, rather than risk drum damage by trying to get the maximum possible life from the linings.

Although some backing plates and brake drums contain inspection holes that allow you to check lining thickness, figure 6-12, you cannot do a complete brake inspection unless you remove the drum. Once the drum is off the car, inspect it as described in Chapter 8. Next, thoroughly clean the friction assembly with a brake washer, or a brake vacuum fitted with a high efficiency particulate air (HEPA) filter. Brake dust and other grit increase lining wear and cause brake noise, and a clean friction assembly is easier to inspect.

When the friction assembly is clean and dry, make a simple three-step inspection. First, check the condition of the brake shoes and linings. Second, make sure there are no fluid leaks from the wheel cylinder or axle, and check the wheel cylinder operation. And finally, confirm that all the brake hardware is mounted properly, in good condition, and operating as designed. These operations are detailed below. Additional checks, made with the friction assembly disassembled, are covered in the brake shoe replacement section later in the chapter.

Lining Inspection

The thickness of the brake linings is the primary factor that determines the need for replacement, so check them first. The linings must be thick enough that the vehicle can be operated safely until the next inspection. As a general rule, replace the brake shoes if the linings are worn to within 1/32 inch (.030" or .75 mm) of the lining table on bonded linings, or the same distance above the rivet heads on riveted linings, figure 6-13. Also inspect the linings for physical damage such as cracks, loose or missing rivets, or separation of a bonded lining from the lining table. If any of these problems are present, replace the shoes.

The linings on both shoes should be worn approximately the same amount. If one lining of a dual-servo brake is worn much more than the other, make sure the primary and secondary shoes are installed in their proper positions; the shoe with the smaller lining should be toward the front of the vehicle. If the brake shoes are applied to serve as the parking brake, and are worn excessively at the ends acted on by the parking brake strut, the parking brake may be adjusted too tight. If the linings are worn only in the center or at one end, the shoes are not properly arced to the curvature of the drum, or standard shoes have been installed in an oversize drum. This is not a major problem on today's lightly loaded rear drum brakes, however, it is a concern on front drum brakes. At least two-thirds of the linings on both shoes of a front drum brake should contact the drum. If not, replace or arc the shoes.

Figure 6-14. A glazed brake shoe lining.

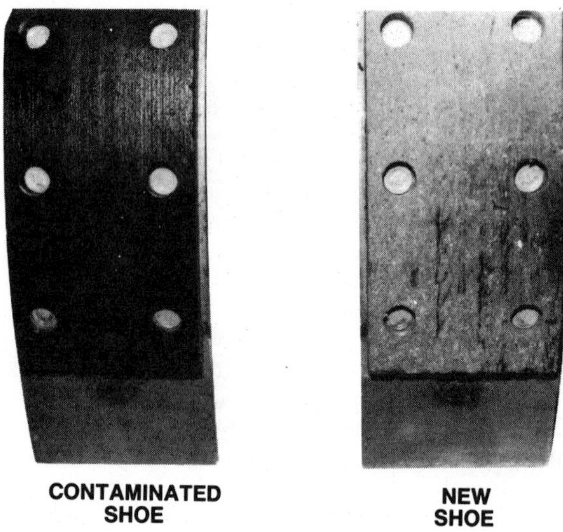

Figure 6-15. A shoe lining suffering from fluid contamination.

Next, inspect the faces of the brake shoe linings. They should be relatively smooth with a dull finish. A hard lining with a shiny or polished surface, figure 6-14, is glazed and must be replaced. If the lining has been contaminated by brake fluid from a leak at the wheel cylinder, or a differential lubricant that has leaked past the axle seals, figure 6-15, the friction surface will be darkened. If the leak is severe, the lining will appear "smeared" and paint will often be stripped from the brake shoe

Figure 6-16. Inspecting a wheel cylinder for fluid leaks.

and friction assembly hardware. Where the contamination is limited to less than 10 percent of the lining surface, you can usually clean and reuse the brake shoes. In most cases, however, you will have to replace the shoes because contamination soaks deep into the lining material and cannot be washed away with brake cleaner.

Wheel Cylinder and Axle Inspection

Detailed inspection of the wheel cylinder is covered in Chapter 5; however, there are some basic checks you should make during a routine brake inspection. First, inspect the outside of the wheel cylinder for signs of leakage; minor stains caused by fluid seepage are considered normal. Next, fold back the cylinder dust boots and look for signs of liquid, figure 6-16; if you find more than a slight amount of dampness, the cylinder must be rebuilt or replaced.

If there are no fluid leaks, check for free movement of the wheel cylinder pistons. With the brake drum from only a single wheel removed, have an assistant gently apply and release the brake pedal while you verify that both brake shoes move outward and return smoothly to their stops. If a shoe does not apply, or returns slowly to its anchor, a piston is frozen or sticking, and the wheel cylinder should be rebuilt or replaced. On brakes without piston stops, use two large screwdrivers to make sure the pistons are not pushed out of the

Figure 6-17. Levering a brake shoe outward should actuate the automatic adjuster.

cylinder bore. Insert the tips of the screwdrivers under the lip at the edge of the backing plate, then lever the screwdriver shafts against the brake shoes to prevent them from moving outward too far.

On cars with a live rear axle, inspect the backing plate where the axle exits its housing. If there is any sign of lubricant leakage, replace the axle seal to prevent contamination of the brake linings.

Hardware Inspection

Drum brake hardware consists of the shoe holddown devices, the shoe return springs, the automatic adjuster linkage, and the parking brake linkage. Replace any hardware that is not in perfect condition, and always replace the same hardware on both friction assemblies of an axle to maintain even braking from side to side.

Make sure the holddown springs, clips, pins, and washers are all present and securely fastened in place. If you pry the shoes slightly away from the backing plate and release them, the holddowns should pull the shoes back into place. While you have the shoes up off the backing plate, inspect the support pads for grooves, notches and any other wear, and make sure the pads are properly lubricated with brake grease.

Inspect for bent, broken, or distorted shoe return springs, and look for nicks in the springs that can weaken them. Also check the springs for discoloration and burnt paint that indicate the springs have been overheated. Excessive heat removes the temper from the springs and causes them to loose tension.

Check the condition of the automatic adjuster hardware. The parts of a link-type adjuster should fit snugly together without excessive play. Adjusting cables should not have any loose or frayed strands, and the cable guide should not be worn. To test the operation of the adjuster, pry the appropriate brake shoe away from its anchor several times, figure 6-17, and check that the adjuster pawl rotates the starwheel to move the brake shoes apart. Be sure to adjust the starwheel back to its original position before you reinstall the brake drum.

Finally, make sure that when the parking brake pedal, lever, or handle is operated, the linkage in the friction assembly moves the shoes apart. The linkage cable should not have any loose or frayed strands, and it should return smoothly when the parking brake is released. If there is an anti-rattle spring on the parking brake strut, it should hold the strut under a constant tension to prevent noise. See Chapter 9 for detailed parking brake inspection and repair information.

BRAKE ADJUSTING

When a drum brake is properly adjusted, the shoes retract only a short distance from the drum when the brakes are released. The exact lining-to-drum clearance varies with the brake design, but it is usually between .006″ and .015″ (.15 and .40 mm). Ideally, you want the smallest clearance possible without brake drag, plus a little extra clearance to allow for heat expansion of the shoes as the brakes are used.

Today, most drum brakes have automatic adjusters that eliminate the need for periodic manual adjustment. As the linings wear, the adjusters automatically reposition the shoes farther outward on the backing plates to maintain the proper lining-to-drum clearance. Manual adjustment of the brakes on cars with automatic adjusters is unnecessary unless scoring makes the brake drum difficult to remove, the brake shoes are replaced, or the automatic adjusters fail.

Drum brakes without automatic adjusters were once used on all cars, and can still be found on the rear axles of some models. These brakes require periodic manual adjustment of the lining-to-drum clearance. The frequency of these adjustments varies with the type of friction material, the size of the brakes in relation to the weight of the car, the type of roads on which the car is driven, and the driving habits of the vehicle operator.

When a drum brake is adjusted properly, a minimum amount of pedal travel is required to apply the brakes, and there is plenty of pedal reserve to compensate for any brake fade that may occur. As the brake linings wear and the lining-to-drum clearance increases, brake pedal travel also increases, and pedal reserve is reduced. As a general rule, you should check and adjust drum brakes with manual adjusters every six months or 7,500 miles (12 000 km), or whenever there is excessive brake pedal travel.

Brake Pedal Travel Test

Although excessive brake pedal travel is the main symptom of brakes that need adjustment, a low pedal can also be caused by other problems such as air in the hydraulic system. A low brake pedal caused by brake lining wear has a firm feel once the brakes are applied. If the pedal has excessive travel but a spongy feel once the brakes are applied, there are other problems in the brake system.

When a car with front disc brakes has a parking brake that applies the shoes of the rear drum service brakes, a simple test can help you isolate the cause of the low pedal, and determine if the rear brakes need adjustment. The test uses the parking brake linkage to apply the brake shoes against the drum; because this eliminates all lining-to-drum clearance, it will also eliminate excessive pedal travel if brake adjustment is the cause. To perform the test:

1. With the parking brake released, apply the service brakes and note the amount of brake pedal travel.
2. Release the service brakes and apply the parking brake.
3. With the parking brake engaged, apply the service brakes several times; this moves the wheel cylinder pistons outward in their bores to take up the slack created when the parking brake was applied.
4. Once the brake pedal travel has stabilized, compare it to the travel you observed in Step 1.

If the pedal travel remains the same, the drum brakes are adjusted properly and the excessive travel is caused by air in the hydraulic system, or a problem with the front disc brakes. If pedal travel returns to normal when the parking brake is applied, adjust the rear brakes.

Parking Brake Caution

On cars where the parking brake applies the rear drum service brake shoes, another symptom of lining wear will be increased travel of the parking brake pedal, lever, or handle. However, unless you check and adjust the rear service brakes *before* you adjust the parking brake linkage, problems will result. Like a brake shoe adjuster, the parking brake linkage moves the shoes outward toward the drum. If you adjust the parking brake without first adjusting the wheel brakes, the parking brake linkage may hold the brake shoes off their anchors. Although this reduces the lining-to-drum clearance and returns the travel of the parking brake control and service brake pedal to normal, it places the parking brake cables under a constant strain that will cause them to stretch and eventually break. As the cables stretch, the brake adjustment will be lost and pedal travel will again increase. Also, because the shoes cannot return to their anchors when the brakes are released, they may not center properly in the drum, resulting in brake drag and increased lining wear.

Whenever you adjust the service brakes, always fully release the parking brake pedal, lever, or handle. If you suspect the parking brake linkage may be improperly adjusted and is partially applying the brake shoes, loosen the parking brake adjustment until there is slack in the cables. Once you complete the service brake adjustment, adjust the parking brake as described in Chapter 9.

BRAKE ADJUSTING PROCEDURE

The exact procedure for adjusting brakes varies with the type of adjuster. However, there are common steps that apply to all brake adjustment. First, raise and properly support the vehicle so the wheels to be adjusted hang free. Unless the brake adjusters or parking brake linkage have been overtightened, the wheels should rotate with no apparent drag. If loosening the adjustments does not free up the wheels, locate and repair the problem before you adjust the brakes.

Next, reduce the lining-to-drum clearance by tightening the adjuster as described in one of the sections below. As you do this, rotate the wheel in the direction that energizes the shoe or shoes being adjusted. This keeps the shoes centered in the drum and properly seated

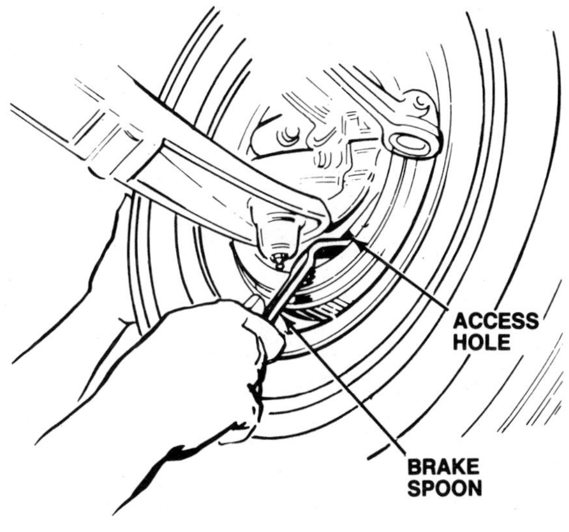

Figure 6-18. Use a brake spoon to turn a starwheel adjuster.

against their anchors. When you adjust a dual-servo or double-leading brake, rotate the wheel in the direction of forward travel. On a leading-trailing brake with a single adjuster, rotate the wheel back and forth as you tighten the adjuster. If a leading-trailing brake has a separate adjuster for each shoe, rotate the wheel in the direction of forward travel to adjust the leading shoe and in the direction of reverse travel to adjust the trailing shoe.

Continue to tighten the adjuster until there is a consistent heavy drag on the wheel, or the brake locks. Then, back off the adjustment the amount specified by the vehicle manufacturer, or until the wheel just turns free. Repeat the adjustment procedure at the other wheels as needed.

Problem Adjustments

On some cars, you may be able to tighten the adjuster a great deal after the linings make contact with the drum. Then, to get free wheel rotation, you will have to back off the adjustment so far that excessive brake pedal travel remains a problem. The pedal may also have a "springy" feel when this occurs. These are all signs that the brake linings are not properly arced to the curvature of the drum (a common problem when standard linings are installed in an oversize drum), or that the shoes are bent or warped. In either case, the lining makes only partial contact when it first touches the drum; further tightening of the adjuster flexes the shoe and brings the lining into greater contact with the drum.

The best repair in these situations is to replace or arc the brake shoes; however, this is not always a realistic option. When the problem is relatively minor, back off the brake adjuster the amount recommended by the manufacturer, and then an additional amount until there is a *very light* drag at the wheel and the brake pedal is reasonably high. A light drag is also acceptable with new brake shoes because it represents only a small area of lining contact with the drum. This portion of the lining will quickly wear away in the first few miles of driving.

SPECIFIC BRAKE ADJUSTERS

Manually adjusted brakes use four different adjuster designs: starwheel, cam, wedge, and anchor. The first three types are used to set the proper clearance between the brake linings and the drum. Anchor adjusters are used on some early brakes to center the brake shoes in the drum, and are always used with another type of adjuster, usually a starwheel, that adjusts the actual lining-to-drum clearance.

Starwheel Brake Adjustment

Starwheel brake adjusters are the most common type, and they are manufactured in both manual and automatic adjusting versions. Starwheel brake adjusters are usually adjusted through an access hole in the brake backing plate. However, some cars with automatic starwheel adjusters have the access hole in the brake drum web, and the tire and wheel must be removed to manually adjust the brakes.

Most brakes that use this adjuster design have a single starwheel adjuster between the shoes that adjusts both shoes at the same time. Some leading-trailing and double-leading brakes use two starwheel adjusters, one at each shoe, to adjust the shoes independently. When you service a brake with dual starwheels, complete the adjustment on one shoe, then proceed to the other.

Manual starwheel adjustment

All starwheel adjusters are first turned in one direction to reduce the lining-to-drum clearance and tighten the adjustment, and then turned in the opposite direction to increase the lining-to-drum clearance and loosen the adjustment. With manual starwheel adjusters, lining-to-drum clearance is usually reduced by levering the starwheel upward as viewed through the access hole in the backing plate. This is not a hard and fast rule, however, and the direction of rotation that tightens or loosens the adjustment will vary from car to car, and sometimes from one side of the vehicle to the other!

Figure 6-19. A lanced area must be removed to reach the brake adjuster on some cars.

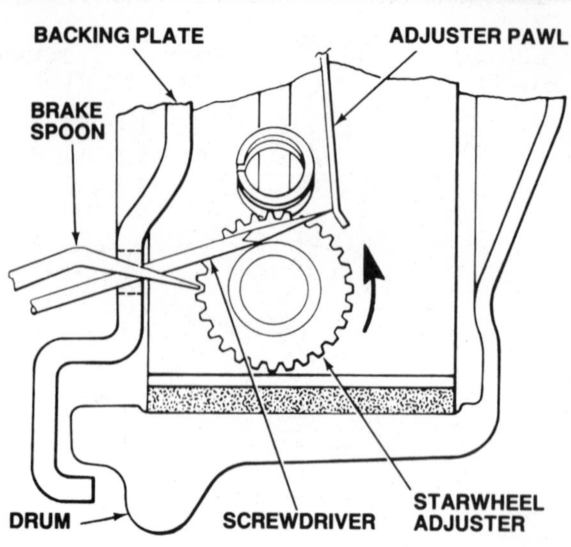

Figure 6-20. Using a screwdriver to push the automatic adjuster pawl out of the way.

To adjust the brakes on a car with manual starwheel adjusters:

1. Remove the plug covering the access hole in the backing plate or brake drum.
2. Insert a brake spoon of the appropriate size and shape through the access hole, figure 6-18, and engage the teeth on the starwheel.
3. Use a lever action to rotate the starwheel and reduce the lining-to-drum clearance. As you rotate the starwheel, the spring holding it in position should snap from one tooth to the next and cause a clicking sound.
4. When there is heavy drag at the wheel, or the brake locks, lever the starwheel in the opposite direction to increase the lining-to-drum clearance. Loosen the adjustment until the wheel just turns free, or you have rotated the starwheel the number of clicks specified by the vehicle manufacturer. As many as 20 to 30 clicks are required in some cases, so check the factory shop manual for the proper number.
5. Reinstall the plug in the adjuster access hole to prevent water and dirt from entering the friction assembly.

Automatic starwheel adjustment

Automatic starwheel adjusters require manual adjustment when scoring makes it difficult to remove the brake drum, when the brake shoes are replaced, or if the automatic adjusters fail. On some cars with automatic starwheel adjusters, the backing plate or drum is lanced for an access hole, figure 6-19, but the metal insert is still in place. To manually adjust the brakes on these cars, remove the insert from the lanced area with a hammer and punch. Take care so that the insert does not fall through the hole into the drum. Once the adjustment is complete, install a metal or rubber plug in the hole to prevent dirt and moisture from affecting braking. The plugs are available from auto dealers and aftermarket brake parts suppliers.

The adjustment procedure is basically the same as for manual starwheel adjusters, with one exception: automatic starwheel adjusters cannot be loosened unless the adjuster ratchet pawl is held away from the starwheel. If you attempt to rotate the starwheel backwards against the pawl, you will round off the edge of the pawl and burr the teeth on the starwheel. If the damage is bad enough, the automatic adjusting mechanism will not operate properly.

To manually reduce the lining-to-drum clearance in a drum brake with automatic starwheel adjusters, use a brake spoon to rotate the starwheel as described above. The adjuster design will allow the starwheel to rotate only in the tightening direction, and the starwheel teeth will ratchet past the pawl at this time. If the brake adjustment loosens as you rotate the starwheel in the only direction it will turn, the starwheel adjuster assembly is installed backwards, or the assemblies have been swapped from side to side on the axle.

To loosen the adjustment, use a standard screwdriver or wire hook to hold the adjuster pawl out of the way. When the access hole is in the backing plate, use the screwdriver to push the lever away from the starwheel, figure 6-20. When the access hole is in the drum, use a wire hook to pull the self-adjuster lever away from

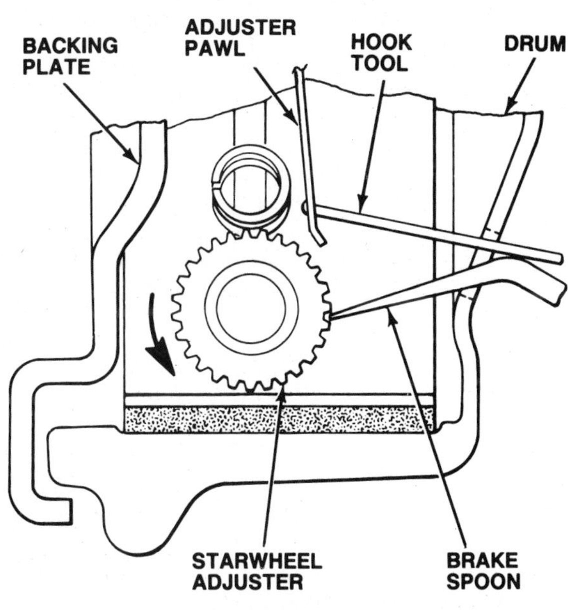

Figure 6-21. Using a wire hook to pull the automatic adjuster pawl out of the way.

Figure 6-22. Cam adjusters are rotated with conventional wrenches.

the starwheel, figure 6-21. In either case, push or pull the pawl just hard enough to disengage it from the starwheel. If you use too much force you may bend parts of the automatic adjuster mechanism.

Cam Brake Adjustment

Cam-type manual adjusters are relatively rare, and are found primarily on older brakes, although a few newer models use them as well.

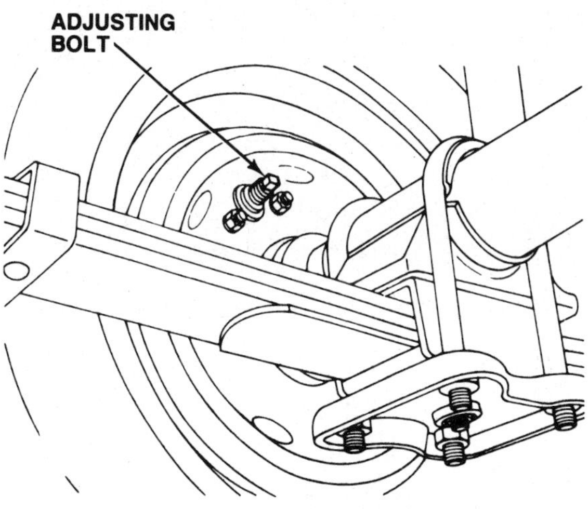

Figure 6-23. The adjusting bolt of a wedge adjuster.

One adjuster is used on each brake shoe, and the shoes are adjusted independently. To adjust a cam-type adjuster, rotate the hex head of the adjuster stud that extends through the backing plate, figure 6-22. In most cases, lining-to-drum clearance is reduced as the wrench on the adjuster is moved downward. Some adjuster studs have arrows on them to indicate the tightening direction.

To adjust a brake equipped with cam-type adjusters, rotate the adjuster stud of the leading shoe to reduce the lining-to-drum clearance. When there is heavy drag at the wheel, or the brake locks, loosen the adjustment so the wheel just turns free. Repeat this process on the trailing shoe.

Wedge Brake Adjustment

Wedge-type manual adjusters are found on some imported cars. With this design, you adjust both brake shoes at the same time by rotating a single adjusting bolt that extends through the backing plate, figure 6-23. Some bolts have a common hex head, but many have a metric-sized square head. Special tools are available to fit these square-head bolts, however, a tight-fitting wrench or an 8-point socket of the proper size can also be used to make the adjustment. In all cases, turning the bolt in a clockwise direction so it threads into the backing plate (travels outward on the vehicle) reduces the brake lining-to-drum clearance.

To adjust a brake with a wedge-type adjuster, rotate the adjusting bolt in a clockwise direction to reduce the lining-to-drum clearance. As the shoes move closer to the brake drum, the flats on the tapered end of the bolt inside the brake

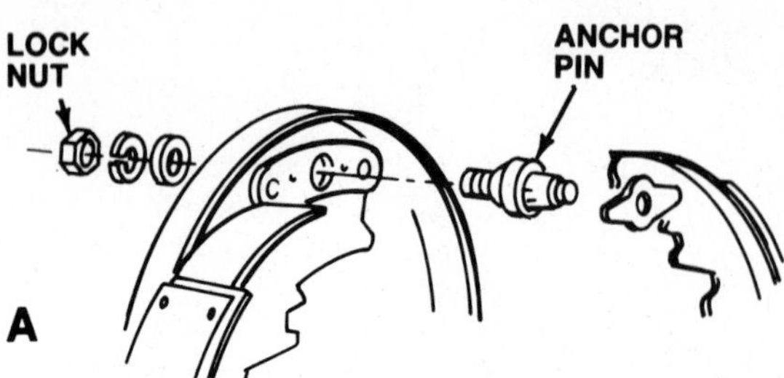

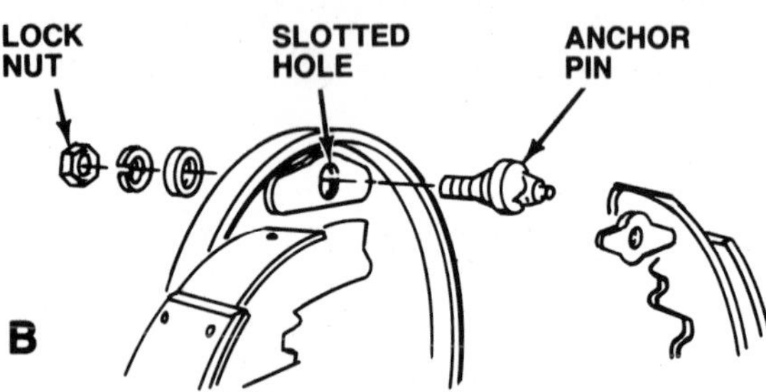

Figure 6-24. The lock nuts for adjustable anchors are on the backside of the backing plate.

will cause the bolt to alternately tighten and loosen, and the brake will grab and release at the same time. When you can tighten the adjusting bolt no further, loosen the bolt the number of flats recommended by the vehicle manufacturer, or until the wheel just turns free.

Anchor Brake Adjustment

Anchor adjusters are used on some older brakes to center the shoes in the drum for better lining contact and more effective braking. If the shoes are not properly centered, it is impossible to accurately adjust the lining-to-drum clearance. Anchor adjusters were dropped from domestic car brakes in the late 1950's when undersize- and cam-ground brake shoes made them unnecessary. Imported cars continued to use anchor adjusters into the 1960's, and some large truck brakes still have adjustable brake shoe anchors.

As mentioned earlier, anchor adjusters are always used with another type of adjuster that sets the lining-to-drum clearance. In normal driving conditions, only the clearance between the brake linings and drum requires periodic attention. This operation is called a minor brake adjustment, and depending on the type of shoe adjuster used, the procedure is identical to one of those described earlier.

Anchor adjustment is normally needed only when the brake shoes are replaced, or at high-mileage intervals when significant lining wear has taken place. This operation is called a major brake adjustment. During a major brake adjustment, you adjust the shoes and the anchor at the same time. The method used to do this varies depending on whether the brake is fitted with an eccentric anchor or a slotted anchor.

Eccentric anchor adjustment

To adjust an eccentric anchor, tighten the shoe adjuster until a medium drag is felt when turning the wheel with one hand. Loosen the anchor pin lock nut one-half turn, figure 6-24A, and slowly rotate the pin in the direction of forward wheel rotation until you feel the least amount of drag as you rotate the wheel. Hold the anchor pin in position and tighten the lock nut, then finish adjusting the lining-to-drum clearance as described earlier.

Slotted anchor adjustment

To adjust a slotted anchor, tighten the shoe adjuster until a heavy drag is felt while turning the wheel with both hands. Loosen the anchor pin lock nut one full turn, figure 6-24B, then tap the anchor pin and backing plate with a plastic hammer to allow the shoes to center in the drum. If the amount of drag on the wheel changes, retighten the shoe adjuster and again tap the anchor pin and backing plate. Repeat this procedure until the drag on the wheel remains constant, then tighten the anchor pin nut to the manufacturer's specification. Finish adjusting the lining-to-drum clearance as described earlier.

BRAKE SHOE REPLACEMENT

Brake shoes are sold and serviced as axle sets. An axle set consists of four shoes, one pair for the friction assembly at each wheel. Shoes from different manufacturers should never be mixed. Although they will fit and may appear the same, the friction coefficients of the linings may be quite different. Even if only one shoe of an axle set is badly worn, the entire set should be replaced after the problem causing uneven wear has been repaired.

The most efficient way to replace the brake shoes on an axle is to disassemble and service both friction assemblies at the same time. However, if you are unfamiliar with a particular brake design, a good practice is to work on only one wheel at a time so the other brake can serve as a model for assembly. This is important because even a small error in assembly can have a big effect on brake operation. If you do disassemble both brakes at the same time, keep the

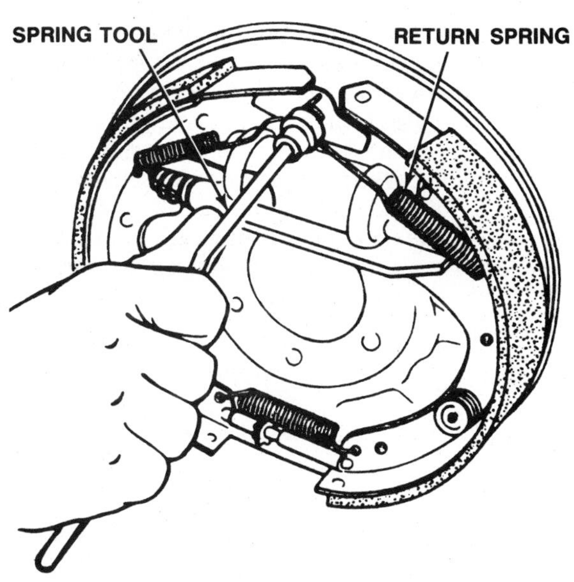

Figure 6-25. Removing a brake return spring.

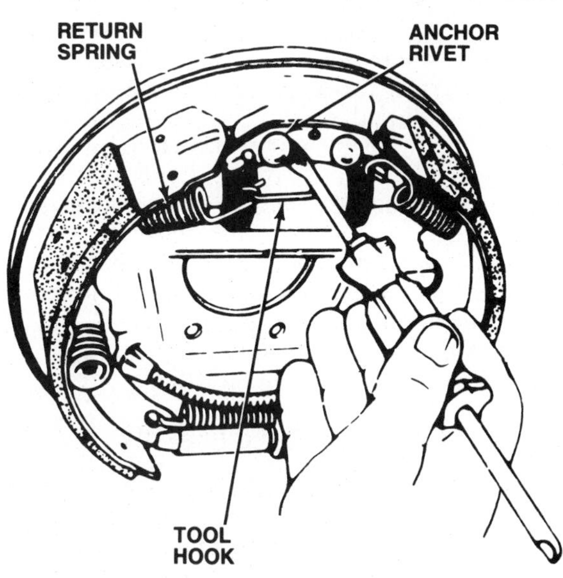

Figure 6-26. Removing a brake return spring on a GM brake.

parts for each wheel separated. Many parts that appear the same on both friction assemblies, starwheel adjusters for example, are actually different from side to side.

The exact procedure used to replace a set of brake shoes varies with the design of the friction assembly. Step-by-step shoe replacement procedures for several typical drum brake friction assemblies are provided at the end of the chapter. The sections below describe a general procedure that includes the common operations

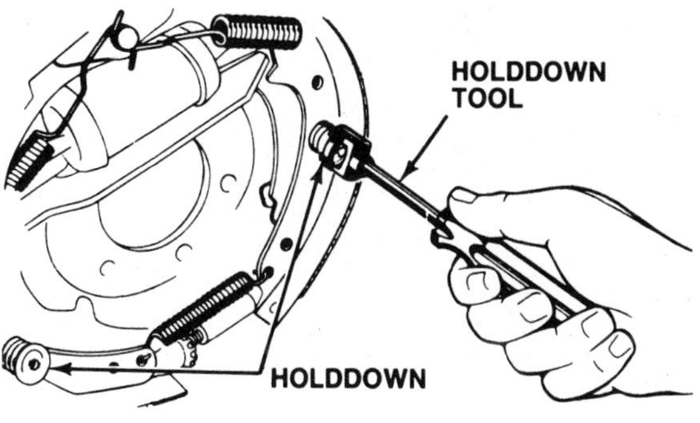

Figure 6-27. Removing a brake shoe holddown.

required to disassemble, inspect, and reassemble drum brakes.

Drum Brake Disassembly

The first step in any shoe replacement job is to remove the brake drum as described earlier in the chapter. After that point, however, the order in which the many parts of the friction assembly are disassembled varies from one brake to another. The shoe return springs are often the first parts removed. Before you remove the springs, note how they are installed, which holes they fit into in the shoe webs, and how the springs overlap on the anchor post (where applicable).

Springs that hook over an anchor post are removed with a special tool as shown in figure 6-25. Place the tool over the post and hook the flange under the end of spring; rotate the tool to lever the spring up and off the anchor. Once the springs are free, remove the anchor plate and adjuster cable or linkage (if fitted) from the anchor post.

The special tool used to remove the return springs from the rear drum brakes on FWD General Motors cars, figure 6-26, works somewhat differently. Attach the hook of the tool to the end of the return spring, then lever the shaft of the tool against the anchor rivet to remove the spring. Return springs that install between two brake shoes can be removed with a pair of brake spring pliers, however, this is often unnecessary because once the shoes are removed from the anchor, they can be collapsed together to release the tension on the shoe-to-shoe spring.

The next step in disassembly is often to remove the brake shoe holddowns. Most holddown devices can be removed by hand or with common hand tools, although there are special tools available to make the job easier. The most common of these is used with spring-and-pin holddowns, figure 6-27. To remove the holddown, place the tool over the spring retaining washer; while holding the pin in place from the backside of the backing plate, push the tool inward and rotate it to release the washer from the flattened end of the pin. When a spring clip is used with a holddown pin, depress the clip by hand and slide it out from under the pin. To remove a coiled spring "beehive" holddown, insert a Phillips screwdriver into the coil, and while holding the retaining clip in place from the backside of the backing plate, push the screwdriver inward and disengage the hook on the spring from the retaining clip.

Once the return springs and shoe holddowns are removed, disassembly of the friction assembly can be completed. In most cases, simply reposition the brake shoes as needed so the shoe-to-shoe return springs, automatic adjusting mechanism, and parking brake linkage, can be disconnected. To prevent problems on assembly, note the location of each part you remove, and how the part is installed.

Drum Brake Inspection

Once the friction assembly is disassembled, inspect the brake drum as described in Chapter 8 and determine whether it needs to be turned or replaced. Even though a drum removed from a car may appear in perfect condition, it often has minor wear and distortion that cannot be spotted visually, but will affect braking performance. Because of this, it is common practice to always turn the drums when replacing brake shoes in order to guarantee a good job. If the drum hub contains the wheel bearings, you should also repack the bearings when you replace the brake shoes. Information on bearing service is located in Chapter 11.

Next, inspect the brake backing plate and its mounting bolts. If the plate is bent or cracked, replace it. If the mounting bolts are loose, torque them to the car manufacturer's specifications. Also, check the shoe support pads for grooves, notches, or any other signs of wear. Minor wear is considered normal; grind or file the pads smooth to repair the problem. If the pads are deeply grooved, replace the backing plate. If a replacement backing plate is unavailable, and you are trained in the proper use of an oxy-acetylene torch, it is acceptable to build up the worn support pads with brazing rod then grind or file them smooth.

Once you are sure the backing plate is in good condition, inspect the wheel cylinder as described in Chapter 5. Just as drums are turned whenever the brake shoes are replaced, it is common practice to always rebuild or replace the wheel cylinders as well. Unless this is done, the thicker linings on the new shoes will force the wheel cylinder pistons back over portions of the bore that may have become rusted or corroded. This will quickly cause a fluid leak that will contaminate the linings, cause an expensive comeback, and result in an unhappy customer. Rebuilding or replacing the wheel cylinders is inexpensive insurance.

When you replace a set of brake shoes, most vehicle manufacturers recommend that you automatically replace the shoe return springs as well. They do this because faulty springs are sometimes hard to spot. Even if a spring appears in perfect condition, it may have lost part of its tension from the hundreds of thousands of times it has been stretched in normal brake operation.

If you are unsure about whether the springs need to be replaced, here are some guidelines and simple tests to follow. Always replace springs that are bent, broken, distorted or have nicks in them. Burnt paint or obvious discoloration on a spring are sure signs that it has been overheated and lost tension, and should be replaced. Most brake shoe return springs have closed-coil construction, and can be visibly tested by holding them up to a light. If you can see light shining between the individual coils of the spring, it has stretched and should be replaced. Closed-coil springs can also be checked using the drop test. Drop the spring from waist height onto a concrete floor; if there is a dull thud as it hits the ground, the spring is probably in good condition; however, if there is a ringing noise as it hits the floor, the spring has open space between its coils and should be replaced.

In addition to the shoe return springs, manufacturers also recommend you always replace shoe holddown hardware and automatic adjuster cables. Like return springs, these parts are difficult to inspect. For example, you cannot tell if a holddown spring or pin is weakened unless it has obvious physical damage; in the same manner, a stretched adjuster cable looks perfectly normal until you compare its length to that of a new one. These parts play a major role in smooth and quiet brake operation; if any of them break loose in operation, severe lining damage and drum scoring will result. The cost of these components is small compared to repairing the problems they will cause if they fail.

The final parts to inspect are those in the automatic adjusting mechanism and parking brake linkage. These differ depending on the design of the brake, however, look for bent components and wear at the points where the

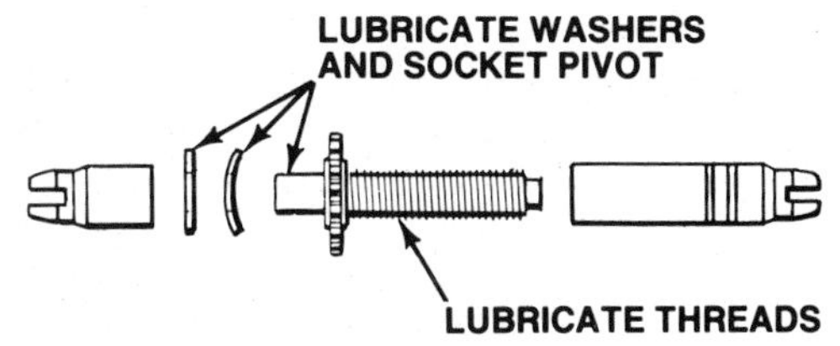

Figure 6-28. Starwheel adjusters must be clean and properly lubricated to work correctly.

parts contact one another. If the car has starwheel adjusters, disassemble the adjuster assembly and clean it thoroughly, figure 6-28. Replace the adjuster if the starwheel teeth are rounded. Use a wire brush to clean any rust and grime from the adjuster threads, then lubricate the threads with brake grease and screw the adjuster together through its full range of travel. If the threads bind at any point, repair the problem or replace the adjuster.

Brake Assembly

Drum brake assembly is essentially the reverse of disassembly, although there are some things to be aware of. Most of these points concern making sure that all parts are reinstalled in their proper locations. There are also a few special techniques used to install specific brake components.

The first step in assembly is to compare the replacement brake shoes to the original equipment parts. The replacement shoes do not have to be absolutely identical to the original parts, however they should have the appropriate holes in the shoe webs, and the linings should be the same basic size and shape as those on the originals. If there are major differences in the construction of the shoes or the sizes of the linings, contact your parts supplier and make sure you have the proper shoes for the application.

Next, determine where the shoes belong on the car. Although they may physically fit in more than one position, most brake shoes are designed to be installed in a particular spot. On a dual-servo brake, for example, the primary shoe generally has a smaller lining than the secondary shoe, and the friction materials used for the two linings may differ as well. Always install the primary shoe so it is pulled away from the anchor when the brakes are applied with the wheel turning in the direction of forward rotation. Some leading-trailing brakes are also designed to use shoes that have different friction characteristics. In these applications, install the replacement shoes in the same relative locations as the original equipment parts. If you have any question about the proper shoe positions, consult the component manufacturer or the shop manual of the vehicle being serviced.

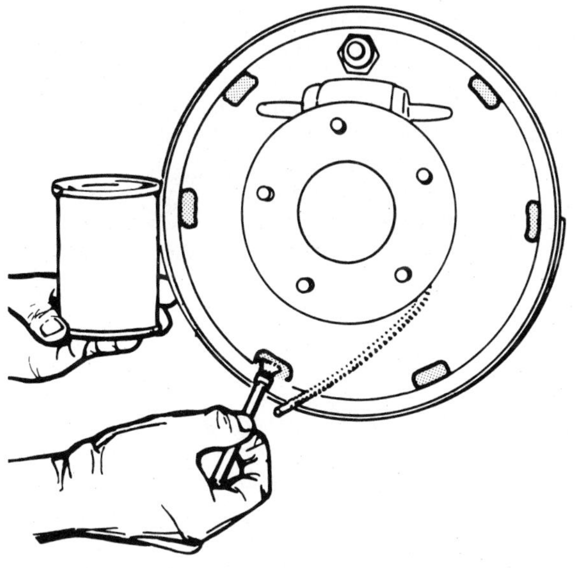

Figure 6-29. Lubricate the shoe contact pads with high-temperature brake grease.

Once you have determined the proper locations for the replacement shoes, remove any parking brake linkage pieces or similar parts from the old shoes. Transfer these parts to the appropriate replacement shoes, and install them using new fasteners. Lubricate the shoe support pads on the backing plate with brake grease, figure 6-29, then assemble the shoes on the backing plate. The procedure required to do this varies with the design of the friction assembly, basically however, you reposition the shoes as needed until the parking brake linkage, brake adjuster, and the shoes themselves are all fitted together in their proper positions.

While assembling a friction assembly, be sure to use only the parts intended for that side of the car. For example, all starwheel adjusters look alike, but right-side adjusters usually have left-hand threads, while left-side adjusters have right-hand threads. If an adjuster is installed on the wrong side, the automatic adjusting mechanism will increase the lining-to-drum clearance rather than reduce it. Some starwheel adjusters are marked L or R to indicate the side of the car on which they belong.

Once the shoes are assembled in position, install the holddowns to secure the shoes in place. On pin and spring holddowns, insert the pin through the holes in the backing plate and

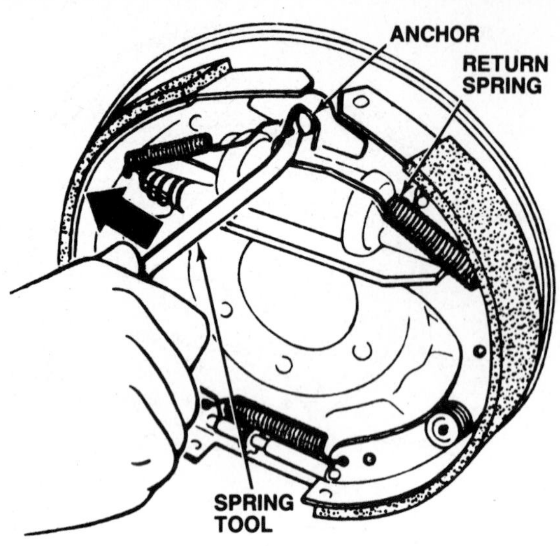

Figure 6-30. Installing a shoe return spring on an anchor post.

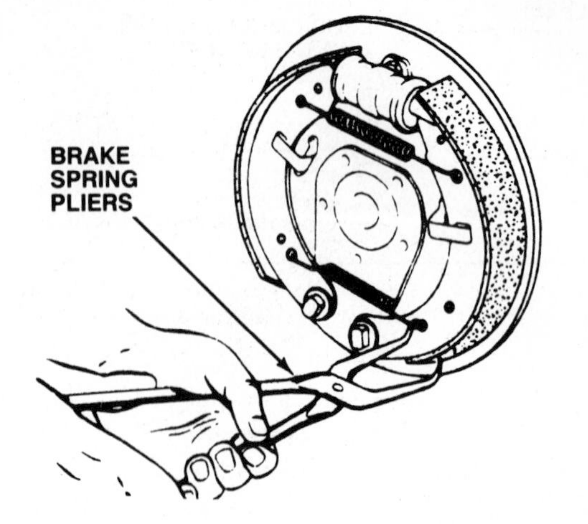

Figure 6-31. Installing a shoe-to-shoe return spring.

brake shoe web. Then, while holding the pin in place from the backside of the backing plate, use the special tool to compress the spring and retaining washer over the end of the pin. Rotate the washer as needed to lock it onto the flattened end of the pin. Where a spring clip is used with a holddown pin, compress the clip by hand and slip it into position under the flattened end of the pin. To install a coil-spring "beehive" holddown, hold the retaining clip in place from the backside of the backing plate, then use a Phillips screwdriver to push the spring inward and engage its hook into the retaining clip.

The final step in drum brake assembly is to install the shoe return springs. It is very important to install the proper spring in the proper direction, in the proper location. Some springs can be installed only one way, and their proper position is easy to identify. Different paint colors are often used to distinguish similar springs that have different tensions. Certain springs have a longer straight section and attachment hook at one end than at the other. These springs must be installed facing a specific direction, or the coiled section of the spring will interfere with another part of the friction assembly. Sometimes, there are several holes in the shoe web where a spring can be attached. If you install a spring in the wrong hole, it will affect the rate at which the brake shoes apply and release. If you failed to note how the springs were installed when you disassembled the friction assembly, consult a shop manual for the proper locations.

Return springs are installed in two basic ways, and both methods require a special tool. To install a spring that fits over an anchor post, attach the appropriate end of the spring into the hole in the shoe web. Then, place the notched end of the spring tool on the anchor post, and drape the hook end of the spring over the tool shaft, figure 6-30. Finally, taking care not to overstretch the spring, lever the tool back so the spring slides down the shaft and into place on the anchor.

Shoe-to-shoe return springs are installed using brake spring pliers, figure 6-31. Install one end of the spring into its hole in the brake shoe web, and place the other end over the hooked arm of the brake spring pliers. Position the pliers over the shoe the spring is to be attached to; the pointed arm of the plier should contact the lining at the same level as the hole the spring is to engage. If possible, position the pointed arm on a lining rivet; otherwise, position it directly on the lining and use extra caution to prevent damage. Squeeze the handle of the pliers to stretch the spring to the appropriate length, then insert the end of the spring into the hole in the web. Remove the plier, and make sure both ends of the spring are fully engaged in the web holes.

INITIAL BRAKE ADJUSTMENT

When you replace a set of brake shoes, you generally bottom the adjuster mechanism so the shoes are at their smallest diameter. This makes it easier to install the return springs, and ensures that the brake drum will easily slip over the shoes. However, once the brake is assembled, and you are sure the drum fits, you must adjust the initial lining-to-drum clearance

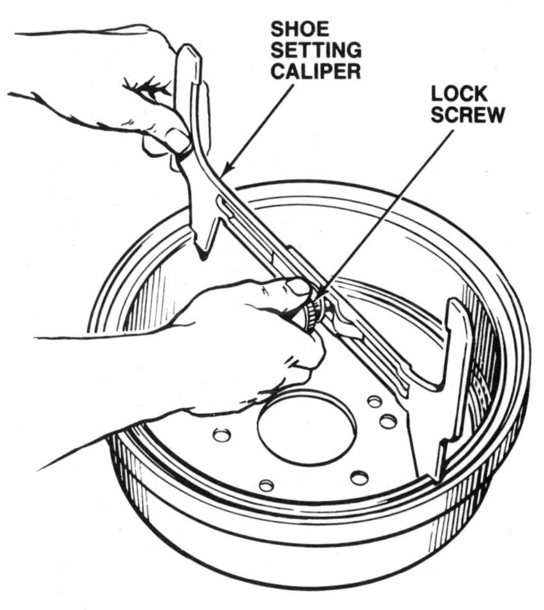

Figure 6-32. Measure the brake drum diameter with a shoe setting caliper.

so the brake pedal travel will be satisfactory. On cars with starwheel brake adjusters, initial adjustment is usually done before the drum is installed. On cars without starwheel brake adjusters, the initial adjustment is performed after the brake drum is installed.

Initial Adjustment — Starwheel Adjusters

Both manual and automatic starwheel brake adjusters make their adjustments in very small increments. If you attempt to adjust the brakes manually after the drum is installed, more than a hundred clicks of the adjuster may be required to get the proper clearance. This is both tiring and time consuming. When you make the initial adjustment before you install the drum, final manual adjustment will be quick and easy, or you can use the automatic adjusters to make the final adjustment during the test drive.

To perform the initial adjustment, place a shoe setting caliper in the brake drum as shown in figure 6-32. Slide the caliper back and forth and spread the jaws until they span the drum at its widest point. Tighten the lock screw to fix the caliper at this setting. Depending on the brand of tool being used, the opening on the opposite side of the caliper is now set to equal the drum diameter, or be approximately .020″ (.50 mm) smaller than the drum diameter to provide a clearance of .010″ (.25 mm) between the drum and each brake shoe.

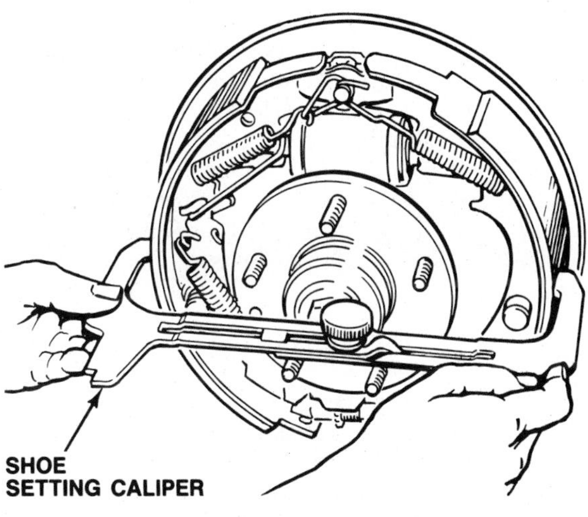

Figure 6-33. Transfer the drum measurement to the shoes with the shoe setting caliper.

Remove the caliper from the drum, and place the open side over the brake shoes as shown in figure 6-33. If the caliper opening matches the drum diameter, rotate the starwheel adjuster as needed until there is approximately .020″ (.50 mm) clearance between the caliper opening and the shoes at their widest point. If the caliper setting includes the desired lining-to-drum clearance, rotate the starwheel adjuster as needed until the caliper just slides over the shoes at their widest point. Hold the automatic adjuster pawl out of the way while turning the starwheel so neither becomes burred. On brakes with dual starwheel adjusters, rotate each starwheel an equal amount.

Once the initial adjustment is completed, install the brake drum and check to make sure the brake pedal is fairly high and firm. On cars with manual starwheel adjusters, perform a final brake adjustment as described earlier in the chapter. On cars with automatic starwheel adjusters, take the car for a test drive and apply the brakes several times while traveling forward or in reverse (depending on the design of the adjuster) to complete the adjustment.

Initial Adjustment — Drum Installed

On cars without starwheel brake adjusters, the initial adjustment is generally made with the drum installed because other adjuster designs makes it quick and easy to set the lining-to-drum clearance. To perform the initial adjustment on brakes with cam- or wedge-type manual adjusters, follow the procedures for manual adjustment described earlier in the chapter.

A few brakes have automatic adjusters that can take up large amounts of lining-to-drum clearance very rapidly. On these brakes, the initial adjustment is made after the brake drum is installed by simply applying the brake pedal or parking brake lever a few times. The parking brake strut adjuster on the Chevrolet Vega, the strut-quadrant adjuster on some Ford Escorts, and the friction washer adjusters on some older imports are examples of these types of brakes.

BRAKE SHOE BURNISH-IN

Whenever new brake shoes are installed, they require a short time of relatively light use called a burnishing-in or bedding-in period. There are two reasons for this. First, new shoes do not make full contact with the drum friction surface. A short period of light braking allows the shoes and drums to become "familiar" with one another and wear into better contact.

The second reason a burnish-in period is needed is because new brake linings are not fully cured when they come from the factory; there are still binding resins and other chemicals that have not completely evaporated from the friction material. Final curing takes place over the first few hundred miles of driving as the shoes are repeatedly heated and cooled.

If new brakes are used hard immediately and become too hot, residual resins boil out of the shoes as gasses and liquids. If these gasses and liquids come out in great enough quantities, they will lubricate the friction material and cause a type of brake fade called green fade. When green fade occurs, the shoes will be glazed when they return to normal operating temperature. Glazed shoes do not have the proper friction coefficient for smooth operation, are usually very noisy, and must be replaced.

Brake Burnishing Procedure

Whenever you replace any of the friction materials in a brake system, take the car for a test drive afterward to make sure the brakes are operating properly. During the test drive, perform the simple procedure below to burnish-in the new shoes and/or pads:

1. Drive the car at approximately 30 mph (50 kph). Check your rear-view mirror to make sure it is safe to do so, then apply the brakes with *moderate to firm* pressure to slow the car to about 5 mph (8 kph).
2. Accelerate back to speed, and drive for approximately 15 seconds to allow the brakes to cool. Make another stop as described in Step 1. Continue to alternate stops with cooldown periods until you have made about a half dozen stops in this manner.
3. Accelerate to approximately 55 mph (88 kph). Check your rear-view mirror to make sure it is safe to do so, then apply the brakes with *moderate to firm* pressure to slow the car to about 20 mph (30 kph).
4. Accelerate back to speed, and drive for approximately 30 seconds to allow the brakes to cool. Make another stop as described in Step 3. Continue to alternate stops with cooldown periods until you have made about a half dozen stops in this manner.
5. When you deliver the car to its owner, caution them that the brakes should not be used hard for approximately 100 miles of in-town driving, or 300 miles of highway driving. This will ensure that the brakes will become properly burnished in.

SHOE REPLACEMENT PROCEDURES

The following pages contain shoe replacement procedures for four different drum brake friction assemblies:

- Chrysler leading-trailing brake with manual adjuster
- Chrysler dual-servo brake with cable-type starwheel automatic adjuster
- General Motors dual-servo brake with lever-type starwheel automatic adjuster
- Ford leading-trailing brake with strut-quadrant automatic adjuster.

While there are many other drum brakes that are not covered, these are popular designs that include many common variations in drum brake construction and overhaul procedures. With the knowledge gained from reading this chapter, and the experience of going through the shoe replacement procedure for each of these brakes, you should be able to understand the various other friction assemblies you will encounter.

Chrysler Leading-Trailing Brake with Manual Adjuster

The Chrysler leading-trailing brake has been used on the rear wheels of all FWD Chrysler Corporation cars since 1978. The early version covered in the shoe replacement procedures has a manual starwheel adjuster; 1983 and later models have automatic adjusters. Chrysler uses a fixed drum with this friction assembly, so the wheel bearings should be repacked and adjusted when the brake shoes are replaced. See Chapter 11 for information on wheel bearing service.

Chrysler Dual-Servo Brake with Cable-Type Starwheel Automatic Adjuster

The Chrysler dual-servo brake with cable-type starwheel automatic adjuster has been used in both front- and rear-wheel applications on all RWD Chrysler Corporation cars since 1969. This brake comes in 9-, 10-, and 11-inch diameters, but the service procedures for all three are the same. The only difference between the front and rear designs is that the front brake does not have a parking brake mechanism. Fixed drums are used on front brakes so the wheel bearings should be repacked and adjusted when the shoes are replaced. See Chapter 11 for information on wheel bearing service.

GM Dual-Servo Brake with Lever-Type Starwheel Automatic Adjuster

The General Motors dual-servo brake with lever-type starwheel automatic adjuster is used as the rear brake on most FWD General Motors cars. This brake has been built in two versions, but both operate the same. On early brakes, the adjuster lever actuating link installs through a hole in the shoe anchor. On later models, the link hooks over a post on the anchor. In all other respects, the operation and servicing of these brakes is identical.

Ford Leading-Trailing Brake with Strut-Quadrant Automatic Adjuster

The Ford seven-inch (180-mm) leading-trailing brake with strut-quadrant automatic adjuster is used on all Escort/Lynx three-door models except those with styled steel wheels. This is a somewhat unusual design that is similar to the brakes on the Merkur XR4Ti. Ford uses a fixed drum with this friction assembly, so the wheel bearings should be repacked and adjusted when the brake shoes are replaced. See Chapter 11 for information on wheel bearing service.

Although the shoe replacement procedure for this brake includes instructions for disassembling the adjuster, this is only necessary if the teeth on the adjuster quadrant or pin are damaged, and a part must be replaced. In most cases, simply pull the quadrant back against spring tension, and reposition it so the pin fits into the third or fourth notch at the narrow end of the quadrant. This shortens the adjuster strut to its minimum length, and makes it easy to install the brake drum.

CHRYSLER LEADING-TRAILING BRAKE WITH MANUAL ADJUSTER

1. Pull back the spring on the parking brake cable and unhook the cable from the parking brake lever.

2. Remove both shoe-to-anchor springs.

3. Remove the shoe hold-down springs, and withdraw the pins from the backing plate.

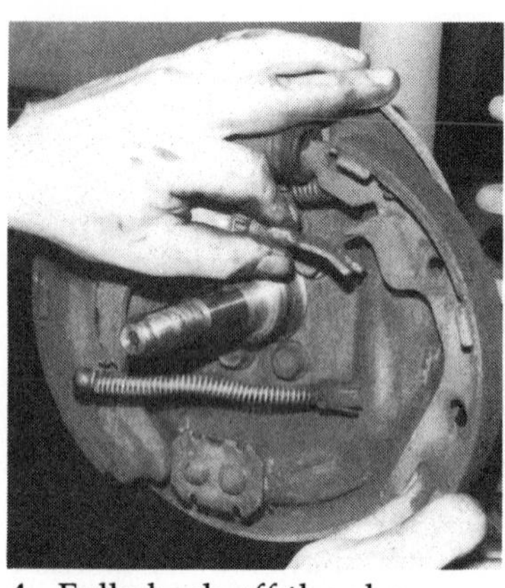

4. Fully back off the shoe adjustment, then spread the shoes and remove the adjuster screw assembly.

5. Raise the parking brake lever, and remove the trailing shoe from under the anchor plate to release tension on the return spring.

6. Unhook the return spring from the backing plate, and remove the trailing shoe assembly.

7. Remove the leading shoe from under the anchor plate to release tension on the return spring.

8. Unhook the return spring from the backing plate, and remove the leading shoe.

9. Lubricate the shoe support pads on the backing plate.

CHRYSLER LEADING-TRAILING BRAKE WITH MANUAL ADJUSTER

10. Hook the return spring of the leading shoe into the backing plate, then move the shoe into position with its end under the anchor plate.

11. Hook the return spring of the trailing shoe into the backing plate, then move the shoe into position with its end under the anchor plate.

12. Spread the shoes as needed to install the adjuster screw assembly. The forked end of the screw must curve downward.

13. Insert the holddown pins through the backing plate and shoes, then install the holddown springs on the pins.

14. Install both shoe-to-anchor springs.

15. Pull back the spring on the parking brake cable, and attach the cable to the parking brake lever. Position the cable washer (if fitted) between the spring and lever.

CHRYSLER DUAL-SERVO BRAKE WITH CABLE-TYPE STARWHEEL AUTOMATIC ADJUSTER

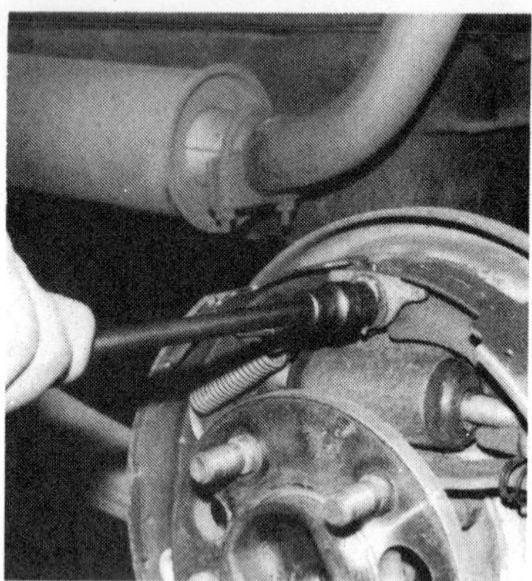

1. Remove the shoe return springs from the anchor post.

2. Lift the eye of the automatic adjuster cable off the anchor, unhook the cable from the adjusting pawl, and remove the cable.

3. Remove the anchor plate from the anchor post.

4. Remove the adjuster cable guide from the secondary shoe.

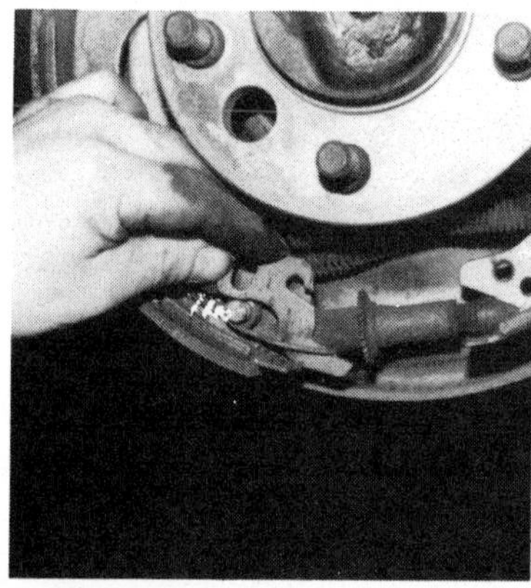

5. Slide the adjuster pawl forward to clear the pivot, and pull it out from under the pawl spring.

6. Remove the pawl spring from the pivot.

7. Remove the shoe-to-shoe return spring.

8. Spread the shoes and remove the starwheel adjuster assembly.

9. Remove the shoe hold-down springs, and withdraw the pins from the backing plate.

CHRYSLER DUAL-SERVO BRAKE WITH CABLE-TYPE STARWHEEL AUTOMATIC ADJUSTER

10. Spread the shoes and remove the parking brake strut and anti-rattle spring.

11. Disconnect the parking brake cable, and lift out the shoes.

12. Remove the parking brake lever from the old secondary shoe, and install it on the new shoe.

13. Connect the parking brake cable to the lever, then position the secondary shoe on the backing plate and install its holddown.

14. Slide the anti-rattle spring over the end of the parking brake strut, and fit the strut into the slot in the parking brake lever.

15. Position the primary shoe on the backing plate so it engages the parking brake strut, then install the shoe holddown.

16. Install the anchor plate and the eye of the adjuster cable over the anchor post.

17. Install the primary shoe return spring through the shoe web and over the anchor post.

18. Install the cable guide onto the secondary shoe so the hole in the guide aligns with the hole in the shoe web.

CHRYSLER DUAL-SERVO BRAKE WITH CABLE-TYPE STARWHEEL AUTOMATIC ADJUSTER

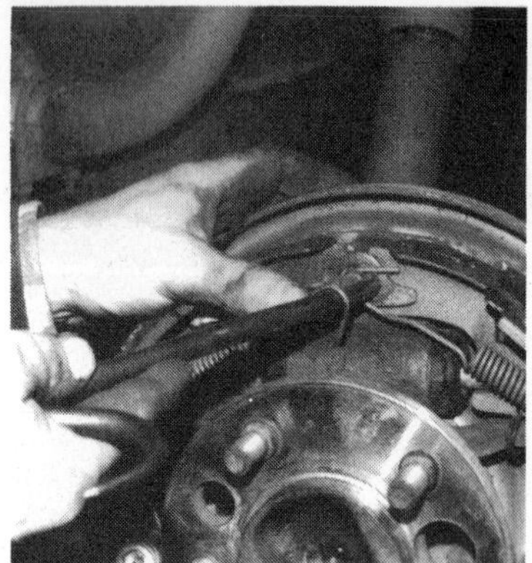

19. Install the secondary shoe return spring through both the cable guide and shoe web, and over the anchor post.

20. Use a pair of pliers to squeeze the end of each spring around the anchor post until it is parallel with the incoming spring wire.

21. Install the starwheel adjuster assembly. The left assembly is cadmium plated and stamped with an L. The right assembly is black and stamped with an R.

22. Install the shoe-to-shoe return spring.

23. Install the adjuster pawl spring over the pivot on the secondary shoe web.

24. Install the adjuster pawl under the spring, route the adjuster cable through the cable guide, and attach the end of the over-travel spring to the pawl.

GM DUAL-SERVO BRAKE WITH LEVER-TYPE AUTOMATIC STARWHEEL ADJUSTER

1. Remove both shoe return springs from the anchor.

2. Remove the shoe hold-down springs, and withdraw the pins from the backing plate.

3. Lift upward on the adjuster lever, and remove the actuating link from the anchor.

4. Remove the adjuster lever, pivot, and lever return spring from the secondary shoe.

5. Spread the shoes and remove the parking brake strut and anti-rattle spring.

6. Disconnect the parking brake cable, and remove the shoes, starwheel adjuster assembly, and shoe-to-shoe return spring.

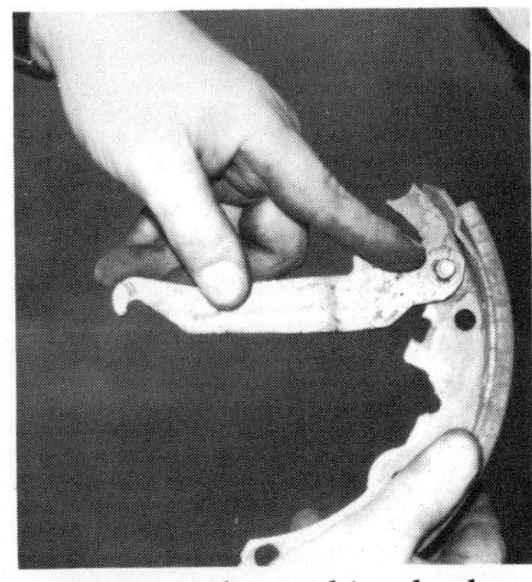

7. Remove the parking brake lever, pin, and retaining clip from the old secondary shoe, and install them on the new shoe.

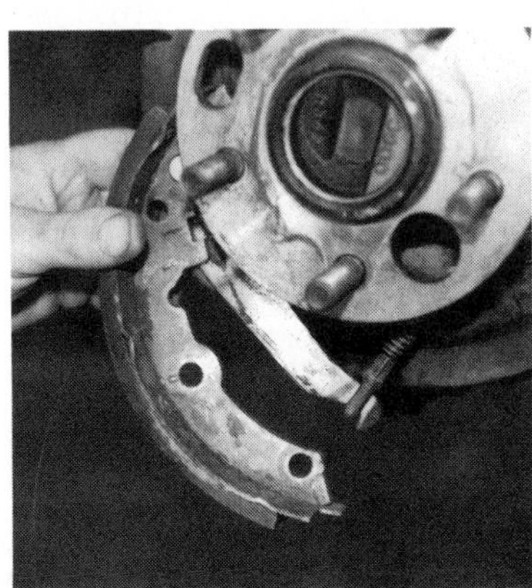

8. Connect the parking brake cable to the lever on the secondary shoe.

9. Install the starwheel adjuster and shoe-to-shoe return spring between the shoes, then position the shoes on the backing plate.

GM DUAL-SERVO BRAKE WITH LEVER-TYPE AUTOMATIC STARWHEEL ADJUSTER

10. Spread the shoes and install the parking brake strut so the end with the anti-rattle spring engages the primary shoe.

11. Install the adjuster lever, pivot, and lever return spring on the secondary shoe.

12. Install the actuating link into the hole or over the post in the anchor.

13. Lift upward on the adjuster lever, and connect the link to the lever.

14. Insert the holddown pins through the backing plate and shoes, then install the holddown springs on the pins.

15. Hook the shoe return springs into their holes in the shoe webs, then install the springs onto the anchor.

FORD LEADING-TRAILING BRAKE WITH STRUT-QUADRANT AUTOMATIC ADJUSTER

1. Remove the shoe hold-down springs, and withdraw the pins from the backing plate.

2. Lift the assembly of the brake shoes, adjuster strut, parking brake lever, and shoe-to-shoe return spring off of the backing plate.

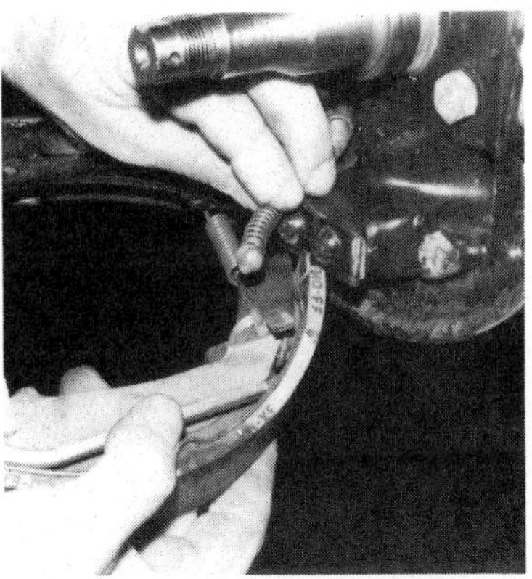

3. Disconnect the parking brake cable from the lever.

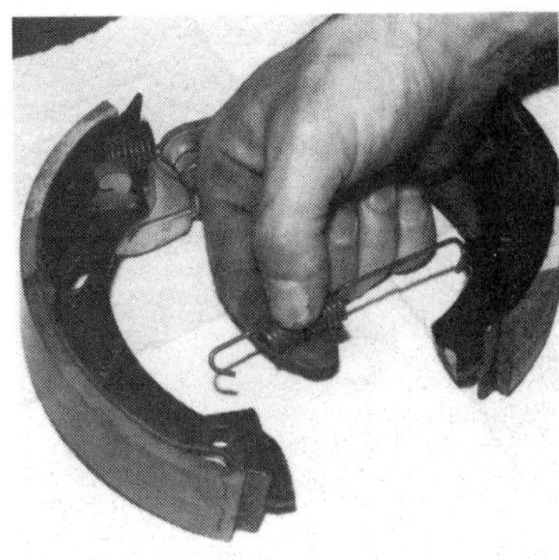

4. Remove the shoe-to-shoe return spring.

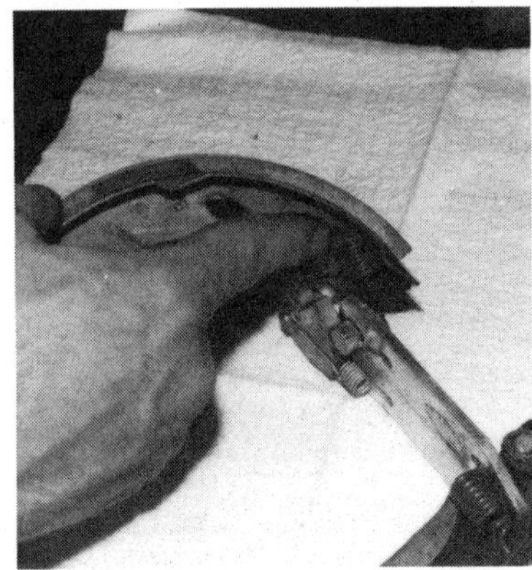

5. Rotate the leading shoe outward to release tension on the return spring, then remove the spring.

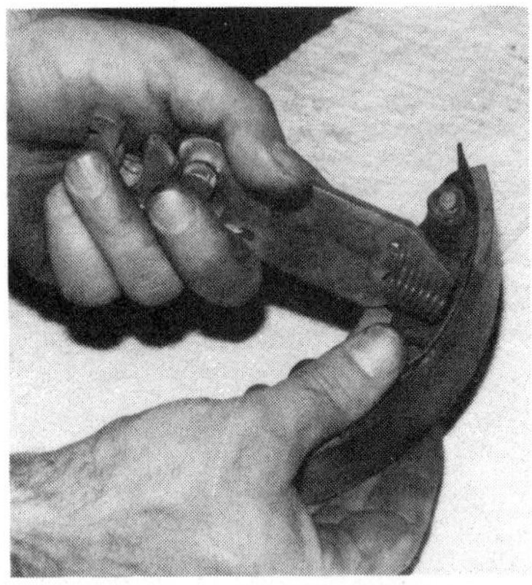

6. Pull the adjuster strut away from the secondary shoe and pivot it downward to release tension on the return spring, then remove the spring.

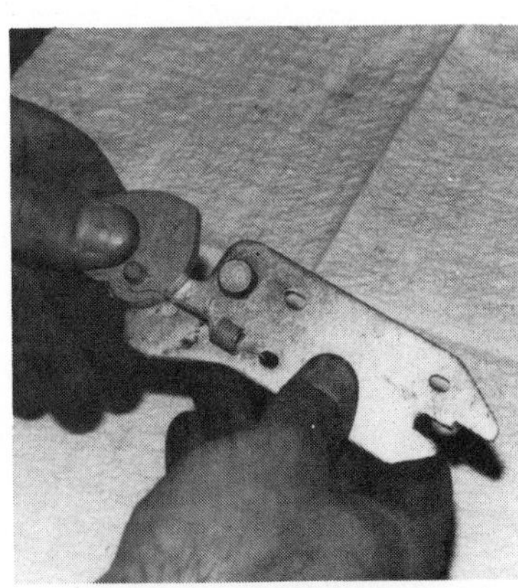

7. Pull the adjuster quadrant away from the knurled pin, and rotate the quadrant until its teeth no longer mesh with the pin.

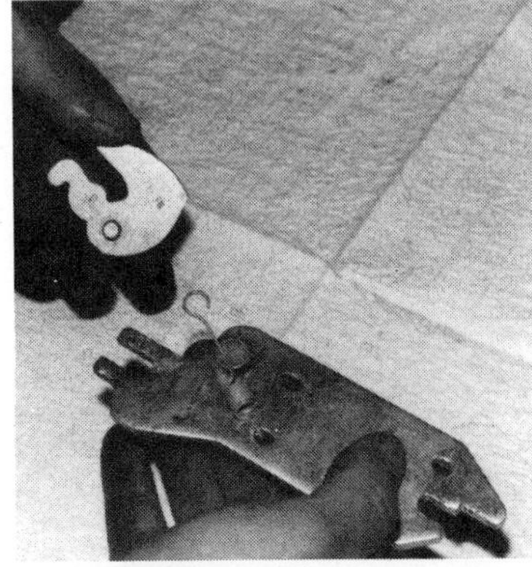

8. Carefully unhook the quadrant spring, taking care not to overstretch it, and remove the quadrant from the slot in the strut.

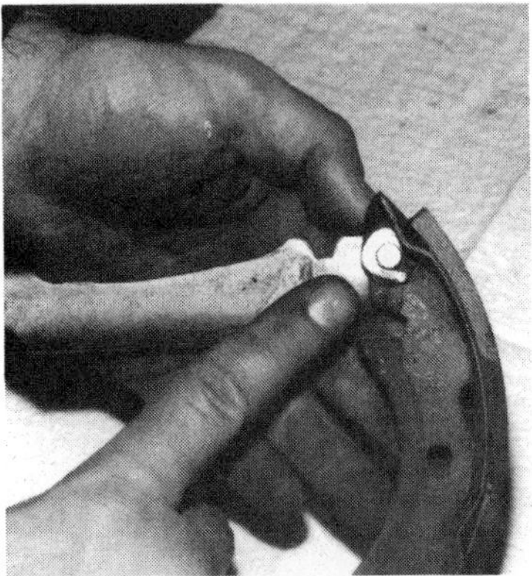

9. Remove the parking brake lever, pin, and retaining clip from the old trailing shoe, and install them on the new shoe.

FORD LEADING-TRAILING BRAKE WITH STRUT-QUADRANT AUTOMATIC ADJUSTER

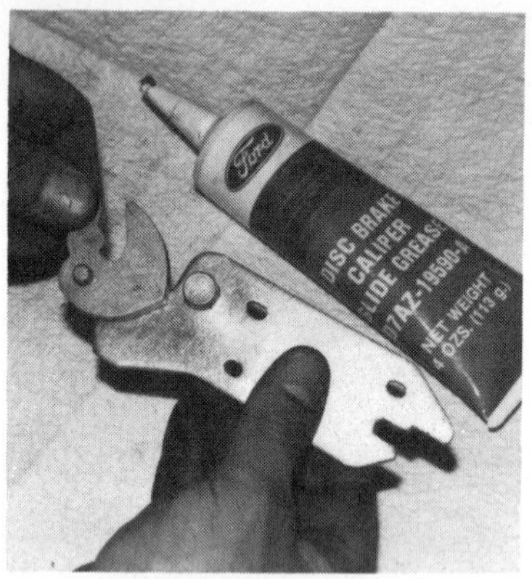

10. Lubricate the contact surface between the adjuster quadrant and strut with brake grease, then install the quadrant into the slot in the strut.

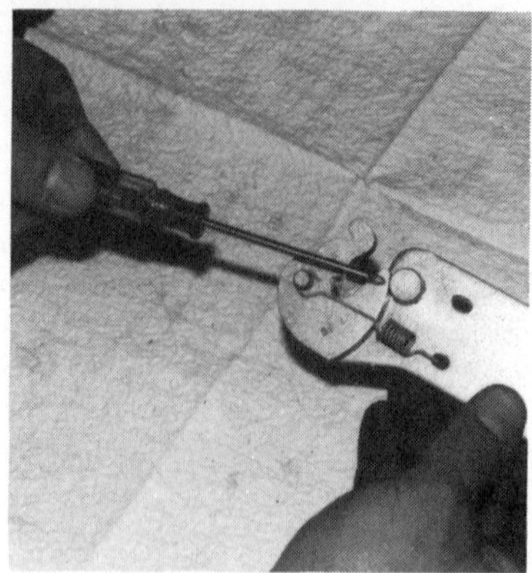

11. Install the quadrant spring, and rotate the quadrant until the third and fourth notch on its narrow end meshes with the pin.

12. Hook the shoe return spring into its holes in the trailing shoe and strut, then pivot the strut upward into position.

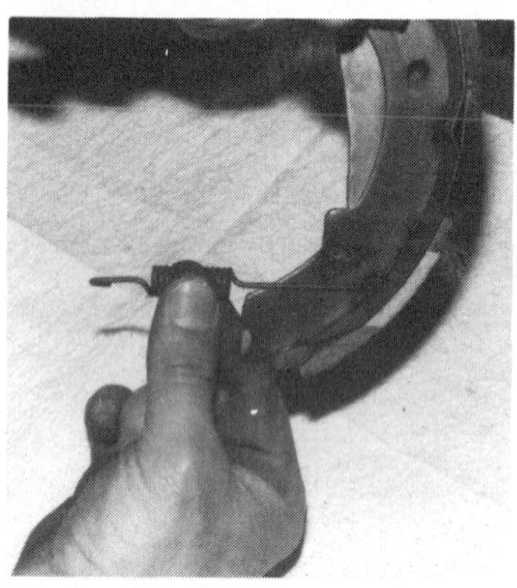

13. Install the shoe-to-shoe return spring so the long, straight section of the spring is attached to the trailing shoe.

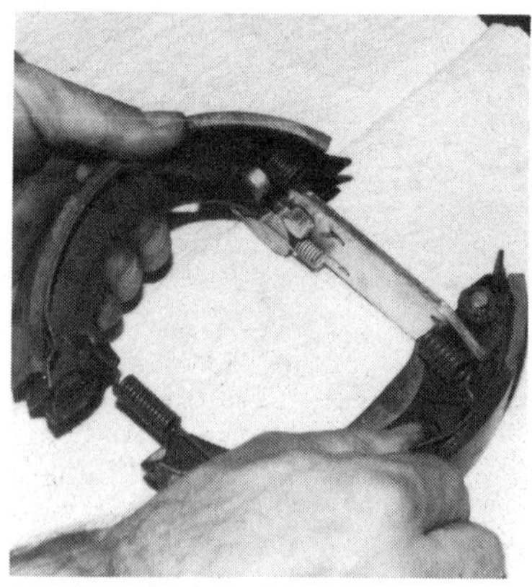

14. Hook the shoe return spring into its holes in the leading shoe and strut, then pivot the shoe rearward into position.

15. Lubricate the shoe support pads on the backing plate.

16. Connect the parking brake cable to the lever.

17. Install the assembly of the brake shoes, adjuster strut, parking brake lever, and shoe-to-shoe return spring onto the backing plate.

18. Insert the holddown pins through the backing plate and shoes, then install the holddown springs on the pins.

Chapter

7

Disc Brake Service

Disc brake service consists of two main operations, pad replacement and caliper overhaul. With the exception of brake fluid level checking, pad replacement is probably the most common form of brake system service. This is because the brake pads are designed to wear as they generate the friction that stops the car.

When replacing brake pads, the caliper is usually removed from the brake rotor, or at least partially disassembled. The caliper is inspected at this time, and then rebuilt or replaced as needed. Caliper overhaul or replacement can also be required because of fluid leaks, or because a caliper piston is frozen in its bore.

This chapter explains the procedures used to inspect and replace brake pads and calipers. It also describes detailed pad replacement and caliper overhaul procedures for several specific brake assemblies.

BRAKE PAD INSPECTION

Front disc brake pads wear more quickly and require more frequent service than either drum brake shoes or disc brake pads on the rear-wheel brakes. This occurs because front disc brakes provide as much as 80 percent of the total vehicle braking. And, compared to drum brake shoes, disc brake pads have less surface area to absorb heat and distribute wear.

Because front brake pads wear at a rapid rate, most vehicle manufacturers recommend that they be inspected approximately every 7,500 miles (12 000 km), or anytime the wheels are removed. Some brake pads are fitted with mechanical wear indicators that contact the rotor and make a high pitched squeal or scraping noise when the pads are worn to the point where they need when to be replaced. Other cars have electrical wear indicators that illuminate a warning light on the instrument panel. The brake pads should be inspected at regular intervals, or whenever any of the above indications of wear are present.

Unfortunately, many car owners wait until they hear a grinding noise from their brakes before having the pads checked. By then it is too late because the noise indicates that the lining material is worn completely away, and a pad backing plate is making metal-to-metal contact with a brake rotor, figure 7-1. Once this happens, the rotor must be machined or replaced as described in Chapter 8. Brake pads are inexpensive compared to rotor service, and it is much better to change a set of brake pads sooner than absolutely necessary, rather than risk rotor damage by trying to get the maximum possible life from the pads.

Figure 7-1. Rotor damage occurs if brake pads are not replaced before the linings are worn away.

There are two types of brake pad inspections. The first is a routine check of lining thickness while the pads are still installed on the car. This inspection is a part of regular vehicle maintenance, and helps determine when the pads should be replaced. Sometimes, a routine pad inspection can also indicate a problem with a caliper. The second type of inspection is made when pads with plenty of lining material are removed from the car in the course of other brake service. This inspection helps determine if the pads can be reused.

On-Car Pad Inspection

The exact method of pad inspection on the car varies with the design of the brake caliper. Some pads can be inspected with the wheels on the car, either directly or by using a small mirror. Other brakes require that the wheels be removed before the pads can be viewed clearly. There are three things to look for when inspecting brake pads: thickness, taper wear, and uneven wear.

The thickness of the lining material is the main factor that determines whether the pads should be replaced. Check the thickness of the lining material, and compare your measurement to the manufacturers specifications. As a general rule, the lining should be at least 1/32 inch (.030" or .75 mm) above the rivet heads on riveted pads, or the same distance above the backing plate on bonded pads, figure 7-2. If the pads are thinner than this, replace them.

Next, inspect the pads for taper wear in which the pads are thinner at one end than at the other. Some pad taper wear is normal in floating calipers because the caliper body tends to flex slightly on its mountings. The leading edges of brake pads may also wear faster than the trailing edges because they operate at a

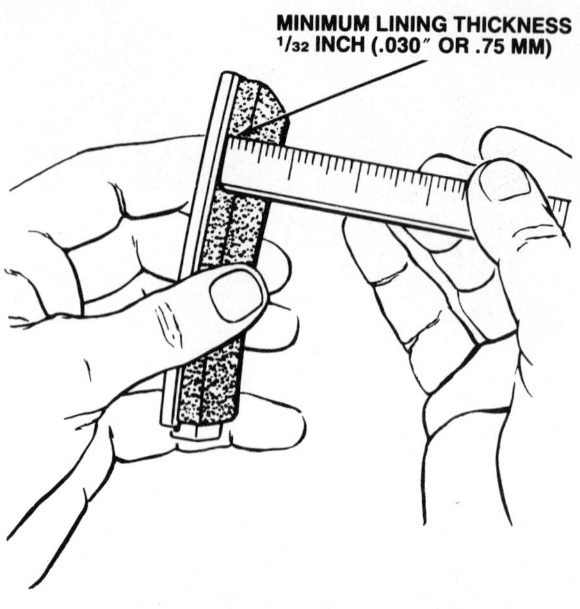

Figure 7-2. Lining thickness determines when a brake pad needs to be replaced.

higher temperatures. However, if there is more than 1/8 inch (3 mm) of taper wear, you should replace the pads and inspect the caliper for possible problems that may be contributing to the taper wear. This test does not apply on certain Datsun vehicles from the early 1970's that use tapered brake pads in a unique pivoting brake caliper.

Next, compare the amount of wear on the two pads in each caliper, then compare the amount of pad wear between the two calipers on the same axle; all of the pads should be worn about the same amount. Uneven wear between pads in the same caliper can be caused if the rotor is rough on one side, causing that pad to wear more rapidly. In a fixed caliper, a frozen piston will cause uneven wear between the two pads. Uneven pad wear occurs in floating and sliding calipers when the mounting hardware rusts or corrodes, causing the caliper to bind where it moves on the anchor plate. A rough rotor can also cause unequal pad wear from one side of the car to the other, but the most common cause is a frozen caliper piston on the side with the least wear.

Whenever you find a caliper with unevenly worn pads, or a car with unequal side-to-side pad wear, locate and repair the cause of the wear problem before you install new pads. Never swap brake pads to the opposite side of the rotor or car in an attempt to balance out wear and extend the life of the brakes. This is not cost effective, and will leave you open to liability problems if there is an accident.

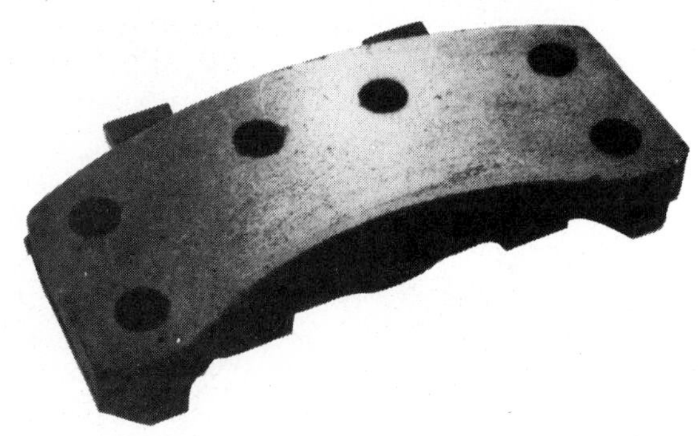

Figure 7-3. A glazed brake pad lining.

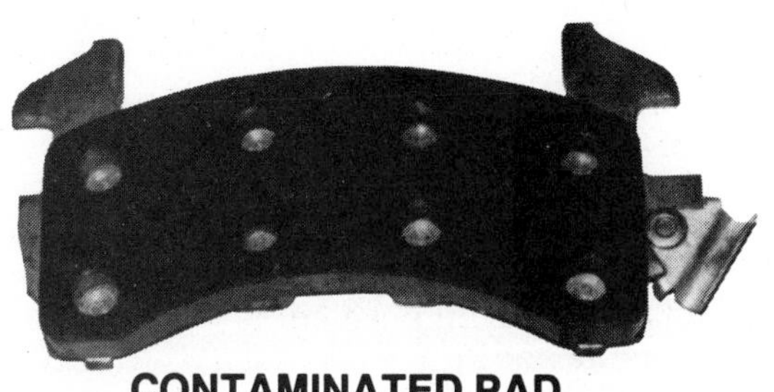

Figure 7-4. A brake pad suffering from fluid contamination.

Off-Car Pad Inspection

Brake pads that have plenty of lining material remaining are often removed in the course of other brake system service, such as rebuilding a leaking or frozen caliper, or turning a brake rotor. These pads can often be reused, however, you should first inspect them for damage and contamination that is difficult to see while the pads are on the car. The major difference in inspecting the brake pads off the car as opposed to checking them on the car is that you can examine the friction surface that contacts the brake rotor. Normally, this surface is relatively smooth with a dull finish. A shiny or polished lining is glazed, figure 7-3, and should be replaced.

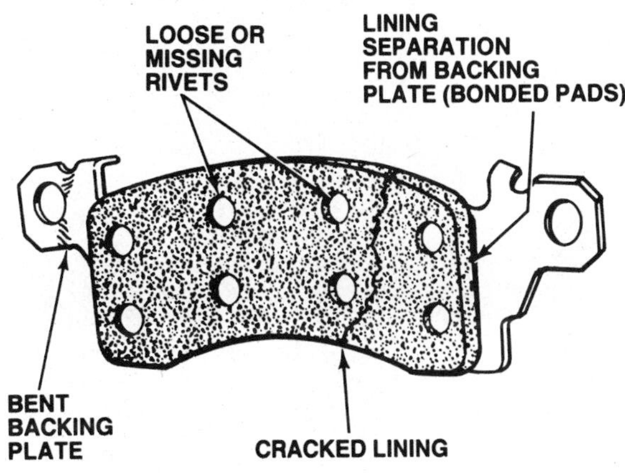

Figure 7-5. Replace any pad that has physical damage.

Also check the lining surface for signs of contamination from brake fluid that has leaked past the piston seal and dust boot. Fluid contamination will cause the lining to darken, and where the leak is severe, the lining will appear "smeared" and the paint will often be stripped from the pad backing plate, figure 7-4. Brake pads that have been soaked with brake fluid must be replaced. However, where the contamination is limited to less than 10 percent of the lining surface, you may be able to clean and reuse the brake pads.

In addition to glazing and contamination, inspect the pads for physical damage. Look for large cracks in the lining, loose or missing rivets, a bent backing plate, or a bonded lining that is separating from the backing plate, figure 7-5. If any of these problems are present, replace the pads.

BRAKE PAD REPLACEMENT

Brake pads are sold and serviced as axle sets. An axle set consists of four pads, the inner and outer pad for the caliper at each wheel. Pads from different manufacturers should never be mixed. Although they will fit and may appear the same, the friction coefficients of the linings may be quite different. Even if only one pad of an axle set is badly worn, the entire set should be replaced after the problem causing the uneven wear has been repaired.

When you replace a set of brake pads, most vehicle manufacturers recommend that you also replace the bushings, O-rings, retaining bolts, retaining clips, and any other caliper mounting hardware unless it is in perfect condition. These parts play a major role in smooth and quiet brake operation, and they are relatively inexpensive compared to the cost of a

Figure 7-6. Lower the fluid level in the master cylinder reservoir before replacing brake pads.

comeback. The necessary parts for all common calipers are sold as a "small parts kit" which is available from the dealer or the aftermarket.

The exact procedure used to replace a set of brake pads varies with the design of the caliper. However, there are a number of basic steps common to any pad replacement job; these are described below. Detailed procedures for replacing the pads in typical brake calipers are included in the photo sequences on caliper overhaul at the end of the chapter.

Pad Removal

The first step in any pad replacement job is to remove some of the brake fluid from the master cylinder reservoir with a brake fluid syringe, figure 7-6. This makes space for the fluid that will be displaced back into the reservoir through the hydraulic system when the caliper pistons are bottomed in their bores to install the new and thicker pads. When removing brake fluid from the reservoir, wear safety glasses to prevent eye injury from accidental fluid spray. Also, take suitable precautions to keep brake fluid from coming in contact with the vehicle finish.

If the hydraulic system of a car has a front/rear split, remove about two-thirds of the fluid from the larger reservoir chamber that supplies the front disc brakes. If the hydraulic system has a diagonal or triagonal split, remove about one-half of the fluid from each of the equally sized reservoir chambers. Be sure to replace the reservoir caps or cover after lowering the fluid

Figure 7-7. A slide hammer will remove frozen brake pads from a fixed caliper.

level; this prevents any damage from brake fluid that may spray out through the compensating ports as the caliper pistons are bottomed in their bores.

Raise the car until the wheels hang free, then support it securely in position. It will be easier to replace the pads if the car is at a convenient working level, so take into account the position you will be in when servicing the brakes. If you will be standing, raise the car so that the brakes are at waist or shoulder level; if you will be seated on a stool, the brakes should be around knee height. Once the car is in position, remove the wheels to gain access to the calipers.

On fixed brake calipers, remove the pad retaining pins, and use a pair of pliers to pull the pads straight out of the caliper. If there is a ridge at the edge of the rotor that prevents a pad from being easily removed, insert a screwdriver between the rotor and pad, and pry the caliper piston back into its bore until there is sufficient clearance to remove the pad. If the pads are rusted or seized in the caliper body, use a screwdriver or a pair of pliers to work the pads back and forth until they can be removed. In severe cases, a hook attached to a slide hammer can be inserted through one of the retaining pin holes in the pad backing plate, figure 7-7; a few blows with the slide hammer should pull the pad out of the caliper.

Removing the pads from a sliding or floating caliper generally requires that the movable portion of the caliper be separated or pivoted away from the anchor plate. Use the appropriate wrench to remove the attaching hardware, and lift or swing the caliper body away from the anchor plate. Take special care when removing rusted or frozen caliper attaching hardware; a broken fastener could result in the need for

Figure 7-8. Hang calipers from the suspension to prevent brake hose damage.

costly caliper replacement. Whenever you remove a caliper during brake pad replacement, hang it from the suspension by a wire, figure 7-8, so there is no strain on the brake hose that might cause internal or external damage.

Once the caliper is secured, remove the brake pads from the caliper and/or anchor plate. Be sure to note the presence and position of any springs, shims, wear sensors, or other parts; detach these items and transfer them to the new pads as necessary. Compare the new pads to the old parts to be sure you have the proper replacements.

Pad Installation

Before you install the new pads, inspect the brake rotor as described in Chapter 8, and inspect the brake caliper as described later in this chapter. These inspections will help you determine if additional service is necessary to these components. The caliper external inspection procedure also describes how to bottom the caliper pistons in their bores so that the new brake pads can be installed.

Because disc brakes are so prone to noise, many shops make it standard practice to apply an anti-squeal coating to the backing plates of new brake pads, figure 7-9. Many new pad sets come with a small tube of this coating, and several brands are available from the aftermarket. Follow the manufacturer's instructions for the product you choose, and allow the coating to set-up properly before you install the pads.

Installing new brake pads is essentially the reverse of removing them, with a few important differences. First, install the pads into the caliper or anchor plate as dictated by the brake design. The pads in fixed calipers slip into place, and spring clips on the retaining pins

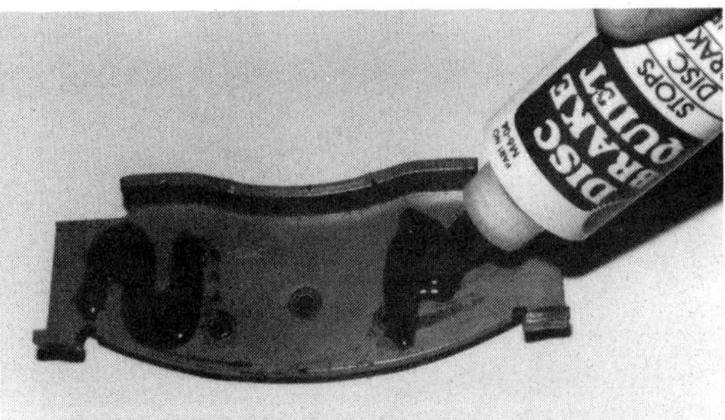

Figure 7-9. Anti-squeal coatings applied to pad backing plates can help reduce brake noise.

and/or shims between the caliper pistons and pad backing plates prevent the pads from vibrating and causing brake noise. The pads in floating and sliding calipers usually have spring clips or bent tabs on the backing plate that lock them securely into the caliper or anchor plate to prevent vibration and brake noise. Bend the springs or tabs as necessary so the pad fits firmly in position.

Once the pads are in position, secure them in the appropriate manner. Once again, the exact method for doing this is determined by the brake design. With a fixed caliper, install the retaining pins. With a sliding caliper, lubricate the ways with high-temperature brake grease, position the caliper body onto the ways over the rotor, and install the retaining hardware. With a floating caliper, lubricate the caliper bushings and retaining bolts with high-temperature brake grease, and position the caliper body over the rotor. Install the retaining bolts and torque them to the manufacturers specifications.

Once the pads are installed and the calipers are secured in place, top up the fluid level in the master cylinder reservoir and apply the brake pedal; it will probably go all the way to the floor. Stroke the pedal as many times as necessary to get a high, firm brake pedal. This will happen when the caliper pistons have moved out of their bores far enough to bring the new brake pads into firm contact with the rotor. If you fail to make this initial brake application, an accident may occur when you drive the car out of the service bay and discover that it has no brakes the first time you apply the pedal.

If you cannot get a high, firm pedal, bleed the brakes as described in Chapter 3. If bleeding fails to restore a good brake pedal, and no other cause can be found for the problem, it is possible that rust and corrosion in the caliper bore have disturbed the piston's seal. When this happens, air continually enters the hydraulic system, and the caliper soon starts to leak fluid. This problem should have been noticed when bottoming the piston, but if not, the calipers should be removed and rebuilt at this time.

Once good pedal height and feel is achieved, adjust the fluid level in the master cylinder to the full mark, and securely install the reservoir cover or caps. Turn the ignition ON, release the parking brake, and make sure the brake system warning light is not illuminated. If the light is on constantly, center the pressure differential switch as described in Chapter 3. Install the wheels, lower the vehicle to the ground, and test drive the car to make sure the brakes operate properly. While on the drive, burnish-in the new pads as described in the next section.

BRAKE PAD BURNISH-IN

Whenever new brake pads are installed, they require a short time of relatively light use called a burnishing-in or bedding-in period. There are two reasons for this. First, regardless of whether the rotors have been refinished or not, new pads do not make full contact with the friction surface. A short period of light braking allows the pads and rotors to become "familiar" with one another and wear into better contact.

The second reason a burnish-in period is needed is because new brake pads are not fully cured when they come from the factory; there are still binding resins and other chemicals that have not completely evaporated from the friction material. Final curing takes place over the first few hundred miles of driving as the pads are repeatedly heated and cooled.

If new brakes are used hard immediately and become too hot, residual resins boil out of the pads as gasses and liquids. When these gasses and liquids come out in great enough quantities, they lubricate the friction material and cause a type of brake fade called green fade. Once green fade occurs, the pads will be glazed when they return to normal operating temperature. Glazed pads do not have the proper friction coefficient for smooth operation, and must be replaced.

Pad Burnishing Procedure

Whenever you replace any of the friction materials in a brake system, take the car for a test drive afterward to make sure the brakes are operating properly. During the test drive, perform the simple procedure below to burnish-in the new shoes and/or pads:

1. Drive the car at approximately 30 mph (50 kph). Check your rear-view mirror to make sure it is safe to do so, then apply the brakes with *moderate to firm* pressure to slow the car to about 5 mph (8 kph).
2. Accelerate back to speed, and drive for approximately 15 seconds to allow the brakes to cool. Make another stop as described in Step 1. Continue to alternate stops with cooldown periods until you have made about a half dozen stops in this manner.
3. Accelerate to approximately 55 mph (88 kph). Check your rear-view mirror to make sure it is safe to do so, then apply the brakes with *moderate to firm* pressure to slow the car to about 20 mph (30 kph).
4. Accelerate back to speed, and drive for approximately 30 seconds to allow the brakes to cool. Make another stop as described in Step 3. Continue to alternate stops with cooldown periods until you have made about a half dozen stops in this manner.
5. When you deliver the car to its owner, caution them that the brakes should not be used hard for approximately 100 miles of in-town driving, or 300 miles of highway driving. This will ensure that the brakes become properly burnished in.

BRAKE CALIPER EXTERNAL INSPECTION

Whenever you replace a set of brake pads, or have the calipers off the car for any reason, you should inspect the calipers to determine if they need an overhaul. Some manufacturers and technicians maintain that the calipers should be rebuilt every time the pads are replaced, others believe that new pads can be installed without overhauling the calipers if the calipers pass a simple inspection. Keep in mind that a set of brake pads usually lasts a minimum of 30,000 miles (48 000 km). If you decide to replace a set of brake pads without rebuilding the calipers, you should be confident the calipers will not leak, seize, or otherwise fail before the new pads are worn out.

If the brake calipers are not rebuilt when the pads are changed, the replacement pads generally do not last as long as the original parts. This happens for two reasons: first, because the rotor friction surface is not as smooth as it was originally; and second, because the rubber used to make fixed piston seals loses flexibility with age, and will not retract the piston and pads as far from the rotor. Both these conditions increase brake drag, which accelerates the rate of lining wear.

Figure 7-10. Rebuild the brake caliper if the dust boot is damaged in any way.

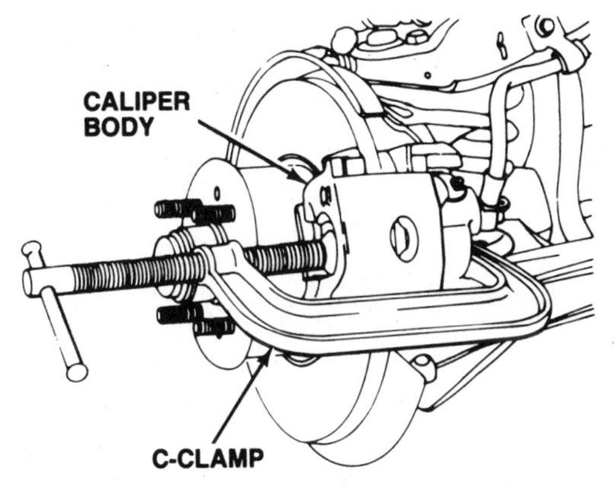

Figure 7-11. Bottoming a caliper piston with a C-clamp.

The next section describes an external caliper inspection procedure, but in addition to the items described, there are a number of other factors that affect the need to rebuild brake calipers. It is impossible to give precise recommendations for every situation, but consider the following items when you make your decision.

A caliper overhaul is more likely to be needed on a car that:

- Has over 50,000 miles (80 000 km) on it and the brake system has never been serviced
- Is more than five years old and the brake system has never been serviced
- Is driven in snowbelt areas where salt is used on the roads
- Is driven in areas with high humidity and rainfall
- Is driven in mountainous regions or in heavy-duty service.

On the other hand, unless there are obvious signs of a problem, a caliper overhaul is less likely to be needed on a car that:

- Has fewer than 50,000 miles (80 000 km) and has had regular brake system service
- Is less than five years old and has had regular brake system service
- Is driven in sunbelt areas where humidity, rainfall, and snowfall are minimal.

Caliper External Inspection Procedure

Once the brake pads are removed, inspect the entire outside of the caliper body; if there are cracks or any other major damage, replace the caliper. Next, visually inspect the dust boot around the piston, figure 7-10; make sure it is fully seated in the caliper body and does not have any holes or tears in it. A leaking dust boot allows moisture and dirt into the caliper bore that will quickly destroy the piston, seal, and bore finish. Also inspect the caliper for brake fluid leakage around the boot, or fluid deposits on the brake pads that can indicate a leaking caliper seal. A leaking seal could result in a total loss of braking in at least half of the split braking system. If the dust boot is unseated or damaged in any way, or there are any signs of fluid leakage, rebuild or replace the caliper.

If the dust boot is in good condition and there are no signs of leaks, bottom the caliper pistons in their bores. On single piston calipers, use a C-clamp, figure 7-11, or a large pair of slip-joint pliers, figure 7-12; take care not to tear or dislodge the dust boot. On multi-piston fixed calipers, insert a screwdriver or pry bar between the rotor and an old brake pad, and lever the pistons back into their bores, figure 7-13. Ford rear disc brake calipers that are applied to serve as the parking brake require a special turning tool to thread the piston back into the bore on the automatic adjuster screw, figure 7-14.

Remember that as you bottom a piston, the fluid inside the caliper is displaced into the hydraulic system under pressure. If other pistons in the hydraulic system, such as those in the wheel cylinders or the caliper on the other side of the car, are not restrained, this pressure may be sufficient to push them out of their bores. Unless these parts are to be rebuilt anyway, take precautions to prevent this from happening. Service one axle or wheel at a time, or in-

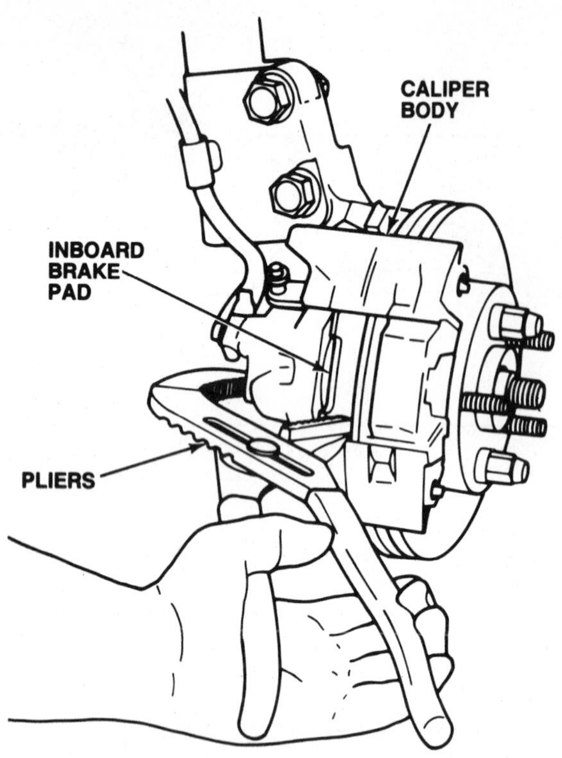

Figure 7-12. Bottoming a caliper piston with slip-joint pliers.

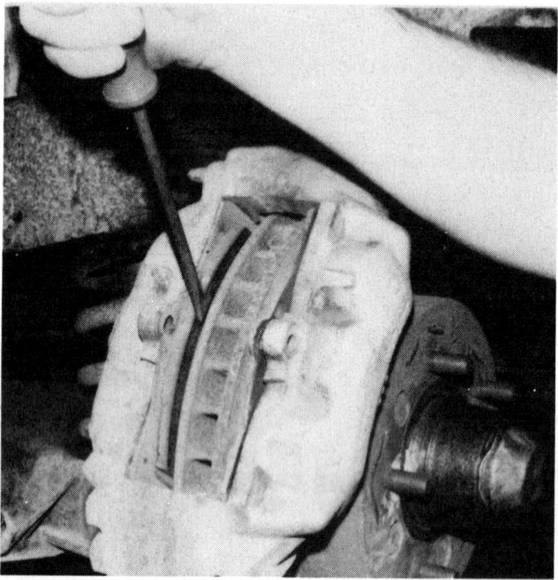

Figure 7-13. Bottoming the caliper pistons in a fixed brake caliper.

sert a block of wood into the opposing caliper to prevent its piston from coming out of the bore.

As you bottom the pistons, note the "feel" of the piston movement. A moderate amount

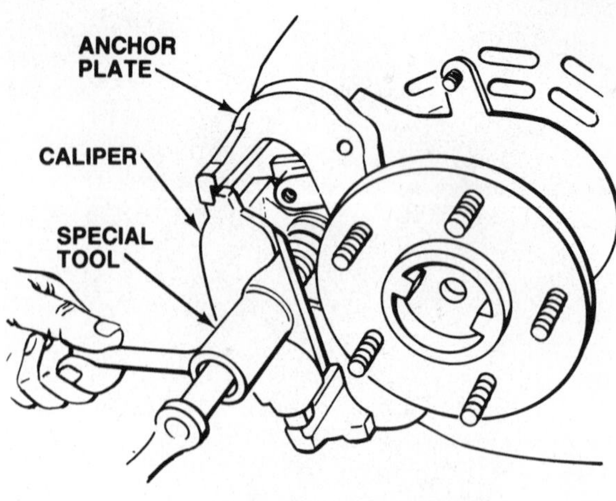

Figure 7-14. A special tool is required to bottom Ford rear disc brake caliper pistons.

of force should be sufficient to move the pistons, and they should slide smoothly into the caliper bores. Furthermore, all of the pistons in a multi-piston caliper, and those in the calipers on both sides of the car, should require about the same amount of force to bottom. If these conditions are met, and the car does not fall into any of the "likely to need an overhaul" catagories discussed above, you can be reasonably safe in assuming that the calipers are in good condition and do not need to be overhauled.

If a piston is frozen or difficult to bottom, the piston and/or bore is rusted or corroded, and the caliper must be overhauled or replaced. Do not return a caliper in marginal condition to service. Even though you may be able to bottom a rusted, corroded or dirty piston in its bore, the piston will not seal properly and the caliper will begin to leak fluid which will contaminate the new pads and cause a comeback.

Just as brake pads are always replaced in axle sets, *both* calipers on an axle should be overhauled if it is determined that one needs to be rebuilt. This ensures that even braking action is maintained, and that the brakes at both wheels have a similar service life.

BRAKE CALIPER OVERHAUL

Every brake caliper requires certain special overhaul procedures, but just as with brake pad replacement, *all* overhaul procedures have some things in common. Step-by-step overhaul procedures for several calipers are provided at the end of the chapter. The sections below describe a general procedure that includes the common operations of bleeder screw removal,

Figure 7-15. Caliper bleeder screws are almost impossible to remove once they become frozen in place.

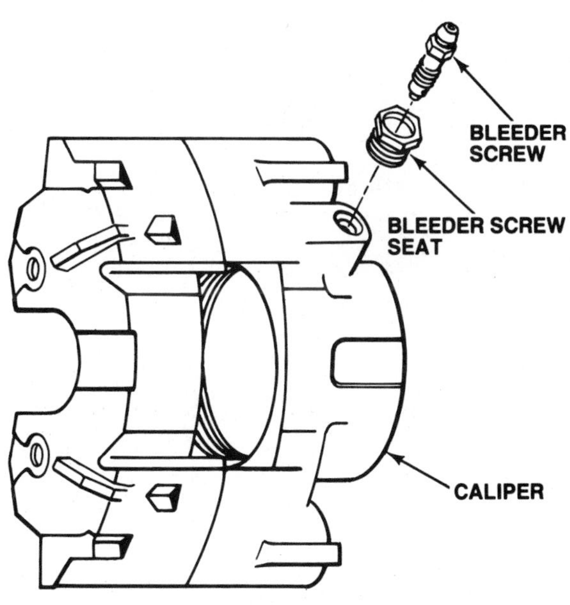

Figure 7-16. A universal replacement bleeder screw assembly.

piston removal, dust boot and piston seal removal, internal inspection, honing, and caliper assembly.

Before you begin an overhaul on any caliper, attempt to loosen the bleeder screw with a bleeder screw wrench. It is not unusual for bleeder screws to become rusted or corroded in place so they cannot be removed. These "frozen" bleeder screws are particularly common in aluminum calipers where the dissimilar metal of the steel screw causes electrolysis. In most cases, calipers with frozen bleeder screws should be replaced. The exceptions are when a replacement caliper is unavailable, or its cost is so high that the time and effort spent on bleeder screw removal can be justified.

Bleeder Screw Removal

If it is necessary to remove a frozen bleeder screw, spray the screw threads with penetrating oil and allow it to soak in for at least 10 minutes. Then, place a deep socket over the bleeder screw so it contacts the body of the caliper around the base of the screw. Wearing safety glasses to prevent injuries, strike the socket several medium-firm blows; this should slightly deform the metal of the caliper and break the surface tension between the bleeder screw threads and those in the caliper opening. Attempt to remove the bleeder screw with the appropriate wrench, but do not apply too much turning force or you will twist the end of the screw off, figure 7-15. Once this happens, it is virtually impossible to remove the portion of the screw that remains in the caliper.

Repeat the procedures with the penetrating oil, socket and hammer, and bleeder wrench as needed until the screw is free. As discussed in the last chapter, special hammer sockets are available from the aftermarket that allow you to apply both hammer blows and turning force at the same time. Once you remove a bleeder screw, buff its threads clean on a wire wheel, and run a drill bit through its passageways to make sure they are clear. In some cases this is not necessary because a new bleeder screw is included in the caliper overhaul kit. Do not reinstall the bleeder screw until after the caliper and screw threads have both been thoroughly cleaned and are ready to be assembled.

When a bleeder screw breaks off, or cannot be removed using any of the techniques above, the caliper should be replaced. If a replacement caliper is unavailable, there are replacement bleeder screw assemblies available from the aftermarket, figure 7-16. These require that you drill out the old bleeder screw, and thread the caliper body to accept the new bleeder screw seat. Because bleeder screws are usually made of very hard steel, this operation can be quite difficult, and is often best done by a machine shop equipped with the proper tooling.

BRAKE CALIPER PISTON REMOVAL

Other than bleeder screw removal, the only potentially difficult part of a brake caliper over-

Figure 7-17. Removing a caliper piston with a mechanical tool.

haul is removing the piston or pistons from the caliper. Pistons fit in their bores with only a few thousandth of an inch clearance, so even a small amount of rust or corrosion can make a piston extremely hard to remove. Unless a piston is free in its bore, or only slightly stuck, a great deal of force will be required to remove it. This force is usually generated in one of three ways: with a mechanical removal tool, or by applying air pressure or hydraulic pressure to the fluid inlet fitting of the caliper.

The problems involved in caliper piston removal are worse with multi-piston calipers. With a single-piston caliper, air or hydraulic pressure only has to act on one piston; when that piston is forced from its bore, you can proceed with the overhaul. However, with a multi-piston caliper, the pressure acts with equal force on all of the pistons, which means that the piston that is least stuck will be forced out of its bore first. Once this happens, pressure is no longer contained within the caliper, and another method must be used to remove the remaining, more securely frozen, caliper pistons.

In some cases, you can work around this difficulty by removing the bridge bolts and splitting the caliper into its component halves. For example, if a two-piston caliper is split, pressure can be applied to each half to remove the pistons individually. This is not always an option, however, because some manufacturers state that their calipers should not be split. And, with four-piston calipers, there will always be two pistons in each caliper half.

One way of dealing with this problem is to use a C-clamp or other means of holding the free pistons in their bores while the frozen pistons are forced out with air or hydraulic pressure. Sometimes, if the caliper is installed on the car, the brake rotor will prevent the pistons from coming entirely out of the bores. In this case, pressure can be applied until all of the pistons contact the rotor, then the caliper is removed from the car. If the rotor does not have sufficient thickness to keep the pistons in their bores, a thin wooden block or a worn-out brake pad can be inserted between the pistons and the rotor. Generally, even a badly frozen piston can be removed manually once it is at least 75 percent of the way out of its bore.

One final caution is necessary before describing the various caliper piston removal procedures. There will always be at least a little brake fluid in the caliper when removing the pistons. Depending on the method used, this fluid will be sprayed or spilled from the caliper when the piston comes out of its bore. When you remove caliper pistons, wear safety glasses to protect your eyes, and take precautions to ensure that brake fluid does not come in contact with the vehicle finish or other painted surfaces.

Mechanical Piston Removal

Special tools are available to mechanically remove brake caliper pistons that are free, or only mildly stuck, in their bores. These tools are not widely used because it is generally faster and easier to use air pressure when removing these types of pistons (see below). However, mechanical removal tools can work well on multi-piston calipers to remove the second piston from a caliper half once another method has been used to remove the first one.

The advantages to removing caliper pistons mechanically are that the tools are simple and relatively inexpensive, and a source of compressed air is not required. The main disadvantage is that mechanical removal tools are human powered, and usually do not provide enough leverage to remove even moderately rusted or corroded pistons.

To remove a caliper piston with a mechanical tool:

1. Remove the caliper from the vehicle and clamp it in a vise.
2. Grip the inner bore of the piston with the removal tool. With a plier type tool, simply squeeze the handles together tightly. With a mechanically-tightened tool, rotate the allen bolt until the jaws of the tool are tightly clamped against the inner bore of the piston.
3. Rotate the piston back and forth until it is free, and pull outward with a twisting motion to remove the piston from its bore, figure 7-17.

If a piston cannot be removed with a manual tool, you will have to use either air pressure or hydraulic pressure, as described below, to remove the piston.

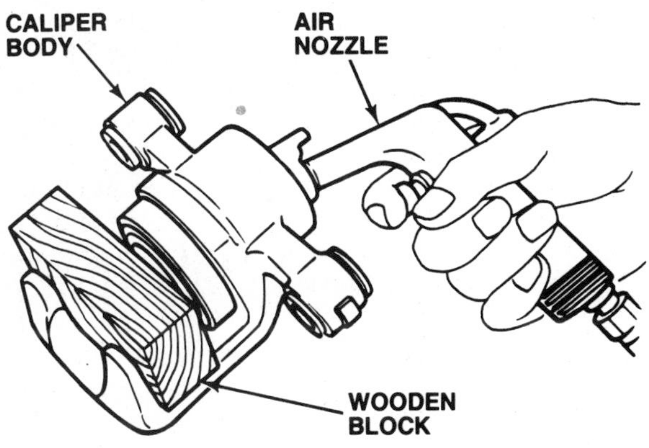

Figure 7-18. Wooden blocks help prevent damage when caliper pistons are removed from their bores.

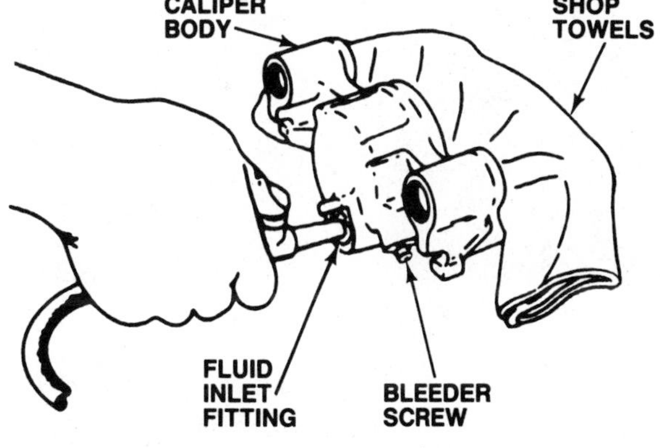

Figure 7-19. Apply air pressure to the caliper fluid inlet to force the piston out of its bore.

Compressed-Air Piston Removal

Using compressed air to remove caliper pistons is a common procedure. This method works well when the piston is free, or somewhat firmly stuck, in the caliper bore. The advantages of removing pistons in this way are that it can be done by a single technician, and sufficient force is available from the compressed air supply in most shops to remove all but severely frozen pistons. For example, 100-psi (690 kPa) of air pressure applied to a 1½-inch (38-mm) diameter caliper piston creates more than 550 pounds (250 kg) of force to drive the piston out of the bore.

The primary disadvantage of using compressed air to remove caliper pistons is that it is very difficult to regulate the amount of pressure applied to the piston. Compressed air contains a great deal of energy that is released very suddenly when the piston pops free. When too much pressure is used, the piston can be damaged as it is driven from the bore with great force, and brake fluid in the caliper may spray out, causing further damage or injury. Use *extreme* caution when you remove caliper pistons in this manner.

To remove a caliper piston with compressed air:

1. Remove the caliper from the vehicle and clamp it in a vise by its mounting flange.
2. Insert a wooden block or bundle of shop towels between the piston and caliper body to prevent damage to the piston when it comes out of the caliper bore, figure 7-18.
3. Cover the caliper body in shop towels to trap any brake fluid spray that may occur when the piston comes out of its bore.
4. Place a rubber-tipped air nozzle into the caliper fluid inlet, and carefully apply air pressure to force the piston out of the bore, figure 7-19. Use the minimum amount of pressure necessary, and *keep fingers and hands away from the piston*.

If full shop air pressure is insufficient to remove the piston from its bore, you will have to use hydraulic pressure to remove the piston.

Hydraulic Piston Removal

Using hydraulic pressure to remove caliper pistons is both the most effective and the safest method. The most common method of hydraulic piston removal is to use the vehicle brake hydraulic system to force the pistons from their bores while the calipers are still mounted on the car. In this manner, both front calipers can be serviced at the same time.

The advantages of this method are that it is relatively fast, does not require special tools, and provides plenty of pressure to do the job. While compressed air can provide several hundred pounds of force to remove a piston, hydraulic pressure can supply well over a thousand pounds! If hydraulic pressure cannot remove a piston from its bore, there is probably so much damage to the piston and bore that the caliper should be replaced anyway. Hydraulic pressure can also be applied with much greater control than air pressure, which reduces the chances of damage caused by flying pistons or brake fluid.

The disadvantages of using hydraulic pressure to remove pistons are that it requires an assistant, and cannot be used if there is air in the hydraulic system or the master cylinder is bypassing internally. In addition, extra fluid may have to be added to the master cylinder reservoir in order to move the pistons all the way out of their bores.

Figure 7-20. Wooden blocks can also prevent caliper pistons from coming entirely out of their bores.

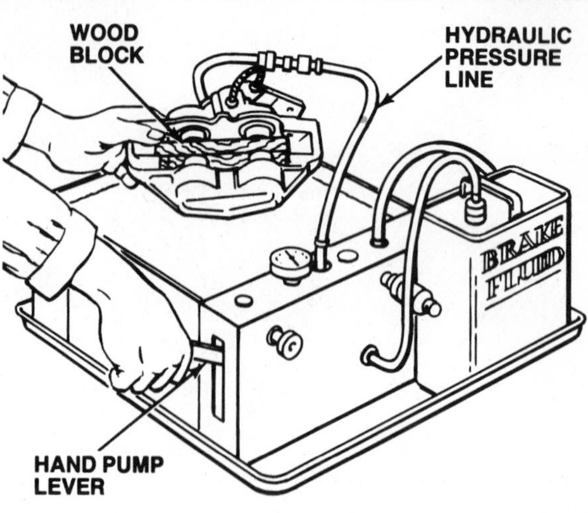

Figure 7-21. Removing a caliper piston with a special hydraulic service bench.

To remove a caliper piston from its bore using the brake hydraulic system:

1. With fixed brake calipers, remove the brake pads from the calipers. If the brake rotor is not thick enough to prevent the pistons from coming out of their bores, insert a set of worn-out brake pads or thin wooden shims between the caliper pistons and the rotor.
2. With floating or sliding calipers, remove the calipers from over the brake rotors and hang them from the suspension to prevent damage to the brake hose. Insert thick wooden blocks into the calipers between the pistons and caliper bodies to prevent the pistons from coming out of their bores, figure 7-20.
3. Place drain pans under the calipers to catch brake fluid in the event that a piston does come out of its bore.
4. Have an assistant *slowly* apply the brake pedal to force the pistons from their bores; several pedal strokes may be needed to move all of the pistons out against the rotors or wooden blocks.
5. Determine which piston was the last to be fully extended; this is the most frozen piston. If the piston is in a fixed caliper, remove worn out pad or wooden shim from between the rotor and the most frozen piston; if no pad or shim was used, remove the caliper from over the rotor, and secure the less frozen pistons in their bores with C-clamps or other suitable tools. If the most frozen piston is in a sliding or floating caliper, remove the wooden block from that caliper.
6. Have your assistant *slowly* apply the brake pedal until the most frozen piston comes entirely out of its bore.
7. Remove both of from the vehicle and finish removing the other pistons with a mechanical tool or compressed air.

Caliper hydraulic service bench
Pistons can also be removed hydraulically with a special caliper service bench. The principle is the same as described above, but rather than using the vehicle brake hydraulic system to do the job, you remove the calipers from the car and attached them to the hydraulic hose of the service bench, figure 7-21. You then pump the hand lever, which is connected to a small master cylinder that provides pressurized fluid to force the piston out of the caliper.

The advantages to this method are that an assistant is not needed, and the condition of the vehicle brake hydraulic system is not a factor in obtaining sufficient pressure to free a frozen piston. The disadvantages are the cost of the specialized caliper service bench, and the fact that only one caliper can be serviced at a time.

Dust Boot and Piston Seal Removal

Once the piston is out of the caliper, remove the dust boot. If the boot is a press fit in the caliper and remains attached to it, use a screwdriver to pry the boot out, figure 7-22. Take care not to scratch the caliper bore as you remove the boot. If the boot is a stretch fit over the piston and remains attached to that part, simply pull it free and set it aside.

Next, remove the piston seals. If the caliper has stroking seals, insert a probe under the seals and lever them off of the pistons. If the caliper has fixed seals, insert a probe under the

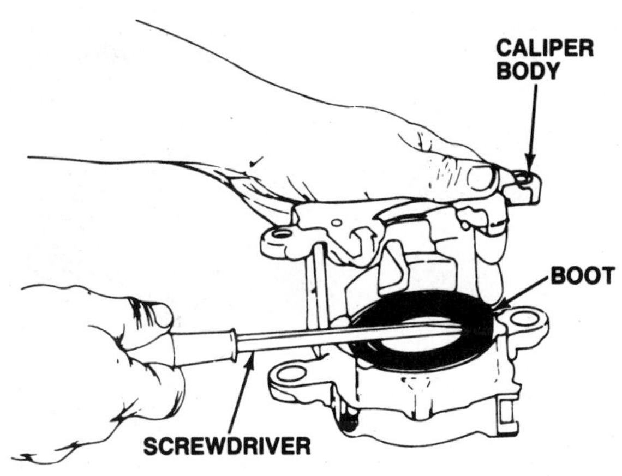

Figure 7-22. Removing a dust boot that is a press fit in the caliper body.

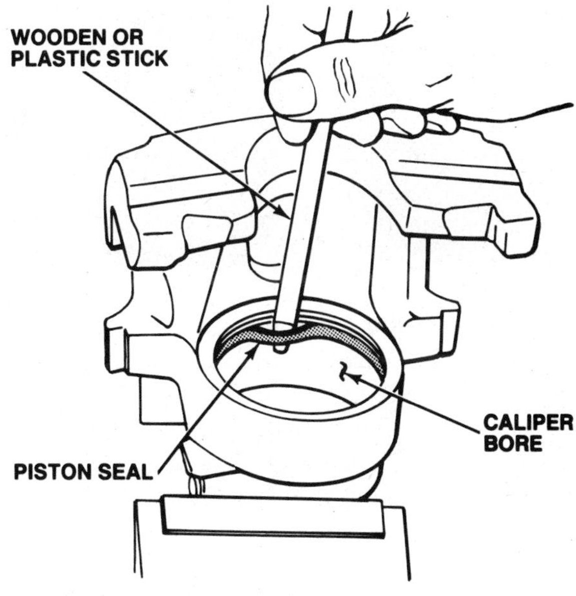

Figure 7-23. The final step in caliper disassembly is to remove the piston seal.

seals and pry them out of the grooves in the caliper bores, figure 7-23. To avoid scratching or otherwise damaging the pistons, caliper bore, or seal groove, always use a wooden or plastic probe to remove seals. Damage in these areas can cause fluid leaks when the caliper is reassembled. If a fixed seal is any shape other than square cut, note how it fits into the groove so that you can install the new seal in the same manner.

Once the caliper is disassembled, clean the body and pistons with brake cleaning solvent. If the parts are particularly dirty, use a soft bristle brush to help remove any contamination. Do not use a wire brush or you may damage the piston sealing surface, the seal groove, or the caliper bore. Rubber parts such as boots and seals do not need to be cleaned because they will be replaced with new parts from the rebuild kit. When the caliper and pistons are clean, allow them to dry and make sure no liquid solvent remains in the fluid passages of the caliper body.

BRAKE CALIPER INTERNAL INSPECTION

Once the caliper body and piston are cleaned, inspect the caliper bore for scoring, rust, corrosion, and wear. This is particularly important on calipers that have stroking seals because the bore provides the sealing surface; even a small imperfection in the bore can create a fluid leak. Minor rust and corrosion can be removed from the bore with crocus cloth or by honing as described later in the chapter. However, if the bore of a caliper with stroking seals is significantly damaged, the caliper must be replaced.

The bore condition of calipers with fixed seals is less critical because the piston provides the sealing surface. Inspect the bores of these calipers for major defects that might prevent the piston from moving freely, and make sure the edges of the seal groove are not rusted or corroded, and do not have any nicks or burrs that can cut the seal or affect its sealing ability. As long as the seal groove is in good condition, and the bore diameter remains within specifications, calipers with fixed seals can be honed clean of even major rust and corrosion. However, if the seal groove is damaged in any way, or honing makes the bore oversize, the caliper must be replaced.

Caliper Piston Inspection

Caliper pistons are made of cast-iron, steel, aluminum, or phenolic resin. As mentioned above, pay particular attention when you inspect pistons used in calipers with fixed seals because the outside diameter of the piston provides the sealing surface for the caliper seal. On pistons used with stroking seals, the important area is the groove in which the seal fits. In both cases, the piston will have to be replaced if the critical surface is not smooth, clean and entirely free of defects.

Inspect cast-iron, steel, and aluminum pistons for rust, corrosion, nicks, and scoring in their sealing areas, figure 7-24. If a cast-iron or steel piston is chrome plated to provide a better

Figure 7-24. Typical caliper piston damage.

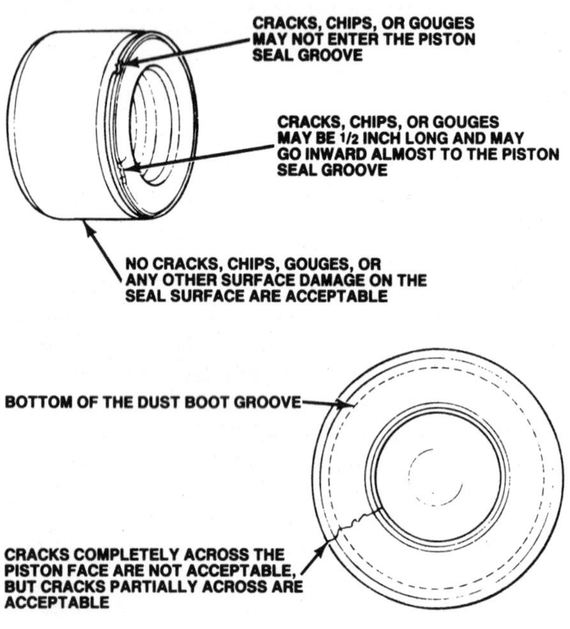

Figure 7-25. Phenolic caliper pistons commonly suffer minor cracking.

sealing surface, make sure the plating is not flaking away anywhere. All aluminum pistons are anodized to provide a durable finish, and this surface should not be damaged in any way. If cast-iron, steel, or aluminum pistons have any of these problems, you should replace them.

Inspect pistons made of phenolic resin for cracks or chips, figures 7-24 and 7-25. Minor cracks or chips that extend partially across the piston face are acceptable, as are minor nicks and gouges at the outer edge of the piston, providing they do not extend into the dust boot groove. Replace any piston if cracks extend across the face of the piston, or nicks and gouges enter the dust boot groove. Finally, inspect the outer diameter of the piston, it should be smooth and even. If the surface is scuffed or scored in any way, replace the piston.

BRAKE CALIPER HONING

In the early years of disc brakes, calipers had stroking seals that would seal against the caliper bore even with a relatively large amount of clearance between the piston and bore. As a result, honing the caliper bore to restore the sealing surface was a routine part of a caliper overhaul. The fixed seals used in virtually all brake calipers today require smaller piston-to-bore clearances in order to seal properly. For this reason, modern calipers are honed rarely, if at all, and only to remove rust, corrosion, or minor scoring that might interfere with the free movement of the piston.

On calipers with stroking seals, honing remains an acceptable rebuilding procedure that can provide a less expensive alternative to caliper replacement. However, just as with master cylinders and wheel cylinders that have been honed, the resulting sealing surface is not nearly as smooth as the finish on the original part. As a result, a honed caliper will not provide the same service life as a new part.

Whether it is used to restore the sealing surface of a caliper with stroking seals, or simply to clean up the bore of a caliper with fixed seals, the honing procedure is the same. Remember to always compress the arms of the hone manually when you move the stones into or out of the caliper bore. Never drag the stones straight out of the bore or you will cause scratches that may result in a leak. To hone a brake caliper:

1. Mount the caliper in a vise so you have clear access to the bore. Do not clamp on the cylinder body or the bore may be distorted.

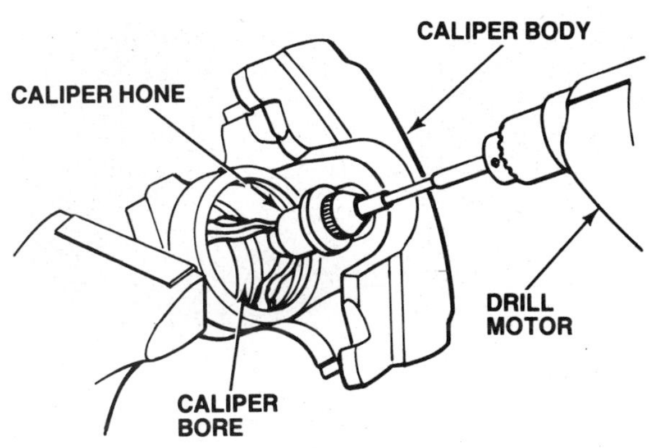

Figure 7-26. Honing a brake caliper bore.

2. Select the proper size caliper hone and chuck it in a drill motor. Use a hone with fine stones.
3. Lubricate the bore and hone with clean brake fluid, then place the hone into the bore. Start the drill motor, and rotate the hone at approximately 500 rpm, figure 7-26. If the caliper uses stroking seals, stroke the hone gently in and out of the bore to obtain a crosshatch pattern for good sealing. If the caliper uses fixed seals, you can simply hold the hone steady in the bore because the honing pattern does not matter.
4. After about 10 seconds, remove the hone, wipe the bore clean, and check the finish. It should be clean, smooth, and free of damage. Repeat the honing sequence as needed. If the bore does not clean up after several attempts, replace the caliper.
5. After you are finished honing, thoroughly clean the caliper bore, seal groove, and fluid inlet passage with brake cleaner. Allow the caliper to dry, and make sure all traces of grit and residue are cleaned away.
6. Measure the caliper bore as described in the next section.

Caliper bore measurement

If the dust boots remain intact and the brake fluid in the system is changed regularly, there is usually little or no caliper piston and bore wear during normal operation. However, whenever you hone a caliper, you must measure the bore to make sure it has not been honed too far oversize. This is particularly true when a great deal of honing is required to clean up the bore, or when honing calipers with fixed seals.

If the clearance is too great in a caliper with stroking seals, the lip of the seal will seat against the caliper bore with less pressure. This reduces the ability of the seal to contain fluid within the caliper, and shortens the service life of the caliper. Most manufacturers specify between .004" and .010" (.10 and .25 mm) piston-to-bore clearance for calipers with stroking seals. Consult the factory shop manual for the car you are servicing to get an exact figure.

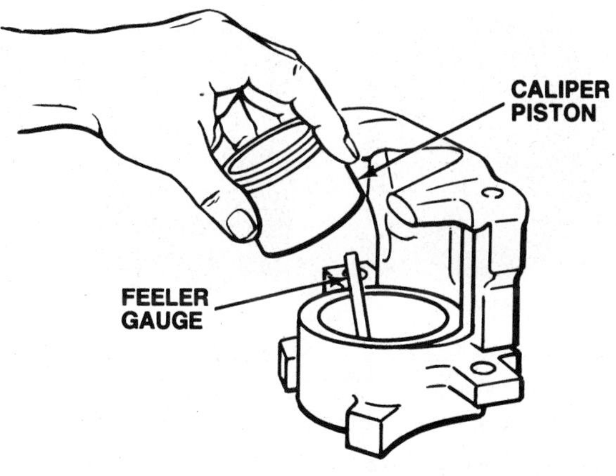

Figure 7-27. Checking the caliper piston-to-bore clearance.

If the clearance is too great in a caliper with fixed seals, seal damage can result, and the self-adjusting action of the caliper may be affected. As a general rule, the maximum piston to bore clearance for fixed seal calipers is .002" to .005" (.06 to .13 mm) when using metal pistons, and .005" to .010" (.13 to .25 mm) when using phenolic pistons. Once again, consult the factory shop manual for the car you are servicing to get an exact figure.

To measure the caliper piston-to-bore clearance, insert a feeler gauge strip whose thickness is equal to the maximum allowed clearance into the caliper bore, figure 7-27. Attempt to insert the piston into the bore alongside the feeler gauge strip. If the piston will enter the bore, repeat the check with a new piston. If a new piston will also enter the bore, replace the caliper.

Sleeved Brake Calipers

Although honing is an acceptable procedure, it is not always the best form of repair. Many calipers with stroking seals are found on older collectible or special-interest cars that get little use. And many of these same calipers have relatively poor dust and water sealing that makes them prone to rust, corrosion, and fluid leaks. To provide a more permanent repair for these cars (the Chevrolet Corvette in particular), several aftermarket firms sell sleeved brake calipers.

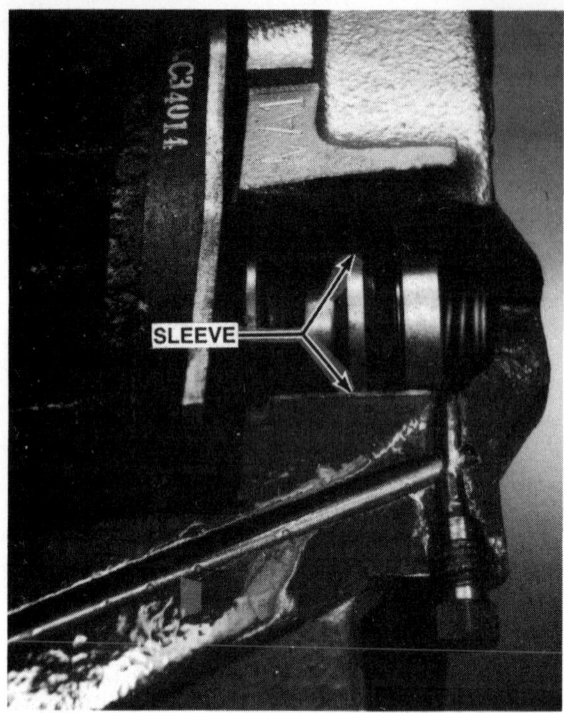

Figure 7-28. Stainless steel sleeves extend the service life of older calipers with stroking seals.

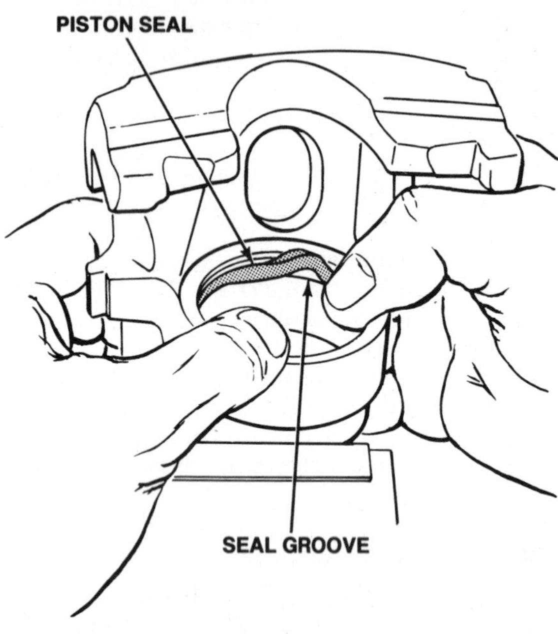

Figure 7-29. Installing a fixed caliper piston seal.

To improve both the sealing and the service life of these calipers, the bores are machined oversize, and stainless steel sleeves are installed, figure 7-28. The sleeve is a press fit into the bore, and provides a very smooth sealing surface. At the same time, stainless steel is far more durable and rust resistant than the original cast-iron caliper bore, which ensures a long service life. If the car you are working on fits into one of the catagories described above, you may want to recommend sleeved calipers to the customer rather than a rebuild of the old parts.

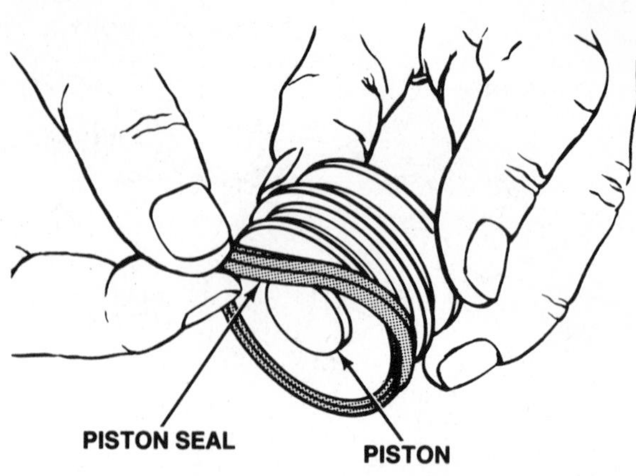

Figure 7-30. Installing a stroking caliper piston seal.

BRAKE CALIPER ASSEMBLY

Once you have disassembled, inspected, cleaned, honed, and measured the caliper and caliper piston, obtain any replacement parts you may need, along with the proper overhaul and small parts kits. Before you begin to reassemble a caliper, make sure your hands are completely clean. Dirty hands will contaminate the caliper with grit or oil, and cause early failure. To help ease assembly, have a container of fresh brake fluid or assembly lube handy.

Piston Seal Installation

The first step in assembling a caliper is to install the piston seals. On calipers with fixed seals, thoroughly lubricate the O-ring seal and the seal groove in the caliper bore. Using your fingers, position one edge of seal in the groove, then work the seal around the bore diameter gently until it is fully seated, figure 7-29. Do not twist or roll the seal. Square-cut O-ring seals can be installed facing either way; however, if the seal is any shape other that square-cut, install it in the same manner as the original part.

On calipers with stroking seals, thoroughly lubricate the lip seal and the seal groove in the caliper piston. Using your fingers only, stretch the seal over the end of the piston and into place in the seal groove, figure 7-30. Take special care not to cut or damage the seal, particularly the edge of the sealing lip, in any way.

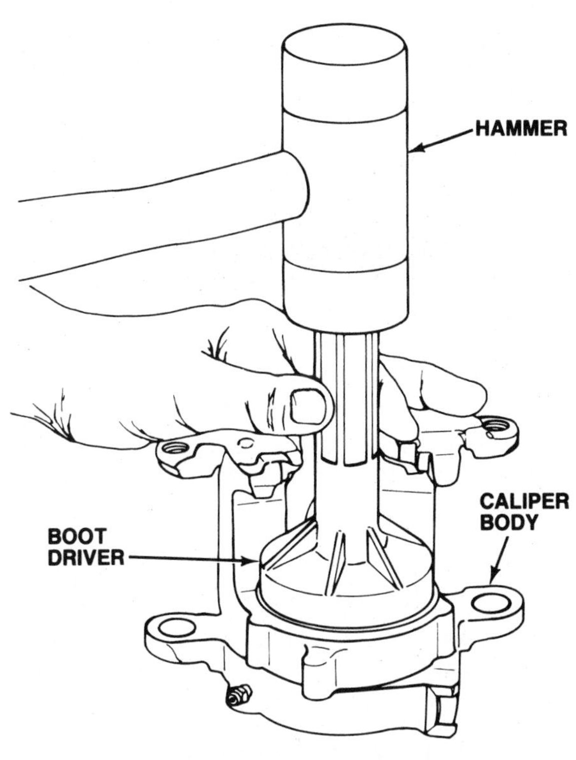

Figure 7-31. Use a hammer and special driver to seat dust boots that are a press fit in the caliper body.

The sealing lip of a stroking seal must face the bottom of the caliper bore when the piston is installed.

Piston and Dust Boot Installation

Once the seals are installed, the next, and essentially final, step in caliper assembly is to install the piston and dust boot. If piston removal is the most potentially difficult part of caliper disassembly, then piston and dust boot installation can be considered the most difficult part of reassembly. The procedure used to do this varies with the design of the dust boot, and all three common methods are described below.

Press-fit dust boot installation

The brake calipers on General Motors cars, late-model Chrysler products (primarily FWD), and some imports use a dust boot with a metal reinforcing ring around the outer edge that is press fit into the caliper body. These boots require a special driver to install properly. To install the pistons and dust boots on these calipers:

1. Lubricate the piston and slip it into the dust boot until the boot snaps into the groove in the piston.

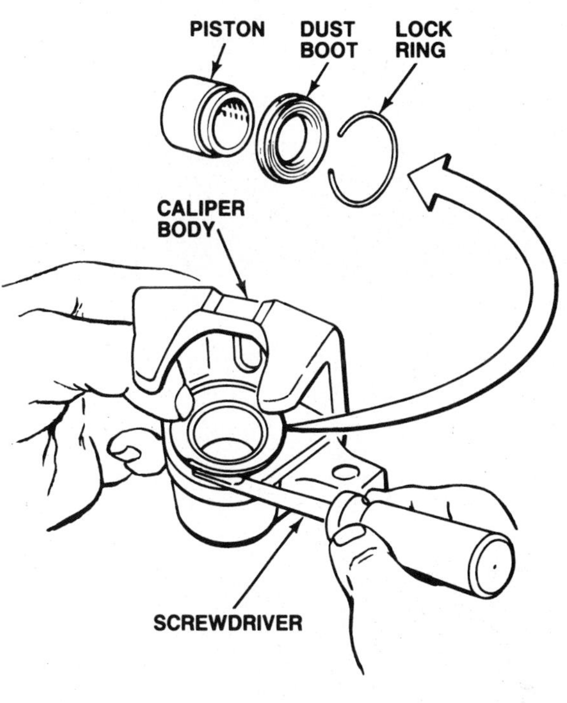

Figure 7-32. Some calipers use a separate retaining ring to lock the dust boot into the caliper body.

2. Using only hand pressure, work the piston into the caliper bore past the O-ring seal. Bottom the piston in the bore.
3. Use a hammer and a dust boot driver of the proper size to seat the metal reinforcing ring at the outer edge of the boot into the groove in the caliper body, figure 7-31.

Retaining ring dust boot installation

The brake calipers on Volkswagens and many Japanese imports use a dust boot with a separate metal ring that secures the outer edge of the boot into the caliper body. No special tools are required to assemble this type of caliper. To install the pistons and dust boots these calipers:

1. Lubricate the piston and slip it into the dust boot until the boot snaps into the groove in the piston.
2. Using only hand pressure, work the piston into the caliper bore past the O-ring seal. Bottom the piston in the bore.
3. Using your fingers, position the outer edge of the boot into the groove in the caliper body; the boot *must* seat properly.
4. Install the metal ring to lock the boot into the groove, figure 7-32.

Figure 7-33. Special rings make it easier to install dust boots that fit into grooves in the caliper bore.

Lip/groove dust boot installation

The brake calipers on Ford Motor Company cars, older Chrysler products (as well as a few newer ones), and domestic American Motors vehicles use a dust boot with a lip around the outer edge that fits into a groove in the caliper bore. Although special tools are not essential to assemble these calipers, a set of metal or plastic rings about one-half inch high will make the job easier. These rings are available from several aftermarket sources. To install the pistons and dust boots on these calipers:

1. Select the ring that just barely fits over the caliper piston.
2. Lubricate the inner edge of the dust boot, and install it over the ring chosen in Step 1.
3. Install the lip at the outer edge of the dust boot into the groove in the caliper bore. Reach through the ring with your fingers and make sure the lip is fully seated in the groove.

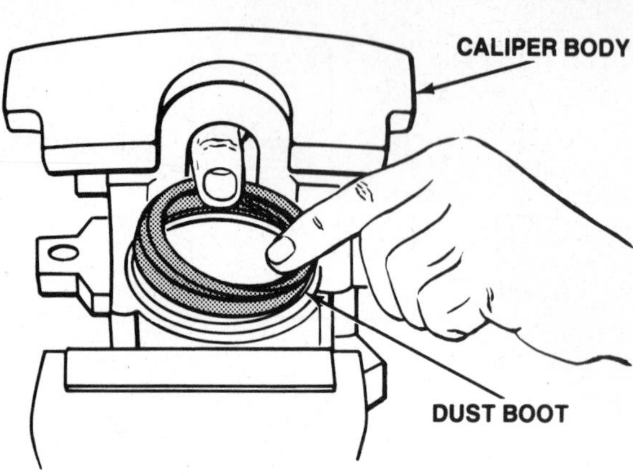

Figure 7-34. Manually installing a lip/groove dust boot.

4. Lubricate the piston and slip it through the ring into the caliper bore, figure 7-33. Using only hand pressure, work the piston past the O-ring seal until it is bottomed in the bore.
5. Carefully remove the ring from the dust boot so that the boot snaps into place on the piston.

There are two common ways to install the pistons and dust boots into these calipers without using the special rings; however, both require a certain "touch," and may require several attempts before the caliper can be assembled properly. The cost of the rings is so low as to make these methods impractical except where the rings are unavailable.

The first method is to install the piston and dust boot entirely by hand:

1. Lubricate the caliper bore, and position the dust boot over the opening.
2. Use your fingers to "pucker" the boot, and carefully work the lip at its outer edge into the groove in the caliper bore, figure 7-34.
3. Once the lip is fully seated, carefully stretch the dust boot and slide the piston through it into the caliper bore and past the O-ring seal. Take care not to dislodge the boot from its groove in the process.

Figure 7-35. Using compressed air to install the piston through a lip/groove dust boot.

The second method of installing the pistons and boots without the special rings uses compressed air. Because water or oil in the compressor storage tank can contaminate the caliper, you should only use this procedure if you know that your compressed air supply is clean and dry.

1. Manually install the lip at the outer edge of the dust boot into the groove in the caliper bore.
2. Lubricate the piston and the inner edge of the dust boot.
3. Center the piston over the caliper bore and lightly rest it against the inner edge of the dust boot to "seal" the bore.
4. Apply a small amount of compressed air at the caliper fluid inlet fitting. As the dust boot begins to inflate, increase the amount of air until you can work the piston through the boot, figure 7-35.
5. Once the boot is over the end of the piston, release the air pressure, and slide the piston into the caliper bore past the O-ring seal. Take care not to dislodge the boot from its groove in the process.

BRAKE PAD REPLACEMENT AND CALIPER OVERHAUL PROCEDURES

The following pages contain pad replacement and overhaul procedures for five different brake calipers:

- General Motors single-piston floating
- Chrysler single-piston sliding
- General Motors four-piston fixed
- Ford single-piston rear
- General Motors single-piston rear.

While there are many other calipers that are not covered, these are popular designs that include most of the common variations in caliper construction and overhaul procedure that you are likely to encounter. With the knowledge gained from reading this chapter, and the experience of going through the overhaul procedure for each of these calipers, you should be able to overhaul the various other calipers you will encounter.

Because every caliper overhaul begins and ends with the removal and installation of the brake pads, the photo sequences are designed so you can use them to either change the brake pads, or completely overhaul the caliper. If only the pads are being changed, simply follow the instructions in the captions and skip the steps indicated.

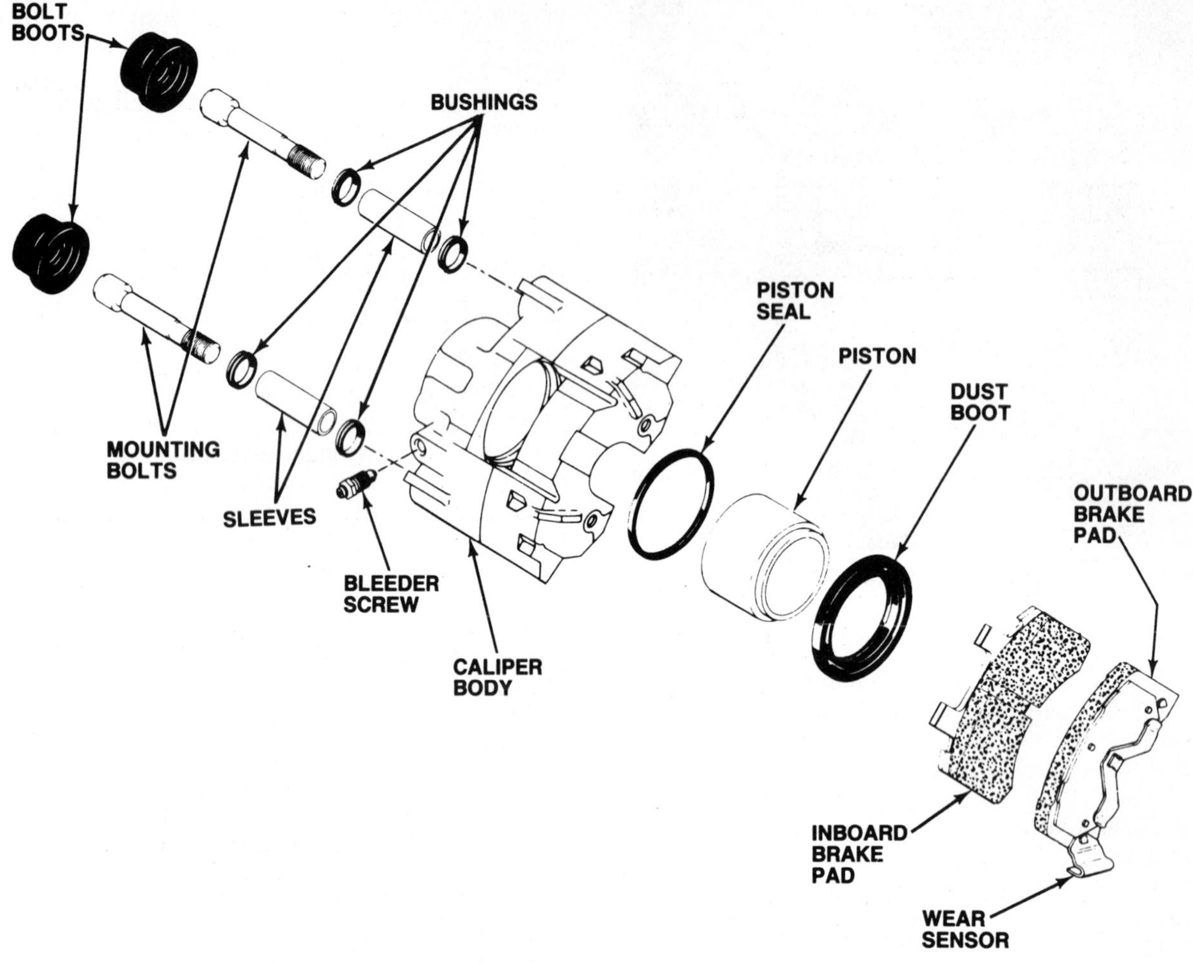

Figure 7-36. The General Motors Series 3200/3300 single-piston, floating caliper.

GM SINGLE-PISTON FLOATING CALIPER SERVICE

The General Motors Series 3200/3300 single-piston, floating caliper, figure 7-36, has been used on a variety of General Motors cars in basically the same form since 1980. This caliper is similar to other front calipers used throughout the General Motors line, and the overhaul procedures for all of them are essentially the same. There are three things to watch for when overhauling these calipers: proper lubrication of the mounting hardware, the tightness of the outboard brake pad in the caliper, and the position of the brake lining wear sensor.

All of these calipers float on rubber bushings that slide along steel sleeves held in place by the caliper mounting bolts. In some cases, the mounting bolt also provides part of the sliding surface. While the exact configuration of the bushings, sleeves, and bolts varies, these parts must be in perfect condition and properly lubricated with high-temperature brake grease, figure 7-37, to ensure smooth caliper operation. To prevent comebacks, replace these parts every time the brake pads are changed or the caliper is rebuilt.

To prevent brake noise, the inboard brake pads on most General Motors calipers are retained by a spring clip that snaps into the caliper piston. Spring clips are used on some outboard pads as well. However, most outboard brake pads are retained by tabs on the backing plate that are bent in some manner to secure the pad tightly in the caliper body. Depending on the caliper design, there are two ways in which this is done. Where the tabs extend over the top of the caliper body, bend them with a hammer off the car, figure 7-38, until the pad snaps into place in the caliper with a slight preload.

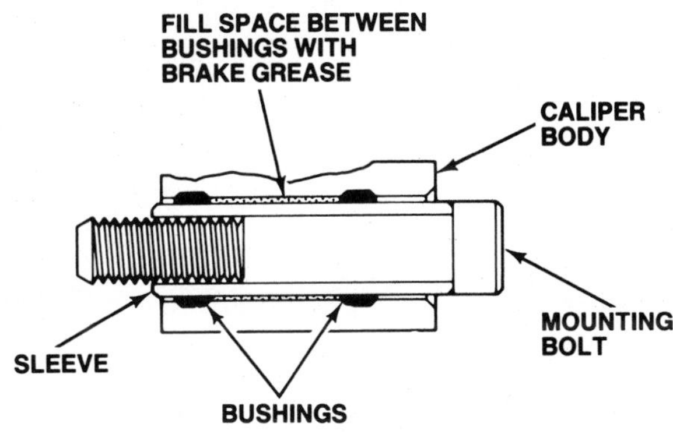

Figure 7-37. Proper lubrication is the key to smooth operation of floating calipers.

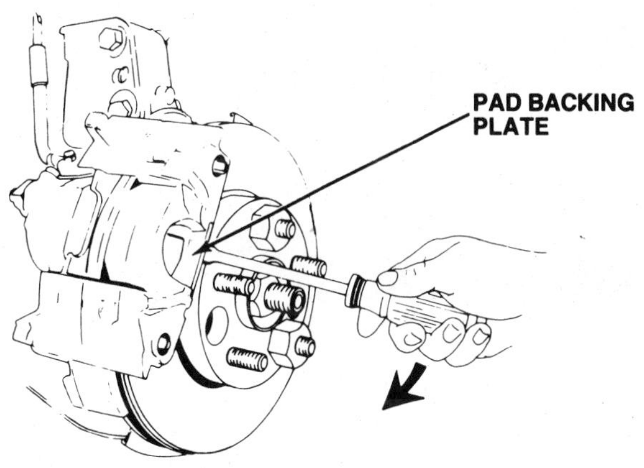

Figure 7-39. Levering a pad into position to clinch the retaining tabs.

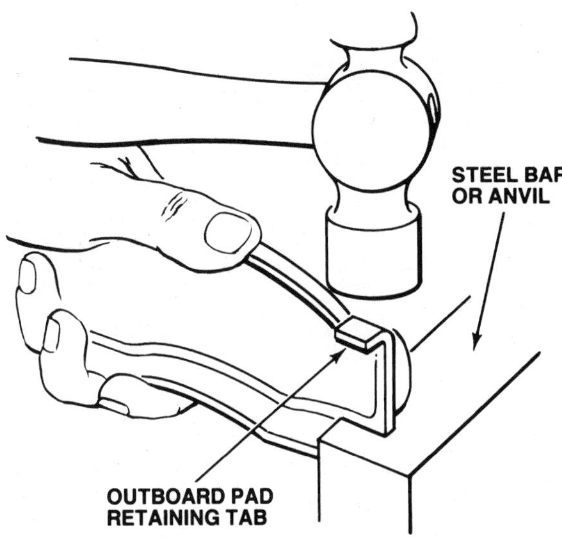

Figure 7-38. Some brake pad retaining tabs can be bent off the car.

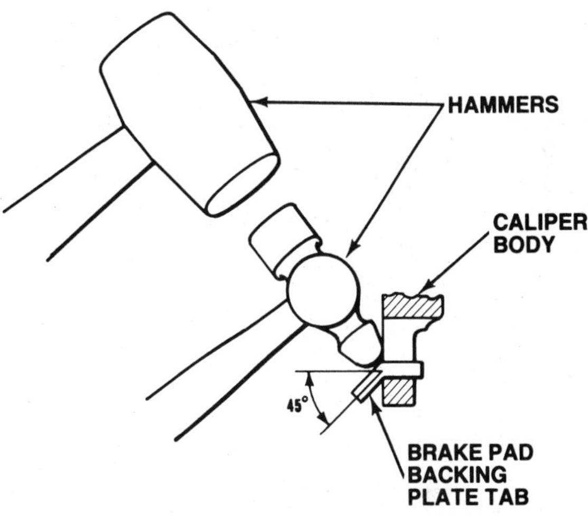

Figure 7-40. Use two hammers to clinch the tab where extends through the hole in the caliper body.

Where the tabs extend through holes in the caliper body, a more complex procedure is called for:

1. Once the new pads are installed, insert a large screwdriver or pry bar between the flange on the bottom of the pad and the rotor hat, figure 7-39.
2. Lever the pad upward so that the flange is firmly seated against the caliper body.
3. Have an assistant apply and hold the brake pedal with firm pressure to lock the pad in position.
4. Use a ball-peen hammer and a larger brass hammer to clinch the tabs against the caliper body as shown in figure 7-40.

The outboard pads on General Motors calipers are usually interchangeable side to side except for the position of the wear sensor. When the caliper is installed on the car, the wear sensor *must* be at the leading edge of the caliper during forward rotor rotation. That is, the rotor should sweep across the sensor before it reaches the brake lining. Make sure you use the proper pad on each side of the car.

GENERAL MOTORS SINGLE-PISTON FLOATING CALIPER SERVICE

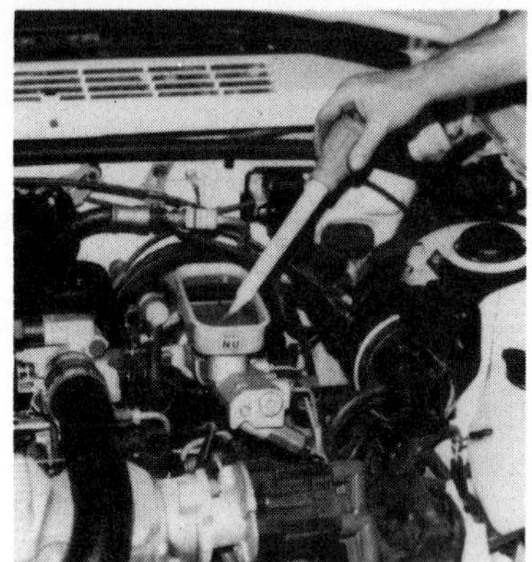

1. Siphon off fluid from the master cylinder reservoir(s) as dictated by the split of the brake hydraulic system.

2. If the caliper is being rebuilt, make sure the bleeder screw is free, then disconnect the brake hose from the caliper fluid inlet.

3. Remove the protective boots, then unscrew and remove the caliper mounting bolts.

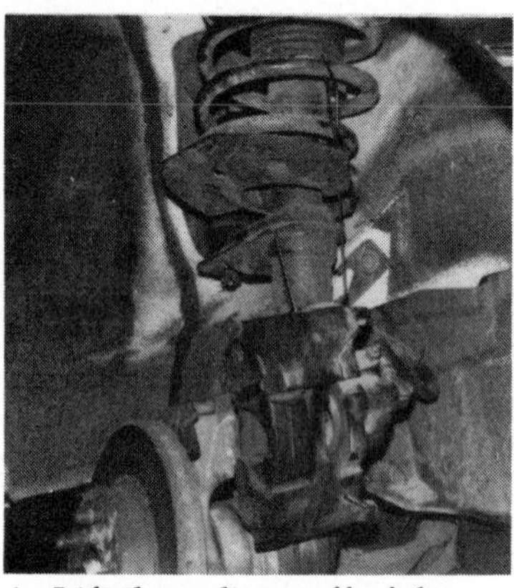

4. Lift the caliper off of the rotor, and hang it from the suspension on a wire if it is not being rebuilt.

5. Lift the inboard brake pad and retaining spring out of the caliper.

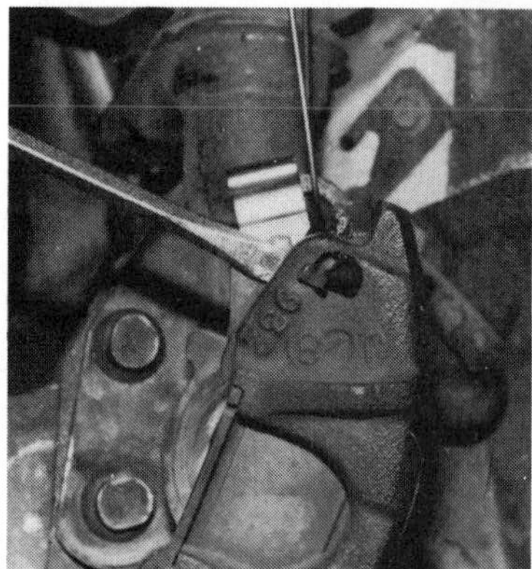

6. With a tab-mounted outboard brake pad, pry the pad out of the caliper body.

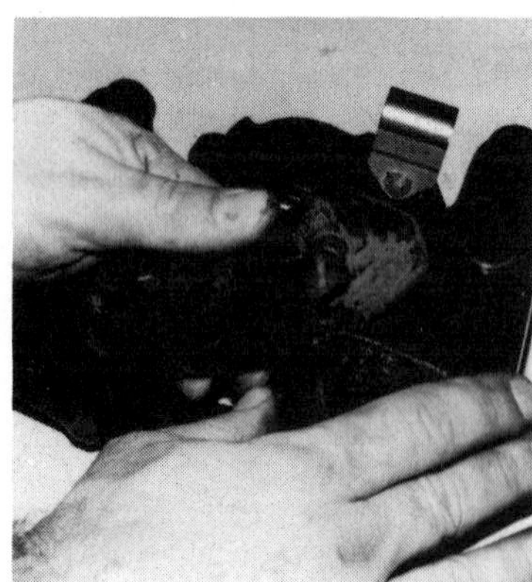

7. With a spring-mounted outboard brake pad, pry the spring clip out of the holes in the caliper body, then slide the pad out.

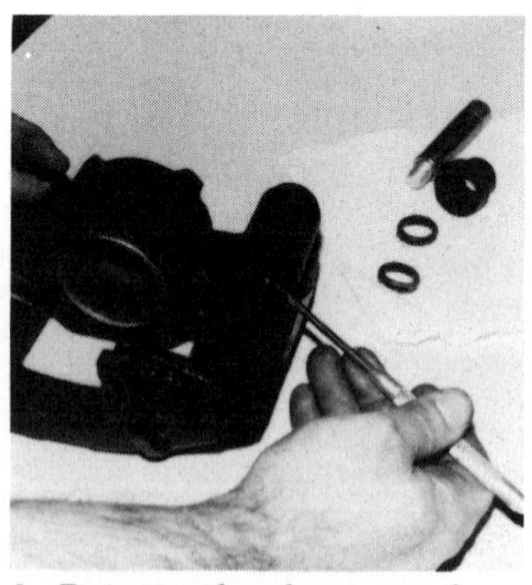

8. Remove the sleeves and bushings from the caliper mounting bolt holes.

9. If the caliper is not being rebuilt, bottom the piston in the caliper bore, then go to Step 19.

GENERAL MOTORS SINGLE-PISTON FLOATING CALIPER SERVICE

10. Remove the caliper piston using one of the methods described earlier, then remove the bleeder screw.

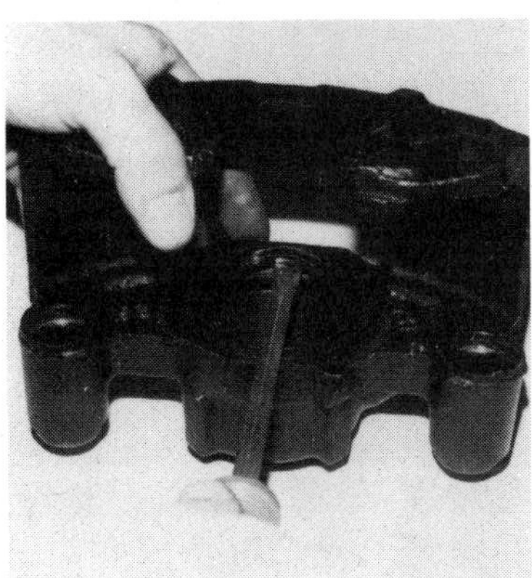

11. Pry the dust boot out of the caliper body with a screwdriver.

12. Remove the piston seal from its groove in the caliper bore.

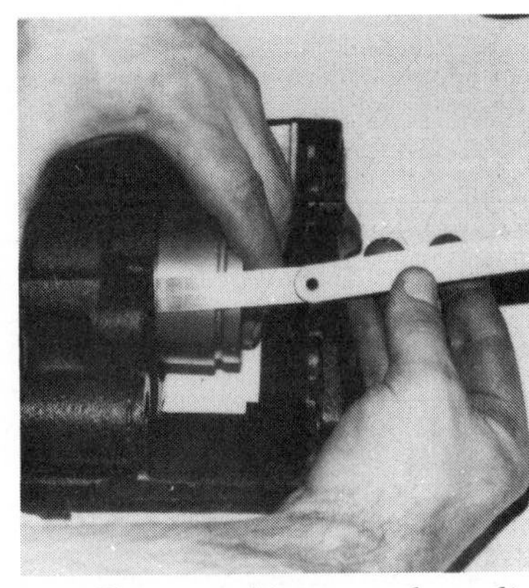

13. Clean and inspect the caliper body and piston. Hone the caliper bore if necessary, and measure the piston-to-bore clearance.

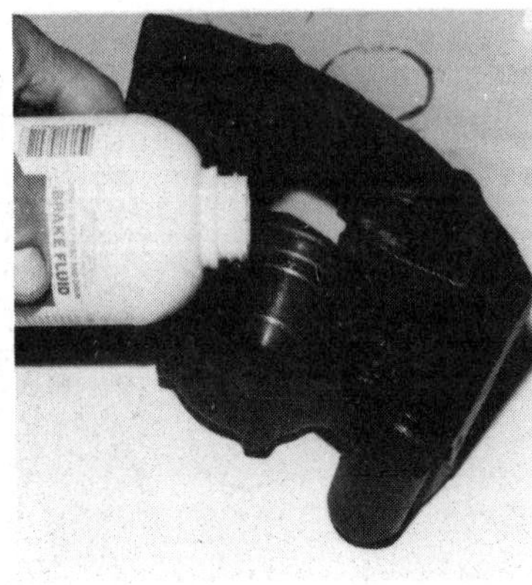

14. Lubricate the caliper bore and the new piston seal. Install the seal into its groove in the caliper bore.

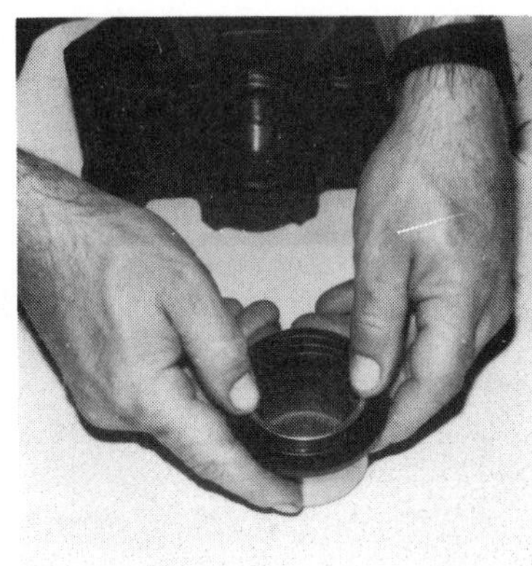

15. Install the new dust boot into its groove on the piston.

16. Lubricate the piston and install it into the caliper bore until it is bottomed.

17. Seat the reinforcing ring at the outer edge of the dust boot into the caliper body with a hammer and boot driver.

18. Install the bleeder screw in the caliper.

GENERAL MOTORS SINGLE-PISTON FLOATING CALIPER SERVICE

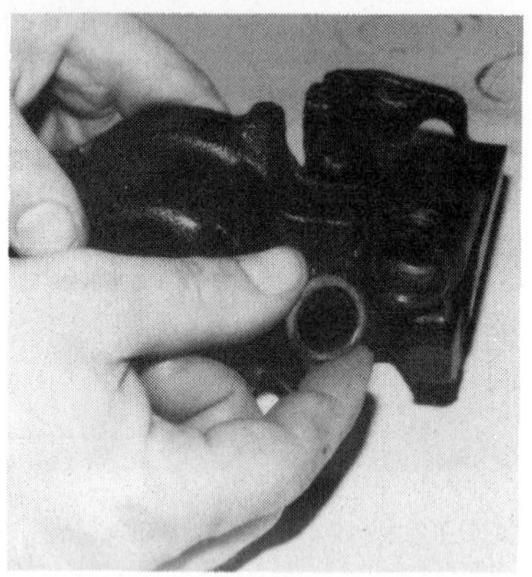

19. Lubricate the new bushings with brake grease and install them into the mounting bolt holes.

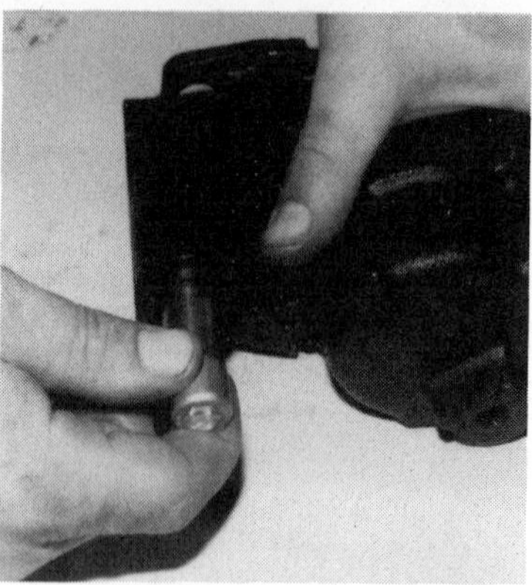

20. Lubricate the outside of the new sleeves with brake grease and install them into the mounting bolt holes.

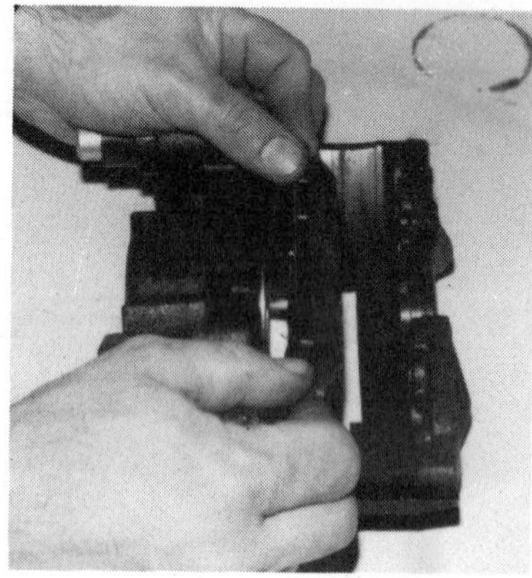

21. Attach the retainer spring to the new inboard brake pad, and install the pad so it seats flat against the piston.

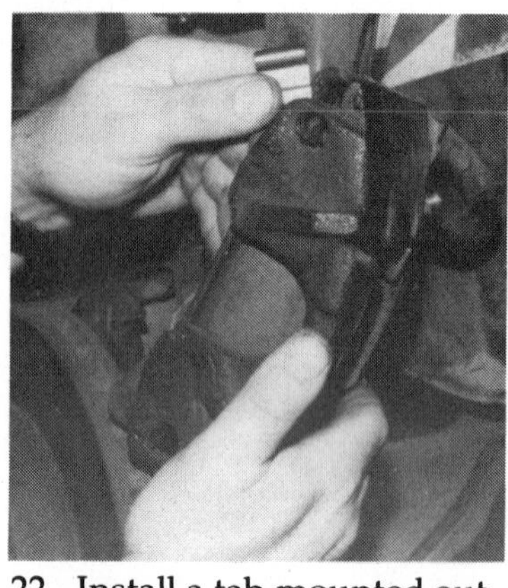

22. Install a tab-mounted outboard brake pad with a slight amount of preload, or so the retaining tabs can be clinched once the caliper is installed.

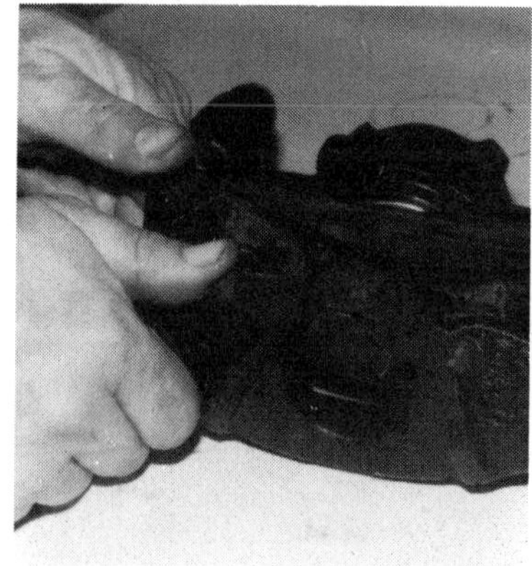

23. Install a spring-mounted outboard pad so the buttons on the spring clip engage the holes in the caliper body.

24. Position the caliper over the rotor, and measure the clearance between the caliper and the anchor plate stops.

25. If the clearance is not .005″ to .012″ (.13 to .30 mm), remove the caliper and file the ends of the anchor plate stops to obtain the correct value.

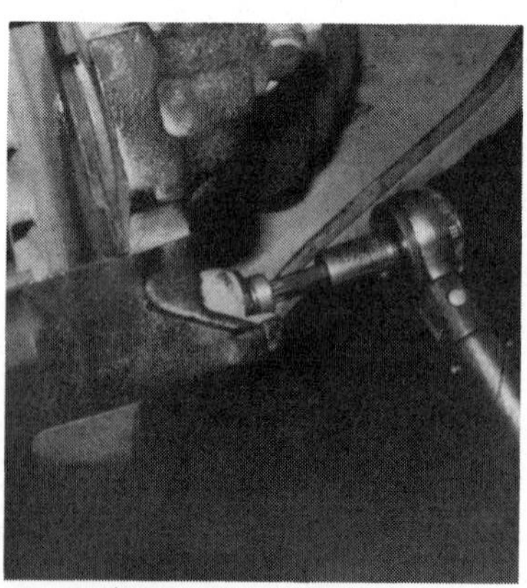

26. Install the caliper mounting bolts and torque them to 38 ft-lb (51 Nm). Install the protective rubber boots over the bolt heads.

27. If the caliper was rebuilt, connect the brake hose to the fluid inlet of the caliper. Torque the bolt to 33 ft-lb (45 Nm). Bleed the brakes as described in Chapter 3.

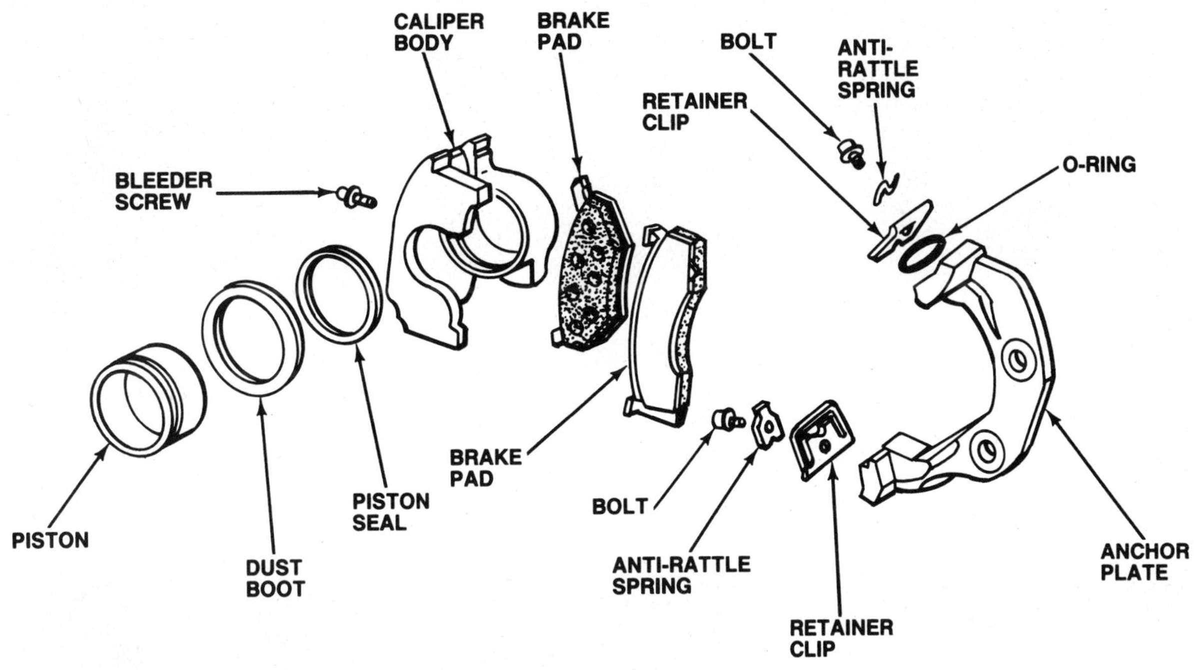

Figure 7-41. The Chrysler single-piston sliding brake caliper.

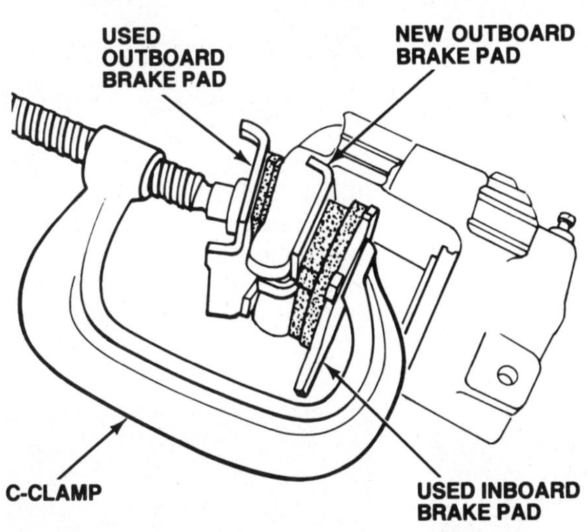

Figure 7-42. Installing a Chrysler outboard brake pad.

CHRYSLER SINGLE-PISTON SLIDING CALIPER SERVICE

The Chrysler single-piston, sliding caliper, figure 7-41, has been used as a front brake caliper on most Chrysler Corporation RWD cars since 1973, and was also used as a rear brake caliper on the 1974 Imperial.

This is a simple brake caliper that requires no special procedures to rebuild. As on many General Motors calipers, the outboard brake pad is retained in the caliper by bent tabs on its backing plate. Install the pad with a slight preload to prevent vibration and brake noise. Chrysler suggests that a C-clamp may be useful when doing this, figure 7-42. If so, use the old brake pads as buffers to avoid damaging the new pad.

CHRYSLER SINGLE-PISTON SLIDING CALIPER SERVICE

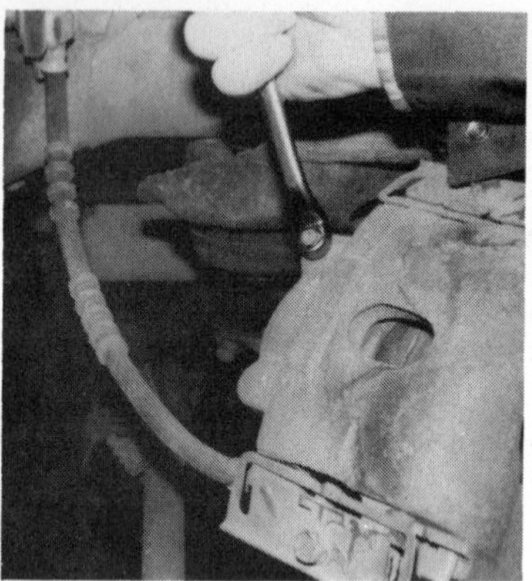

1. If the caliper is being rebuilt, make sure the bleeder screw is free, then disconnect the brake hose from the caliper fluid inlet.

2. Unbolt the caliper retaining clips, and remove the anti-rattle springs.

3. Lift the caliper off of the rotor, and hang it from the suspension on a wire if it is not being rebuilt.

4. Remove the O-ring from the upper way on the anchor plate.

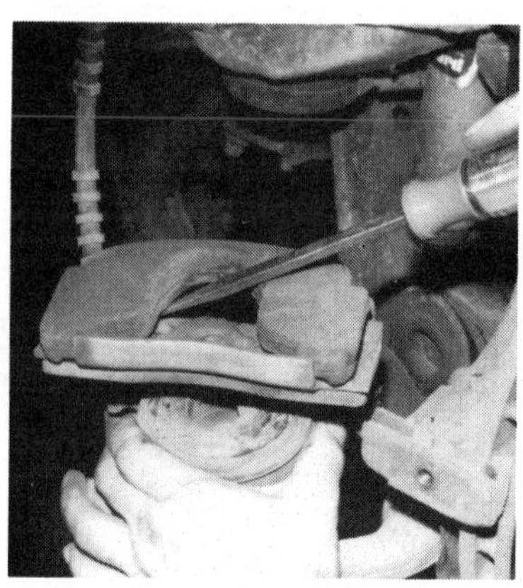

5. Pry the outboard brake pad from the caliper with a screwdriver.

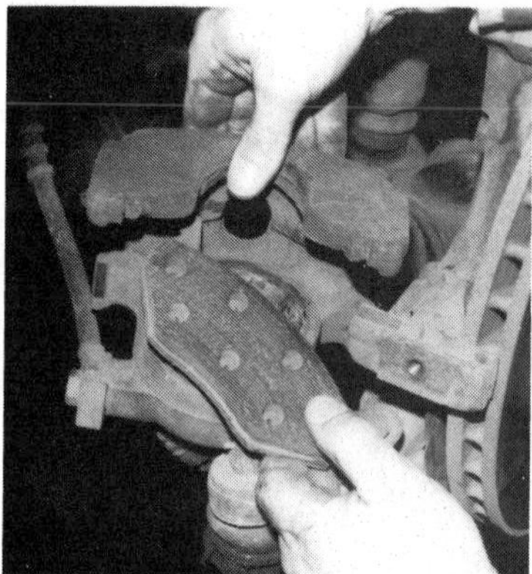

6. Lift the inboard brake pad out of the anchor plate.

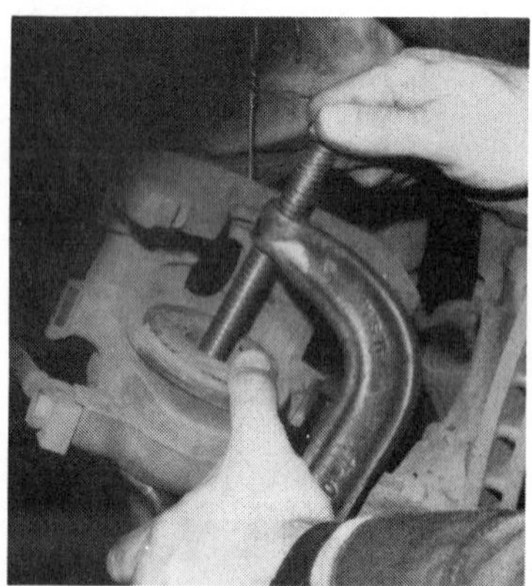

7. If the caliper is not being rebuilt, bottom the piston in the caliper bore then go to Step 17.

8. Remove the caliper piston using one of the methods described earlier, then remove the bleeder screw.

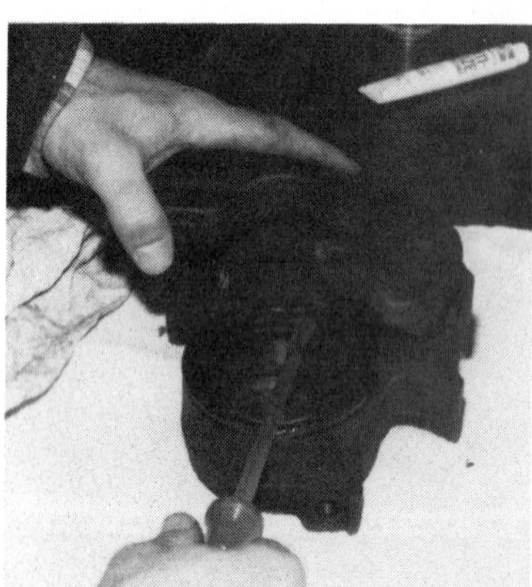

9. Pry the dust boot out of the caliper body with a screwdriver.

CHRYSLER SINGLE-PISTON SLIDING CALIPER SERVICE

10. Remove the piston seal from its groove in the caliper bore.

11. Clean and inspect the caliper body and piston. Hone the caliper bore if necessary, and measure the piston-to-bore clearance.

12. Lubricate the caliper bore and the new piston seal.

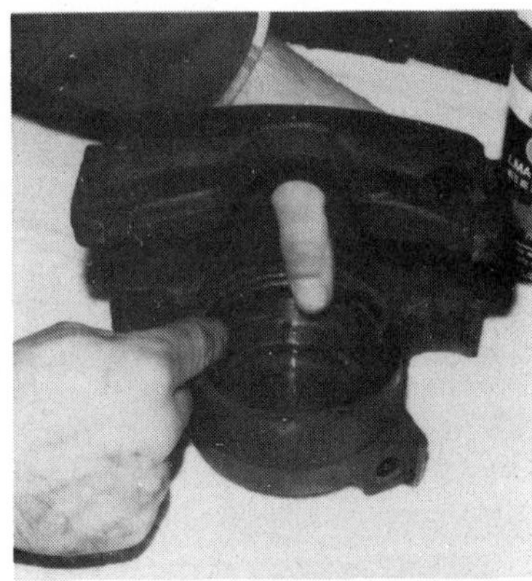

13. Install the piston seal into its groove in the caliper bore.

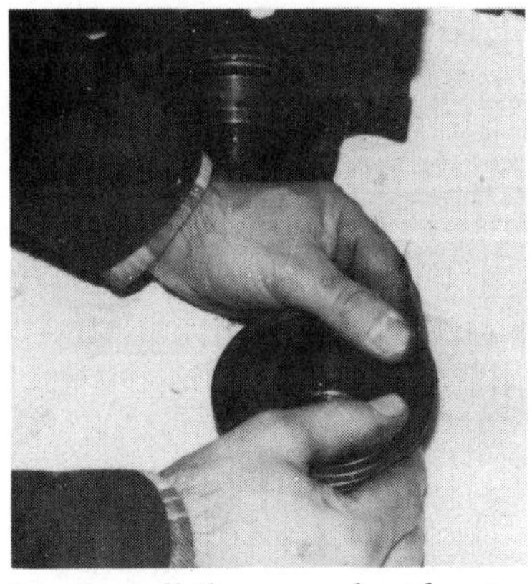

14. Install the new dust boot into its groove on the piston.

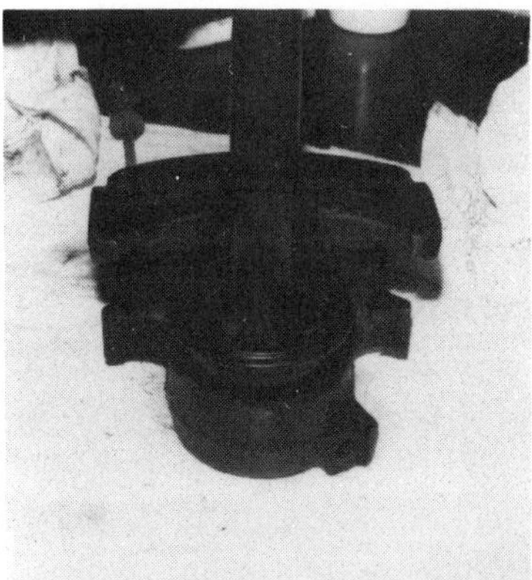

15. Lubricate the piston and install it into the caliper bore until it is bottomed.

16. Seat the reinforcing ring at the outer edge of the dust boot into the caliper body with a hammer and boot driver.

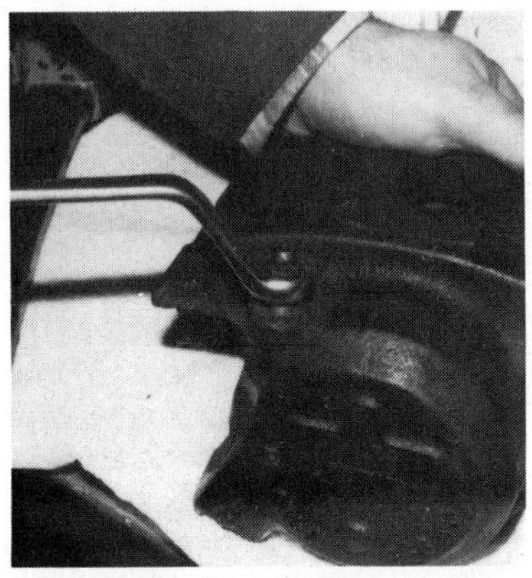

17. Install the bleeder screw in the caliper.

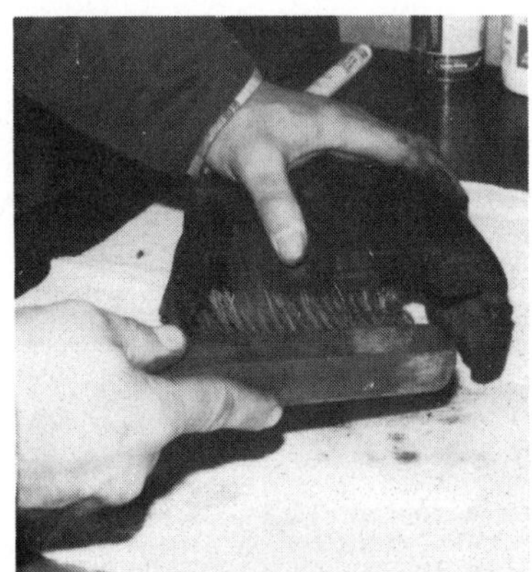

18. Clean the machined ways on the caliper and anchor plate with a wire brush.

CHRYSLER SINGLE-PISTON SLIDING CALIPER SERVICE

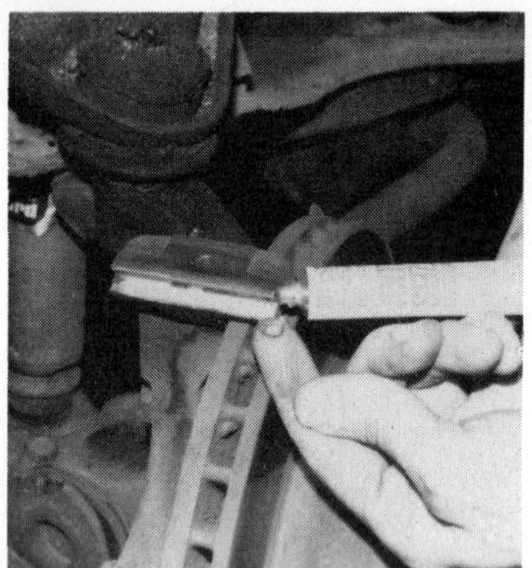

19. Lubricate the anchor plate ways with brake grease, and install the new O-ring on the upper way.

20. Install the inboard brake pad on the anchor plate ways.

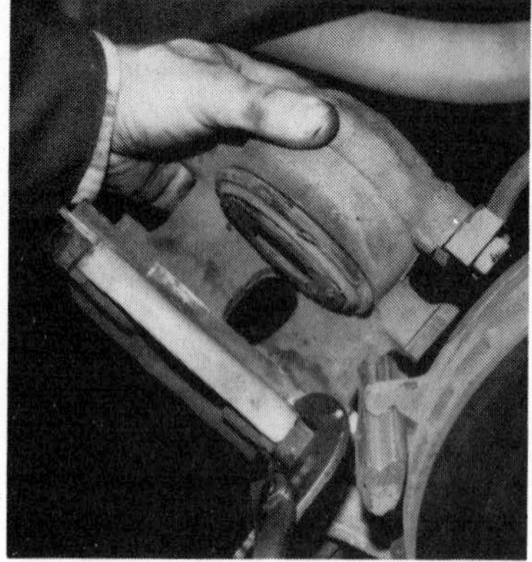

21. Install the outboard brake pad into the caliper with a slight preload.

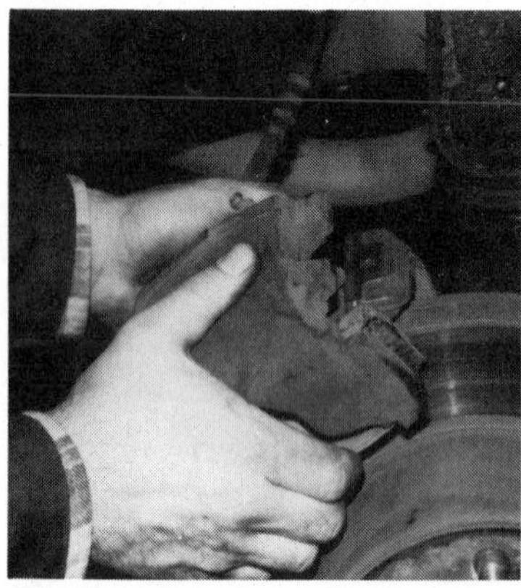

22. Position the caliper over the rotor and align it on the adapter ways.

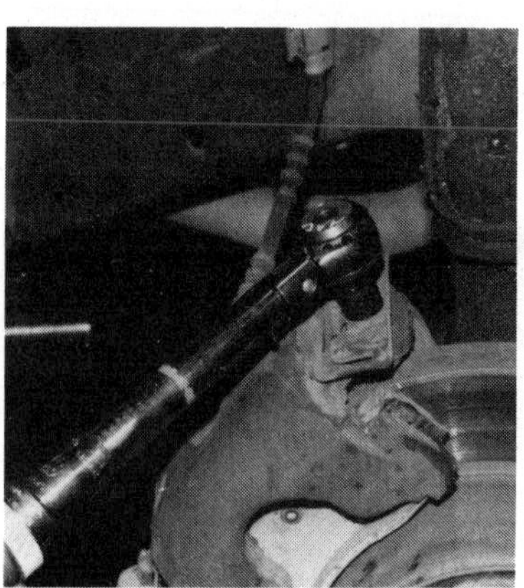

23. Install the anti-rattle springs and retaining clips. Torque the retaining bolts to 15 ft-lb (20 Nm).

24. If the caliper was rebuilt, connect the brake hose to the fluid inlet of the caliper. Torque the bolt to 25 ft-lb (34 Nm). Bleed the brakes as described in Chapter 3.

GM FOUR-PISTON FIXED CALIPER SERVICE

While they are still used on a few high-performance imported cars, four-piston fixed brake calipers are primarily found on older domestic models. The General Motors four-piston caliper with stroking piston seals shown in figure 7-43 is typical of older fixed caliper designs. Introduced in 1965, this caliper was used on all four wheels of the Corvette through 1982, and saw service on other General Motors products in the 1967-69 model years. The caliper shown is the version used from 1965-66. The 1967 and later calipers use the same service procedures, but have a one piece piston design without the plastic insulator.

There are several special procedures involved in working on these calipers. First, because of the piston shape, the outer edges of the dust boots should be removed from the caliper body before the pistons are removed from their bores. To do this, insert a flat-blade screwdriver under the edge of the dust boot reinforcing ring, and using the piston as a fulcrum, lever the ring out of its groove in the caliper body, figure 7-44.

When you install the pistons into their bores, use extra care to avoid damaging the sealing lip of the stroking seal. Special tools are available to ease seal installation, however, a little caution and a thin feeler gauge strip are usually all that is required to work the lip past the edge of the caliper bore, figure 7-45.

Finally, this caliper design is known for its poor sealing which makes it prone to rust, corrosion, and fluid leaks. This often causes severe damage to the caliper bore, and rebuilt calipers with stainless steel sleeves are common in this

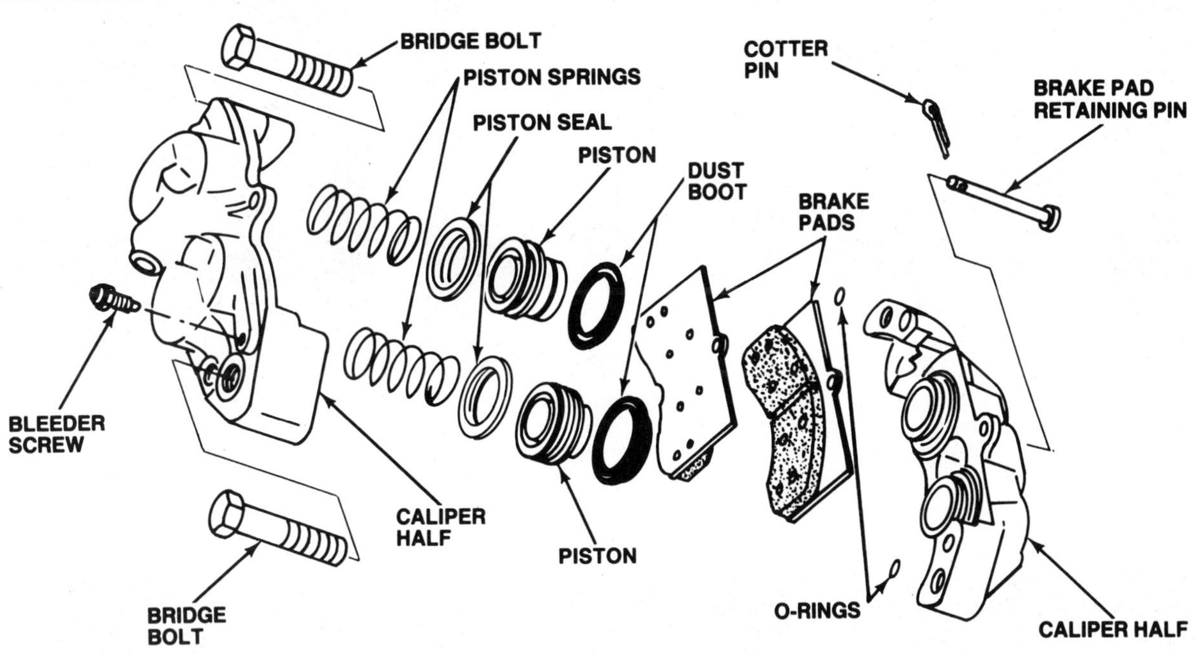

Figure 7-43. The General Motors 4-piston fixed brake caliper.

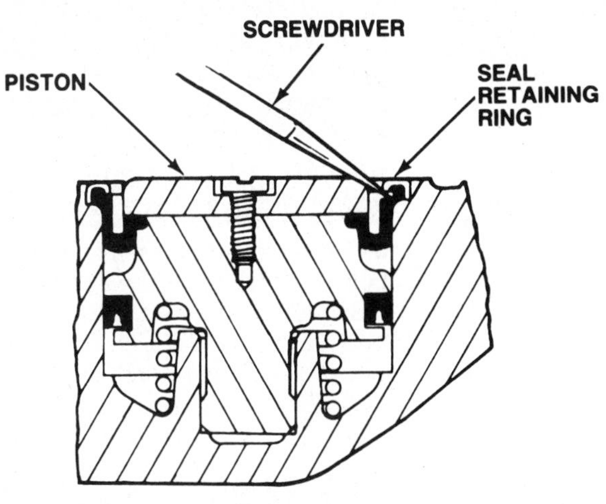

Figure 7-44. Removing a dust boot from the GM fixed brake caliper.

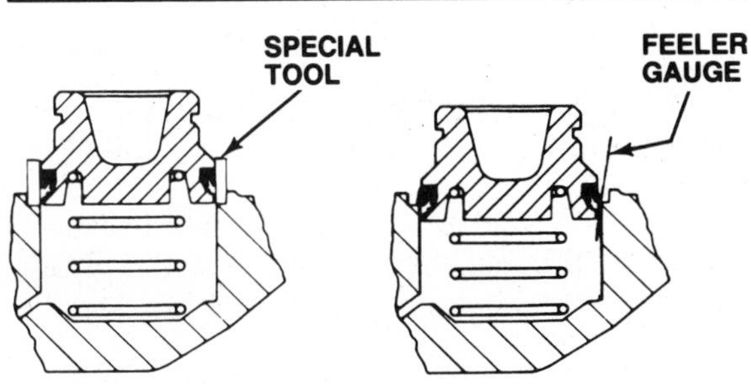

Figure 7-45. Stroking seals require special care during installation.

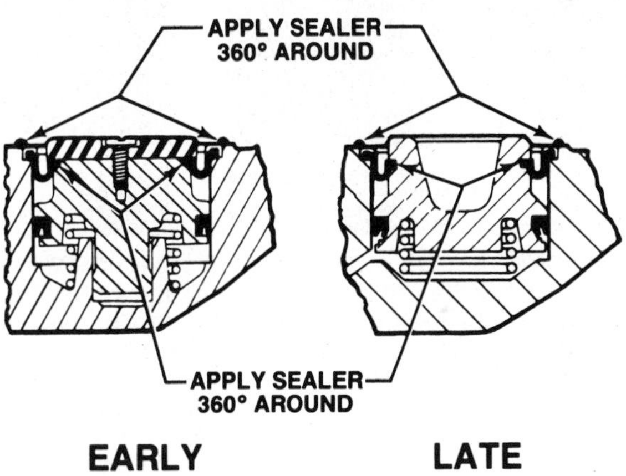

Figure 7-46. Silicone sealer is required to improve dust boot sealing on the GM fixed brake caliper.

application. To ensure a long service life, special sealing procedures are required. Before you install the dust boot, apply a bead of silicone sealer to the dust boot groove in the piston. Once the piston is installed in the bore and the boot is driven into its groove in the caliper body, apply a second bead of sealer around the outer edge of the boot, figure 7-46.

Because this is a fixed caliper, it is not necessary to remove the caliper body from the car when you replace the brake pads. On the other hand, if the caliper is going to be rebuilt, it can be removed from the car with the brake pads still in place.

GENERAL MOTORS FOUR-PISTON FIXED CALIPER SERVICE

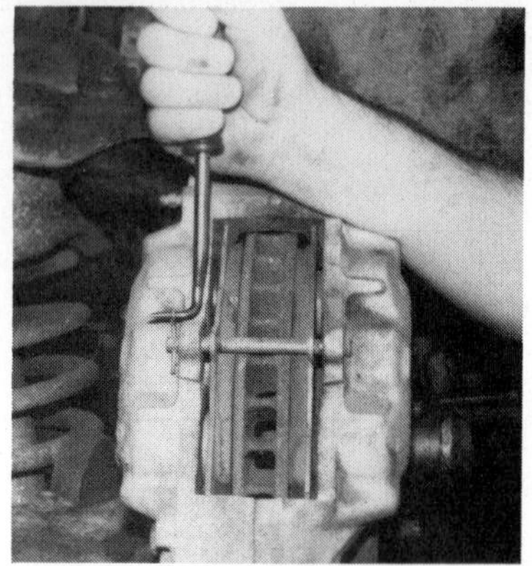
1. Remove the cotter pin from the brake pad retaining pin, and pull the retaining pin out of the caliper.

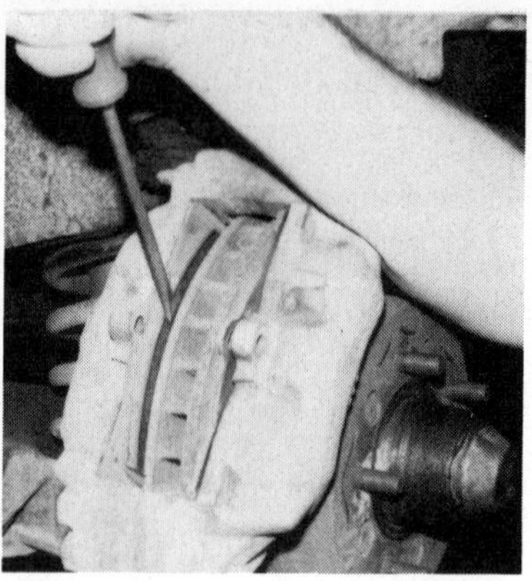
2. Bottom the pistons in their bores.

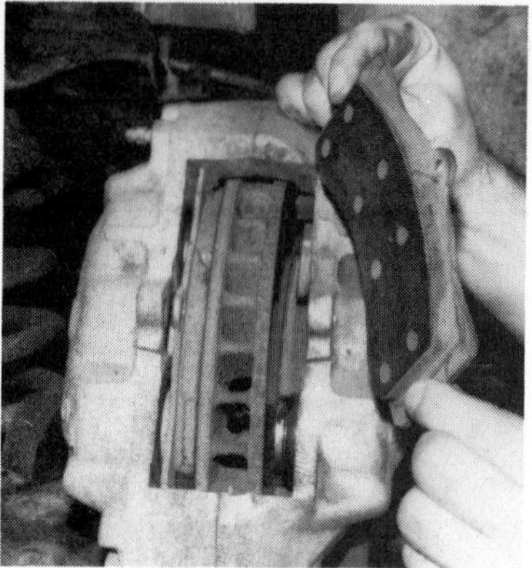
3. Remove the brake pads. If the caliper is not being rebuilt, go to Step 23.

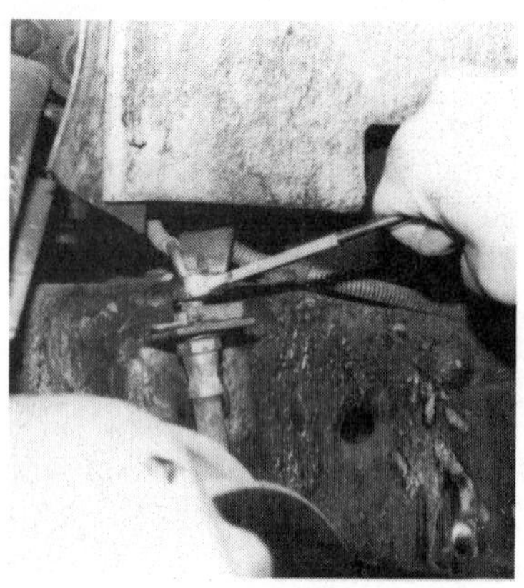
4. Disconnect the brake hose from its bracket, or the tube fitting from the caliper inlet.

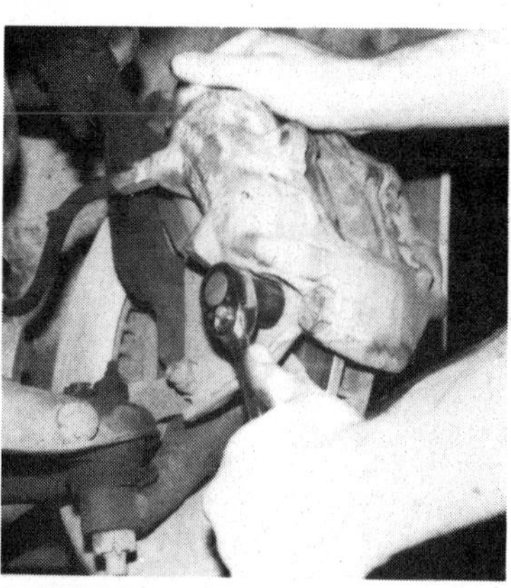
5. Remove the caliper mounting bolts and lift the caliper off of the rotor.

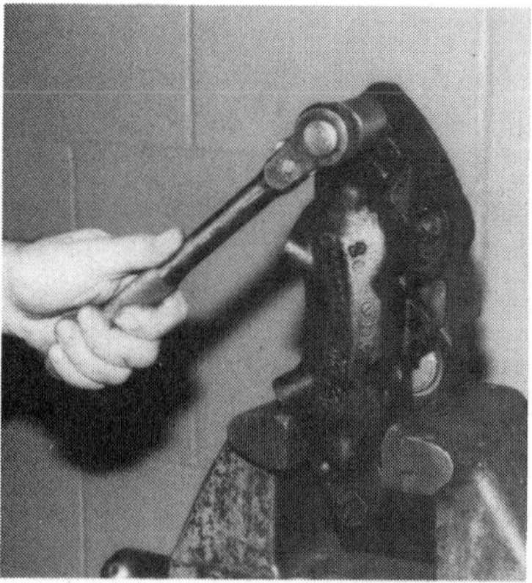
6. Remove the caliper bridge bolts and separate the caliper halves.

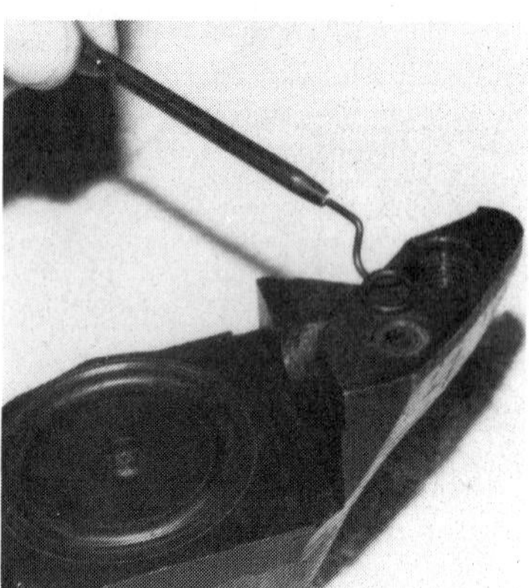
7. Remove the O-rings from the caliper fluid transfer passages.

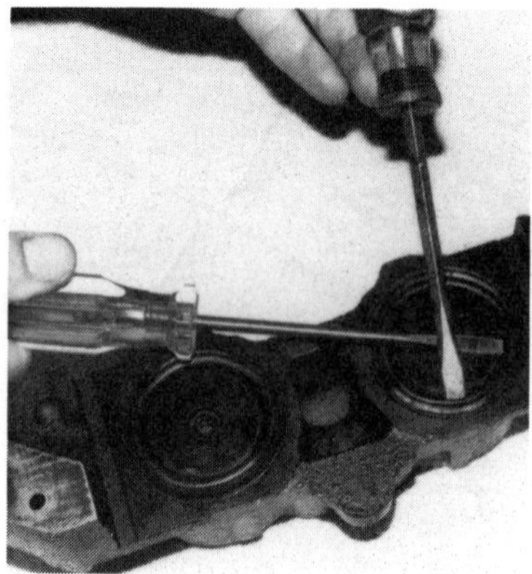
8. Lever the dust boot reinforcing rings out of their grooves in the caliper body.

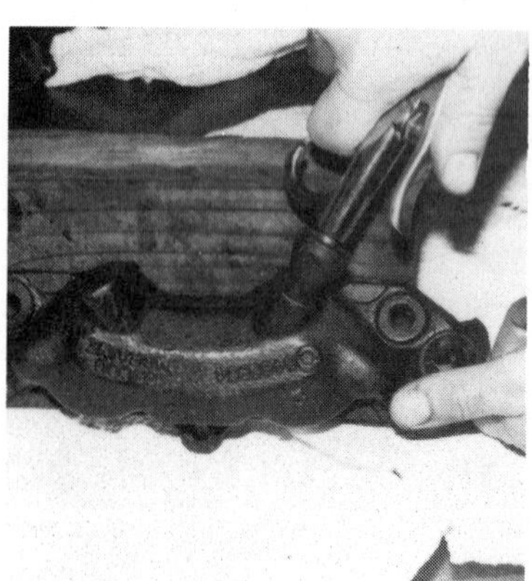
9. Remove the caliper pistons using one of the methods described earlier, then remove the bleeder screw.

GENERAL MOTORS FOUR-PISTON FIXED CALIPER SERVICE

10. Remove the dust boots and seals from their grooves in the pistons.

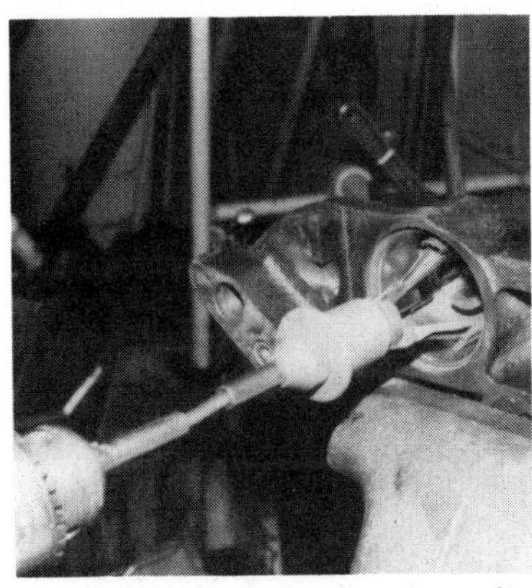

11. Clean and inspect the caliper body and pistons. Hone the caliper bore if necessary, and measure the piston-to-bore clearance.

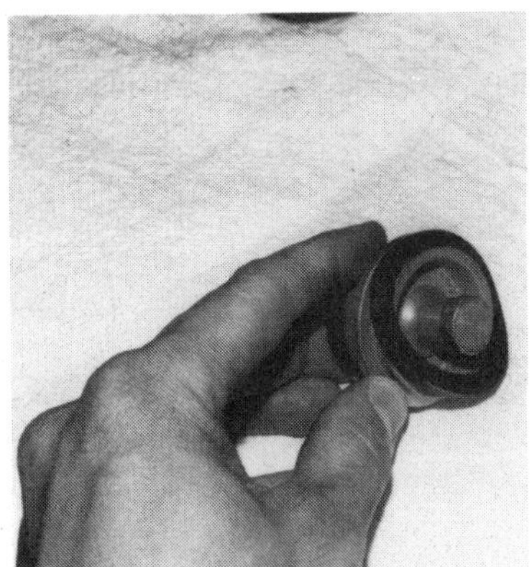

12. Install the seals so the sealing lips face the small ends of the pistons.

13. Install the springs in the caliper bores.

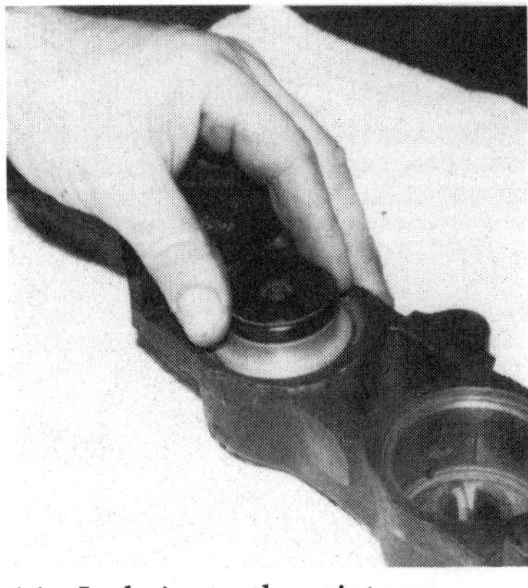

14. Lubricate the pistons, seals, and caliper bores, and install the pistons; make sure they travel smoothly through the bores.

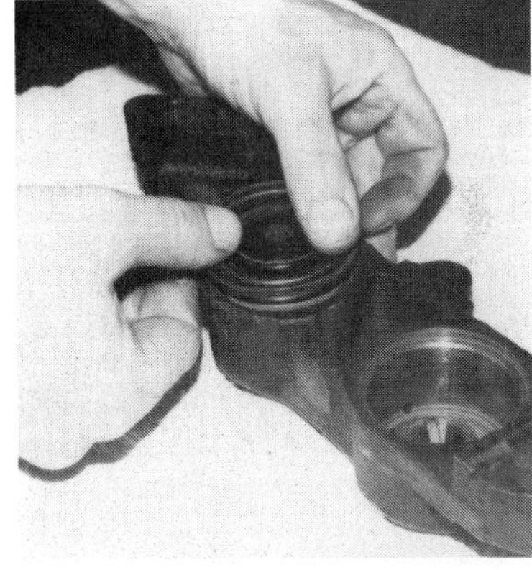

15. Apply a bead of silicone sealer to the dust boot groove in each piston, and install the boots so the folds face the seals.

16. Seat the reinforcing rings of the dust boots into the caliper body. The rings must be flush with or below the machined faces of the caliper.

17. Apply a bead of silicone sealer around the outer edge of the boot retaining rings.

18. Install the new O-rings into the fluid transfer passages, and assemble the caliper halves.

GENERAL MOTORS FOUR-PISTON FIXED CALIPER SERVICE

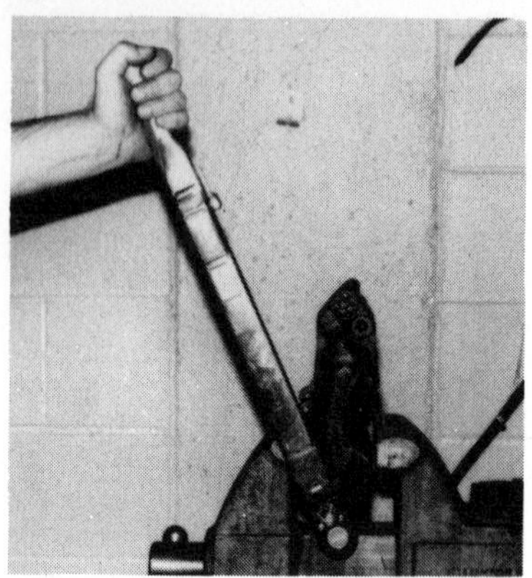

19. Lubricate the caliper bridge bolts, and torque them to 130 ft-lbs (175 Nm) on front brakes, and 60 ft-lbs (80 Nm) on rear brakes.

20. Using a brake pad spreader to hold the pistons out of the way, position the caliper over the rotor.

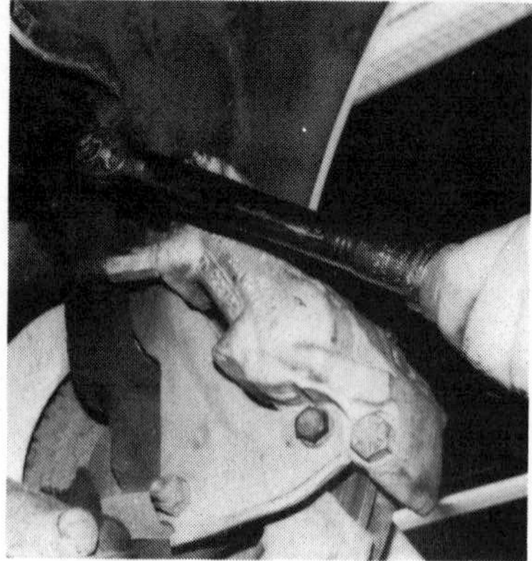

21. Install the caliper mounting bolts; torque them to 70 ft-lbs (95 Nm).

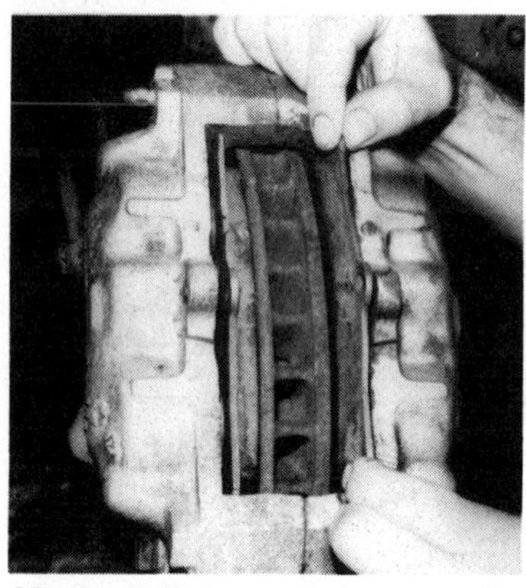

22. Install the new brake pads into the caliper.

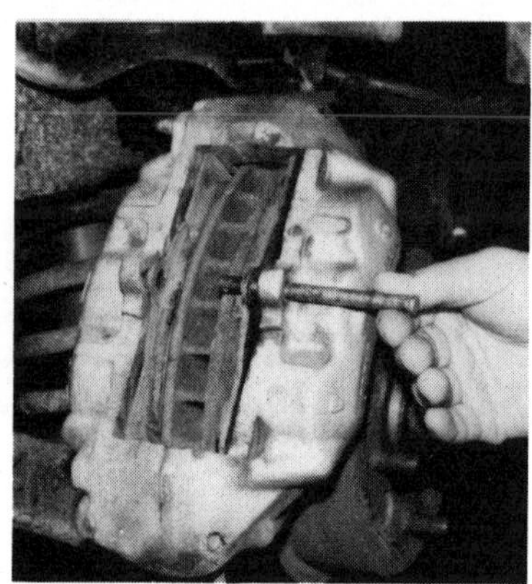

23. Install the brake pad retaining pin and secure it with a new cotter pin.

24. Attach the brake hose or tube fitting, if it was removed. Bleed the brakes as described in Chapter 3.

FORD SINGLE-PISTON REAR CALIPER SERVICE

The Ford single-piston, sliding rear brake caliper with integral parking brake mechanism, figure 7-47, has been used in its present form since 1982, and is similar in most respects to the rear disc brake used by Ford beginning in 1975. There are several special considerations you should be aware of when overhauling these calipers. First, they are specific left- and right-side parts with unique parking brake levers, pistons, and internal thrust screws. Do not mix calipers or their components from side-to-side, and make sure any replacement parts are for the proper caliper. Many of the caliper parts are marked to indicate the side of the car on which they belong.

Second, the design of the self-adjuster mechanism in this caliper prevents the piston from being forced back into the bore in the usual manner. A special piston turning tool, figure 7-14, is required to thread the piston back into its bore once the brake pads are removed.

Because of the adjuster design, it may be difficult to remove one of these calipers if there is a ridge at the outer edge of the rotor, or the brake pads are badly worn and interlock with scoring on the rotor. If you encounter this problem, remove the parking brake cable and lever, and loosen the parking brake end retainer up to one-half turn; this allows the piston to move slightly back into the bore, and makes it possible to remove the caliper from the rotor. Take care when removing the caliper in this manner; if the end retainer is loosened more than one-half turn, the seal between the caliper body and thrust screw may be broken. Brake fluid will then leak into the parking brake mechanism, and the caliper will have to be disassembled to

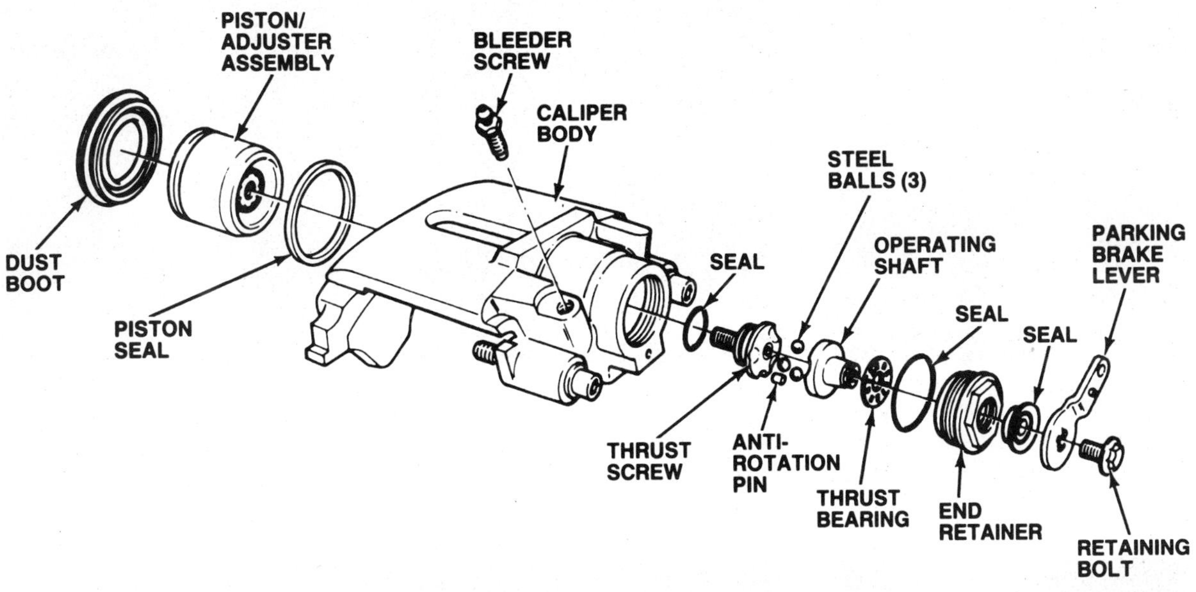

Figure 7-47. The Ford single-piston, sliding rear brake caliper with integral parking brake mechanism.

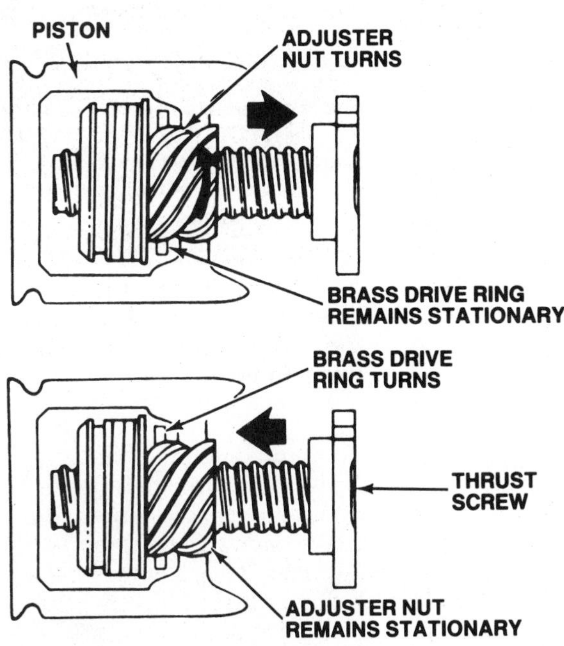

Figure 7-48. Testing the Ford rear brake caliper piston/adjuster mechanism.

clean and relubricate the mechanism. If the caliper is being overhauled, this is not a concern, and the retainer can be unscrewed as necessary to remove the caliper.

Once the caliper is disassembled, clean and inspect the caliper body and all of the other parts except the piston. Hone the caliper bore if necessary, and measure the piston-to-bore clearance. Inspect the thrust screw bore; it must be smooth and free of pits. The thrust screw threads, seal groove, ball pockets, and bearing surface must show no signs of wear or pitting. Inspect the thrust bearing for wear or corrosion, and also the bearing surface inside the end retainer. Replace any of these parts that are not in perfect condition.

Next, wipe the caliper piston/adjuster assembly clean; do not use liquid cleaners because they may remain inside the piston and contaminate the hydraulic system. Inspect the piston sealing surface for damage, and flush any old brake fluid from the piston with new fluid. Inspect the adjuster assembly for a loose fit in the piston, and check the adjuster operation. As shown in figure 7-48, thread the thrust screw into the piston/adjuster assembly. Pull the two pieces apart about ¼ inch (6 mm), the brass drive ring should remain stationary and the adjuster nut should turn. When the two pieces are released, the nut should remain stationary and the drive ring should rotate. The piston/adjuster assembly cannot be taken apart for service. Replace the entire assembly if the adjuster is loose in the piston, or does not function as described above.

Finally, when you have completed the overhaul of one of these calipers, adjust the parking brake mechanism. With the engine running, apply the brake pedal lightly 40 times, allowing a one second pause between applications. With the engine not running, apply the pedal with moderate pressure 30 times. Once you have finished this procedure, adjust the parking brake linkage as described in Chapter 9.

FORD SINGLE-PISTON REAR CALIPER SERVICE

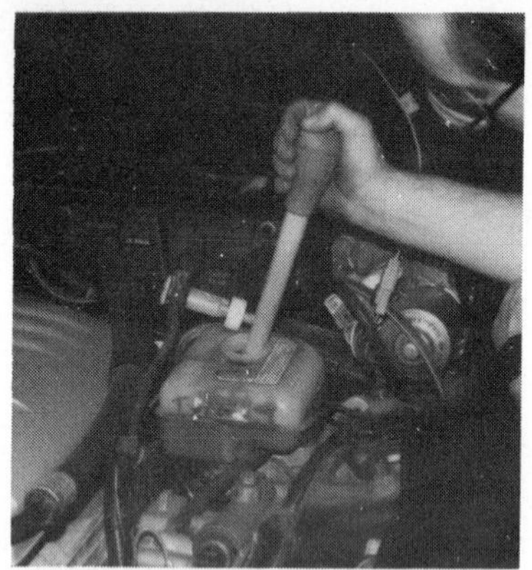

1. Siphon off two-thirds of the fluid in the front brake section of the master cylinder.

2. Remove the clip and pin, and disconnect the parking brake cable from the caliper lever.

3. If the caliper is being rebuilt, make sure the bleeder screw is free, then disconnect the brake hose from the fluid inlet.

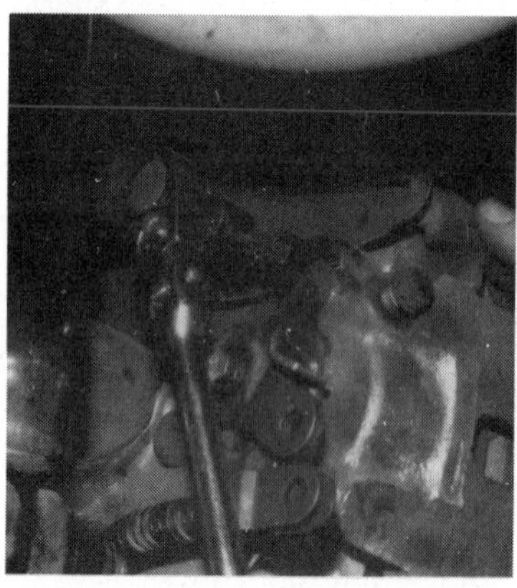

4. Remove the caliper mounting pins and pin bushings.

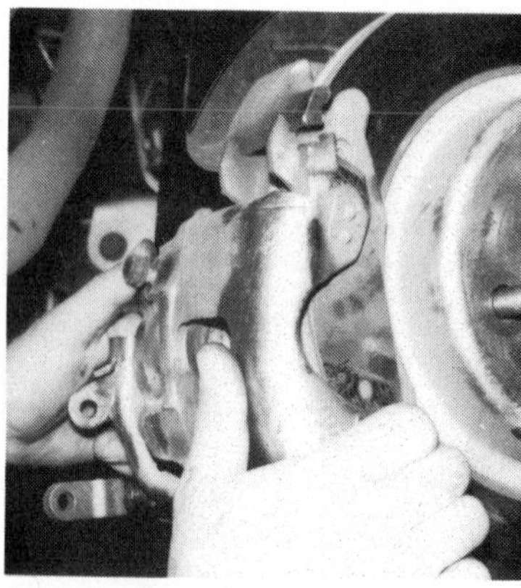

5. Lift the caliper off of the rotor and hang it from the suspension on a wire if it is not being rebuilt. If the caliper is free, go to Step 8; otherwise, go to Step 6.

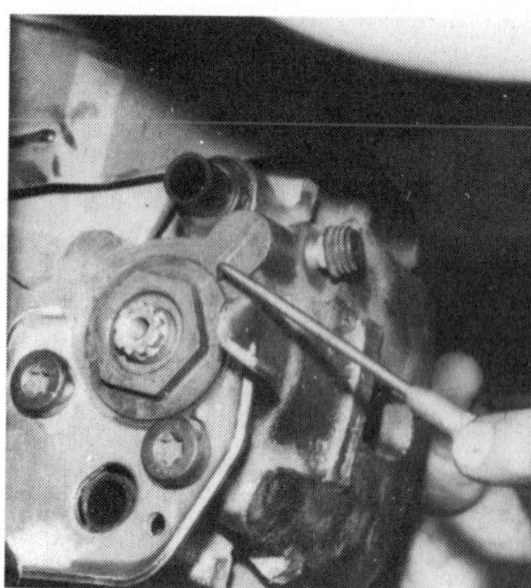

6. Remove the parking brake lever. Scribe the end retainer for reference, and loosen it one-half turn.

7. Pry the piston into its bore, and lift the caliper off of the rotor; hang it from the suspension on a wire if it is not being rebuilt.

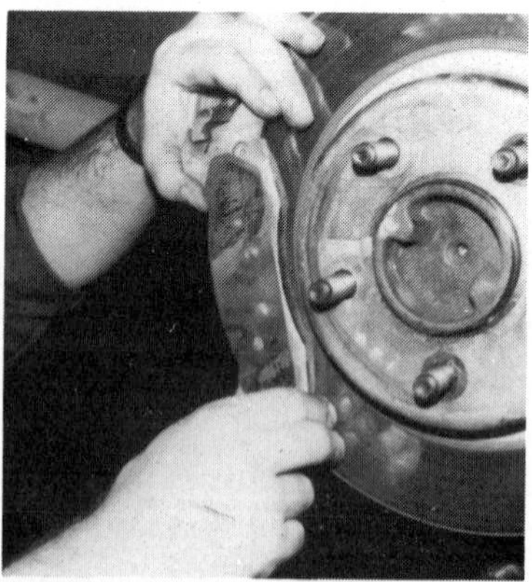

8. Remove the outer brake pad from the anchor plate.

9. Unscrew the two retaining nuts, and remove the brake rotor.

FORD SINGLE-PISTON REAR CALIPER SERVICE

10. Remove the inner brake pad from the anchor plate. If the caliper is not being overhauled, go to Step 37.

11. Remove the bleeder screw and parking brake lever, then unscrew the caliper end retainer.

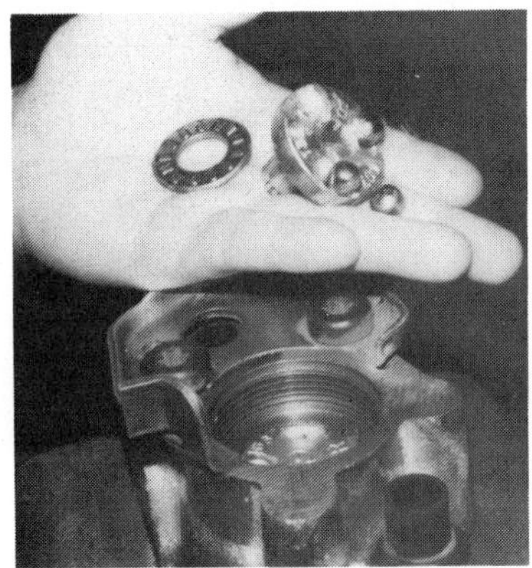

12. Lift the thrust bearing, operating shaft, and three steel balls out of the caliper.

13. Remove the anti-rotation pin with a magnet or tweezers. If the pin is free, go to Step 16; otherwise, go to Step 14.

14. Bottom the caliper piston in the caliper bore with a pair of slip-joint pliers (this is possible with the end retainer removed).

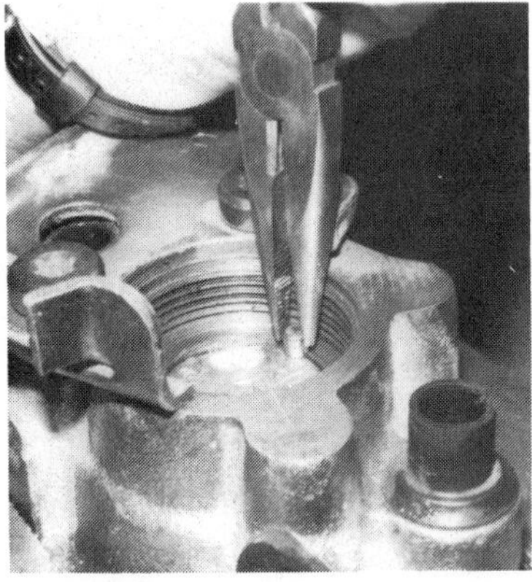

15. Tap the thrust screw into the caliper until the anti-rotation pin stands proud, then remove the pin with a pair of pliers.

16. Remove the thrust screw by rotating it counterclockwise with a ¼-inch Allen wrench.

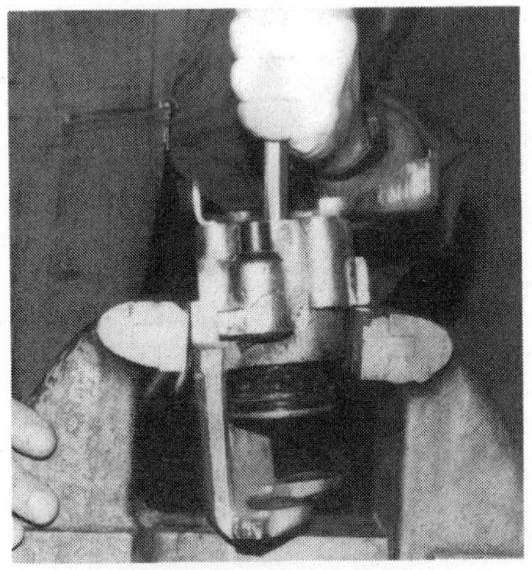

17. Push the caliper piston out of its bore using a punch inserted into the opening in the backside of the piston.

18. Remove and discard the dust boot, piston seal, thrust screw O-ring seal, end retainer O-ring seal, and end retainer lip seal.

FORD SINGLE-PISTON REAR CALIPER SERVICE

19. Install and tighten the bleeder screw.

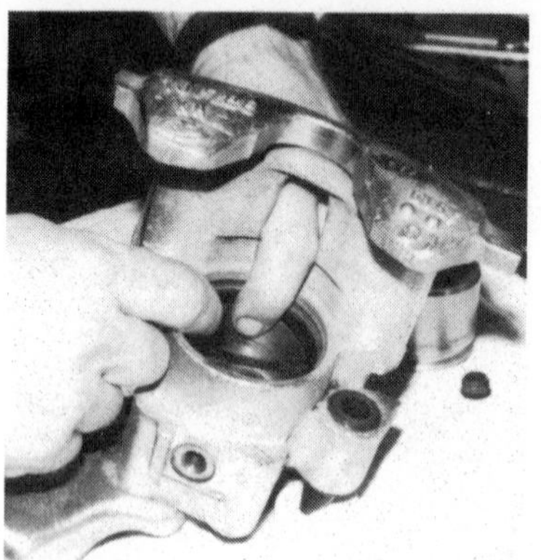

20. Lubricate the caliper bore and the new piston seal. Install the seal into its groove in the caliper bore.

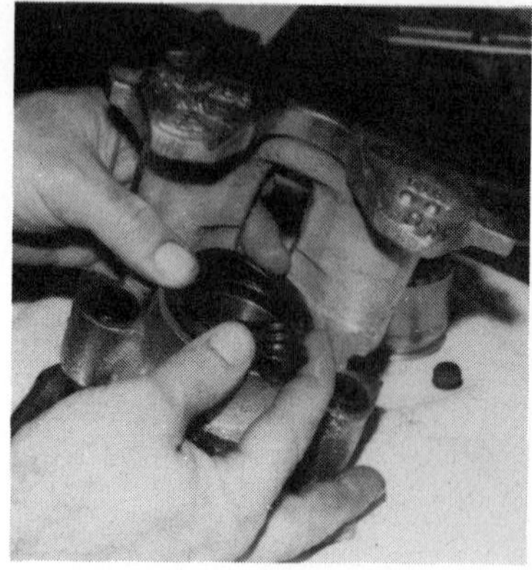

21. Install the outer edge of the new dust boot into its groove in the caliper bore.

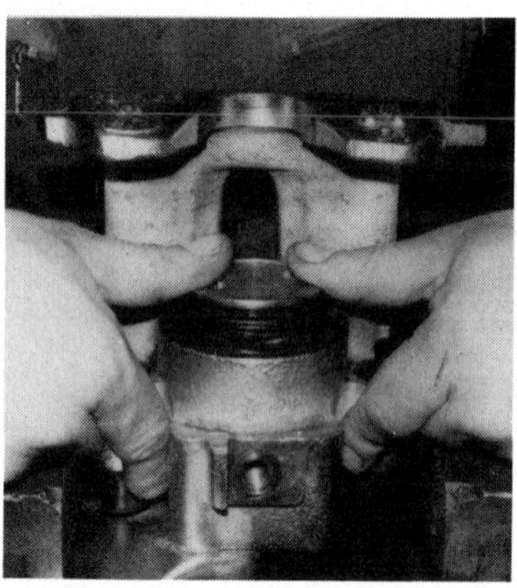

22. Coat the piston with brake fluid and install it through the dust boot into the caliper bore.

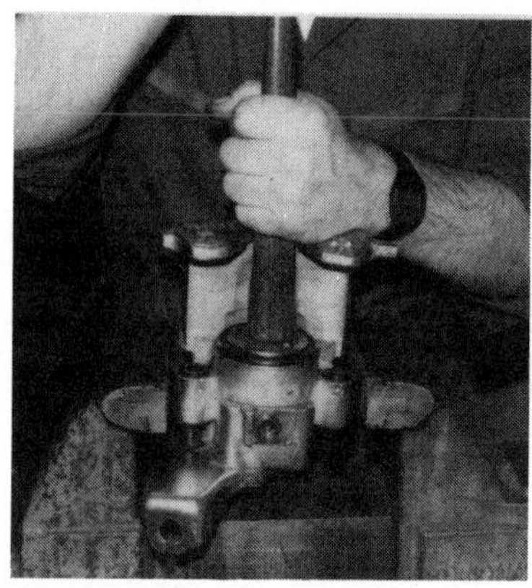

23. Seat the inner edge of the dust boot into its groove on the piston, and bottom the piston in the caliper bore.

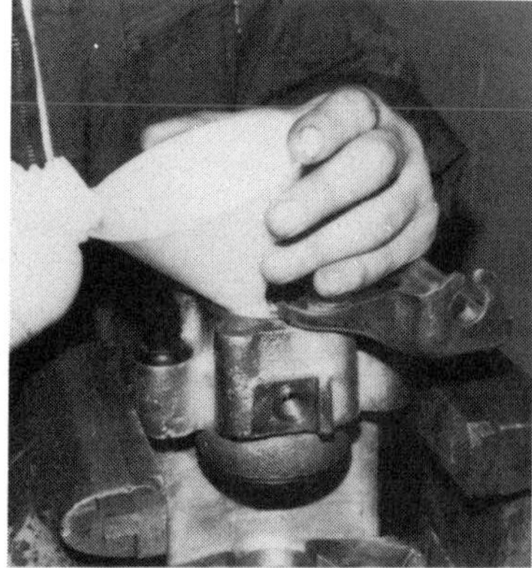

24. Clamp the caliper in a vise with the end retainer side up. Fill the piston with fresh brake fluid to the bottom edge of the thrust screw bore.

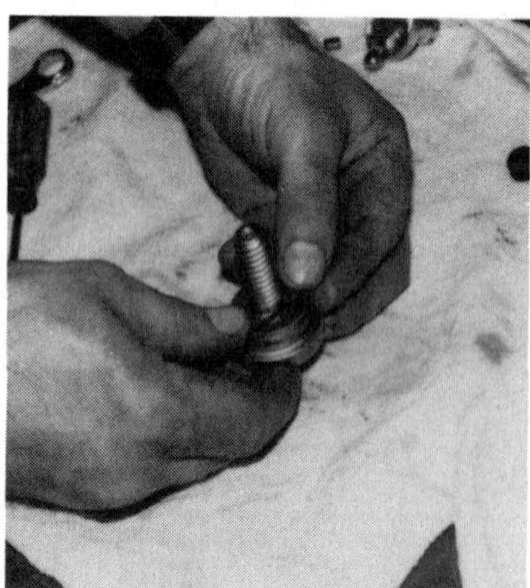

25. Lubricate the new O-ring seal and install it on the thrust screw.

26. Thread the thrust screw into the piston with a 1/4-inch Allen wrench. The top of the thrust screw should be flush with the bottom of the threaded bore.

27. Align the notch in the thrust screw with the notch in the caliper housing, then install the anti-rotation pin.

FORD SINGLE-PISTON REAR CALIPER SERVICE

28. Coat the end of the thrust screw with high-temperature brake grease, and install a greased ball in each of the three pockets.

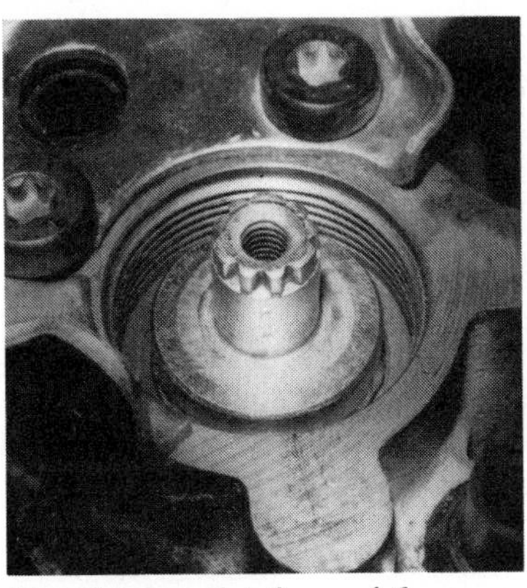
29. Grease the face of the operating shaft, and install it over the balls.

30. Grease the thrust bearing and install it on the operating shaft.

31. Grease the bearing surface inside the end retainer, then grease and install a new O-ring and lip seal on the end retainer.

32. Thread the end retainer into the caliper body while holding a downward pressure on the operating shaft to prevent dislocating the balls.

33. Tighten the end retainer to 75 to 95 ft-lb (102 to 128 Nm).

34. Install the parking brake lever and tighten the retaining screw to 16 to 22 ft-lb (22 to 29 Nm). The lever should rotate freely.

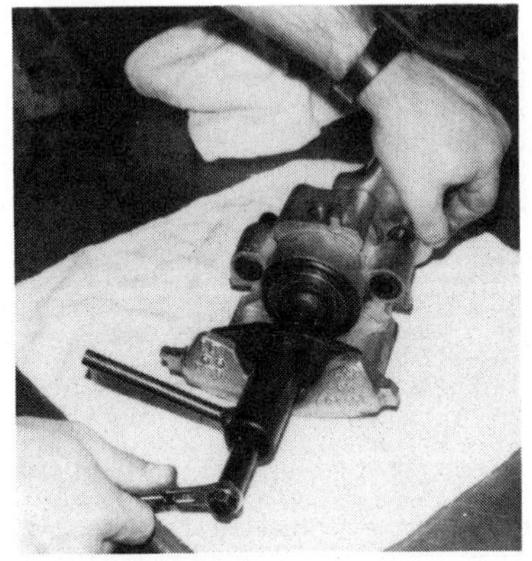
35. Rotate the piston in a clockwise direction with the turning tool. Though it will continue to turn, the piston is bottomed when there is no more inward movement.

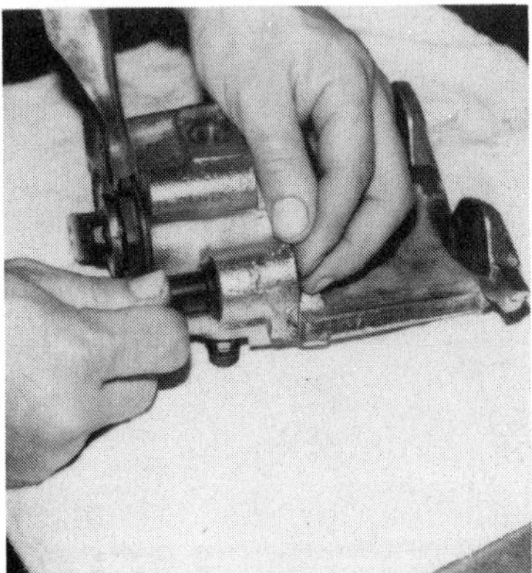
36. Sand or wire brush any rust out of the bushing bores of the caliper, then install new mounting pin bushings.

FORD SINGLE-PISTON REAR CALIPER SERVICE

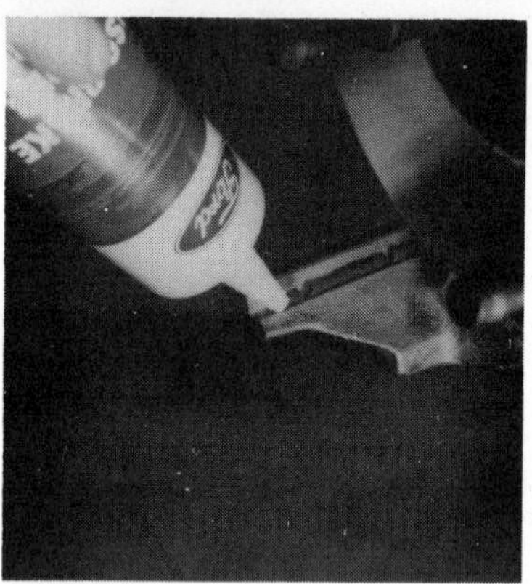

37. Grease the anchor plate sliding ways and install the inner brake pad.

38. Install the brake rotor and its two retaining nuts.

39. Install the outer brake pad with the wear indicator (if fitted) at the top of the caliper.

40. Position the caliper on the car and pull it outward until the inner pad is in firm contact with the rotor.

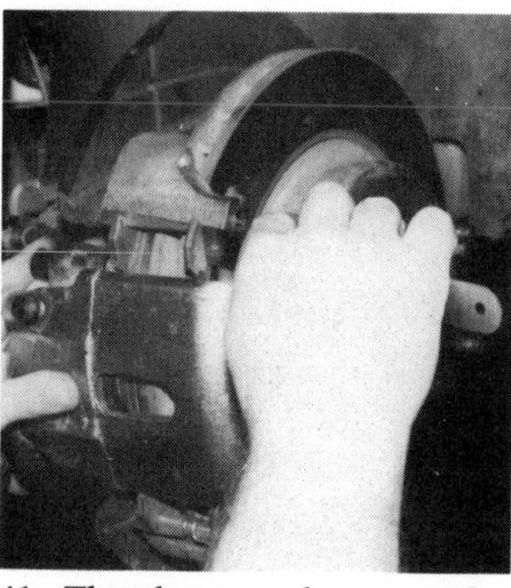

41. The clearance between the outer pad and caliper must be less than 3/32 inch (2.4 mm) or the piston/adjuster assembly will be damaged when the service brakes are applied.

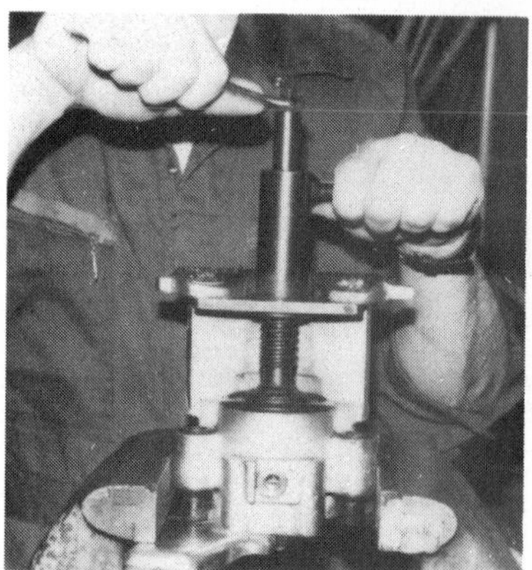

42. Using the turning tool, unscrew the piston as needed to obtain the proper clearance; 1/4 turn equals approximately 1/16 inch (1.6 mm).

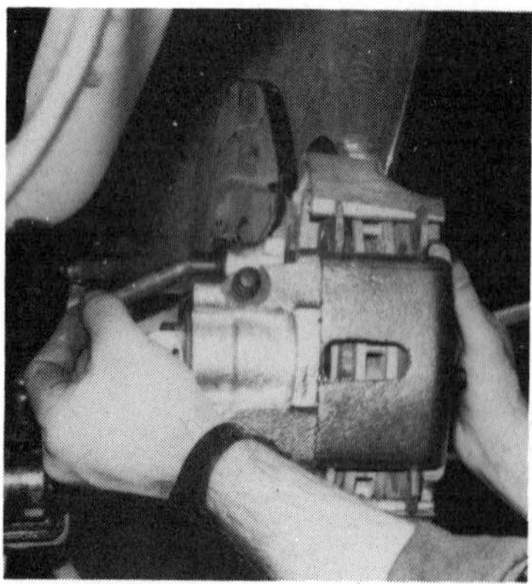

43. Grease the mounting pins and install them in the caliper. Tighten them to 29 to 37 ft-lb (39 to 50 Nm).

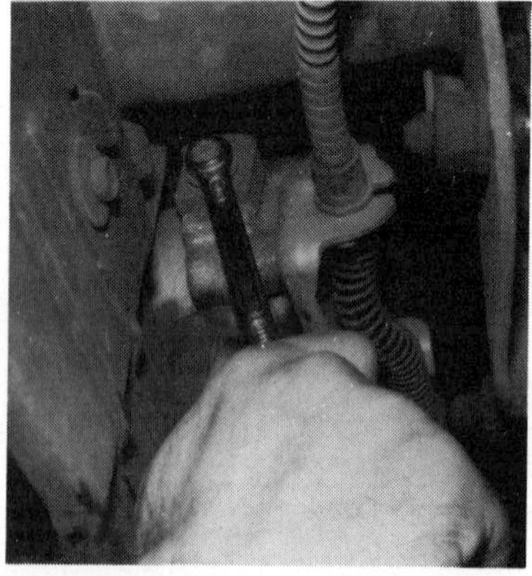

44. If the caliper was rebuilt, connect the brake hose to the fluid inlet of the caliper. Bleed the brakes as described in Chapter 3.

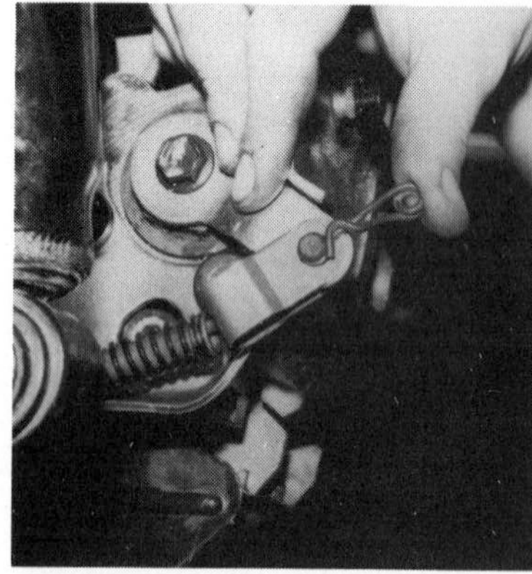

45. Connect the parking brake cable and adjust the caliper mechanism and linkage as described earlier.

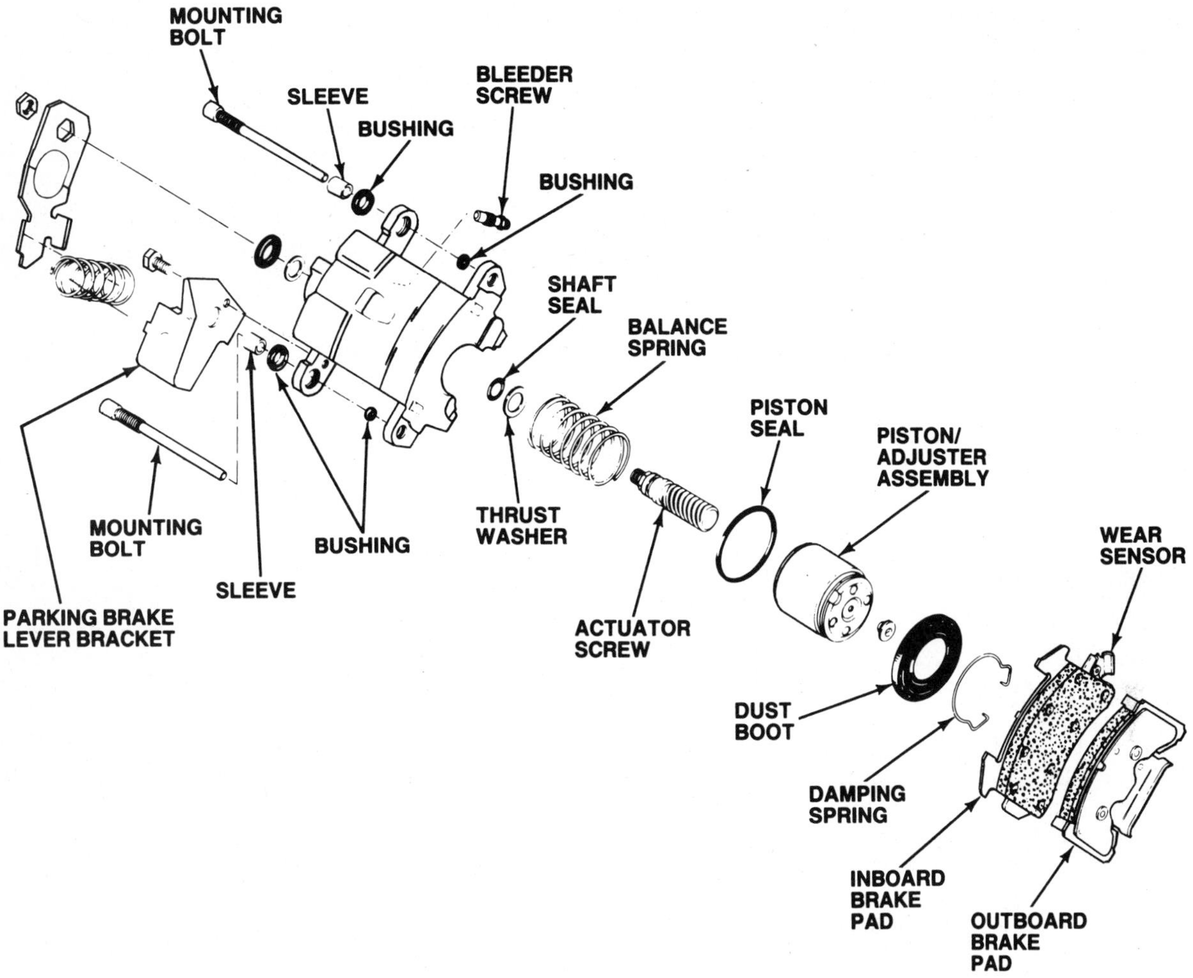

Figure 7-49. The GM single-piston, floating rear brake caliper with integral parking brake mechanism.

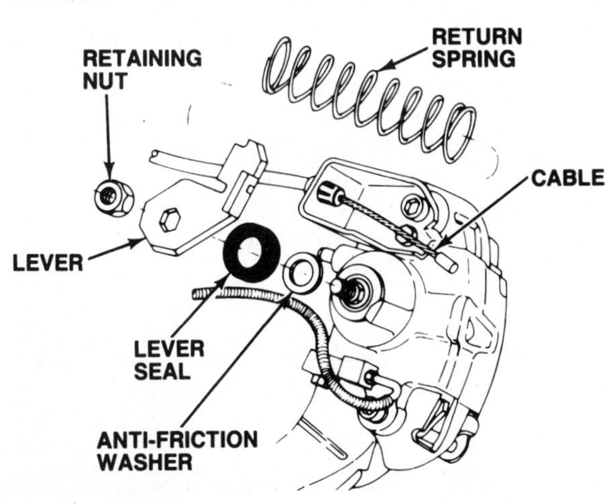

Figure 7-50. The parking brake lever components of a General Motors rear brake caliper.

GM SINGLE-PISTON REAR CALIPER SERVICE

The General Motors 3500 Series single-piston, floating rear brake caliper with integral parking brake mechanism, figure 7-49, is used on all GM rear disc brake applications through 1986, with the exception of the Chevrolet Corvette, and Pontiac Fiero and 6000STE. The construction and operation of this caliper are similar to that of the 3200/3300 front brake caliper described earlier. The major external difference is the parking brake actuating lever, bracket, and associated hardware. Several different lever designs are used, but all consist of essentially the same parts; a typical installation is shown in figure 7-50.

Because this caliper is similar to other General Motors designs, all of the service tips, precautions, and special instructions described earlier for the 3200/3300 calipers apply to this unit

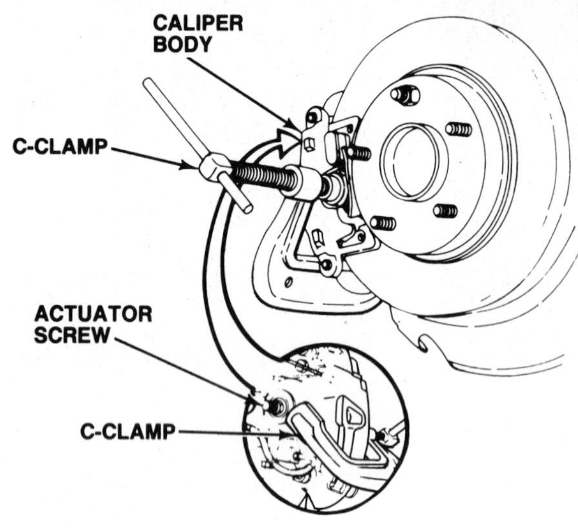

Figure 7-51. Bottoming the piston in a GM rear disc brake caliper.

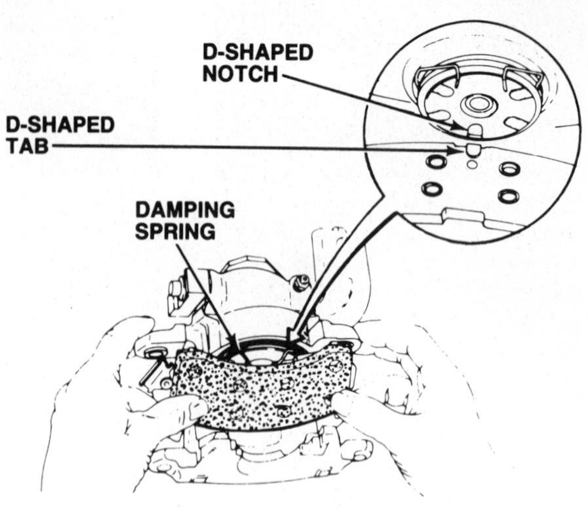

Figure 7-52. Proper inboard brake pad mounting on the GM rear brake caliper.

as well. In addition, there are several other things to be aware of when you overhaul one of these calipers.

First, General Motors rear calipers, like Ford designs, are specific left- and right-side units that cannot be swapped from side to side. Also like the Ford rear caliper, the piston in the General Motors caliper contains the self-adjusting mechanism. Never clean the piston/adjuster assembly with liquid cleaners that can be trapped inside and contaminate the brake hydraulic system.

The self-adjusting mechanism of the General Motors caliper cannot be tested when the caliper is disassembled, so before you remove the caliper from the car, check the range of movement of the parking brake lever. If the lever is frozen, or rotates more than approximately 45 degrees, the self-adjusting mechanism is damaged and the piston/adjuster assembly must be replaced.

Second, and unlike the Ford design, the piston in the General Motors rear caliper can be bottomed in its bore using conventional methods once the parking brake lever has been removed. General Motors recommends a C-clamp for this purpose, figure 7-51. With the parking brake lever removed, place the C-clamp across the inboard side of the caliper body and the backing plate of the outboard brake pad. Do not allow the C-clamp to contact the actuator screw or the self-adjusting mechanism in the piston may be damaged.

Third, when you install the inboard brake pad, the D-shaped tabs on the pad backing plate must engage the notches in the caliper piston, figure 7-52; rotate the piston as needed to obtain the proper alignment. To install the pad, hold it at an angle in the caliper, and slide the edge of the backing plate under the ends of the damping spring. Then, lift the pad up against spring tension, and engage the D-shaped tabs into the piston notches. When installed properly, the pad will seat flat against the piston.

Finally, whenever new pads are installed in this caliper, or an overhaul is completed, adjust the parking brake cables as described in Chapter 9. Then, set the clearance between the brake pads and the rotor by applying the brake pedal at least three times with about 175 lb (778 N) of force. This allows the self-adjusting mechanism in the piston to take up the slack. When the clearance is correct, the parking brake pedal will have a high, firm feel.

GENERAL MOTORS SINGLE-PISTON REAR CALIPER SERVICE

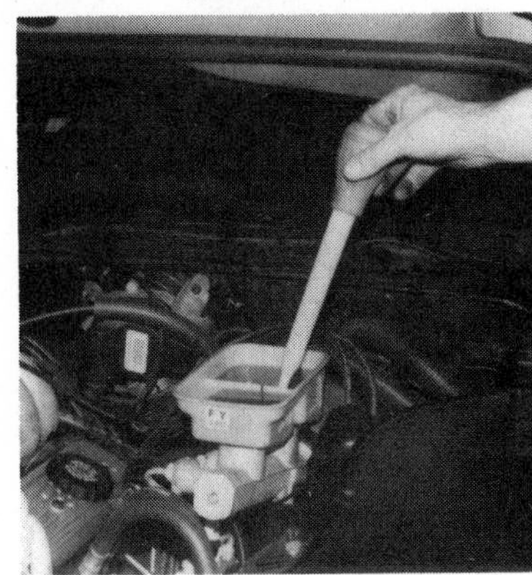
1. Siphon off fluid from the master cylinder reservoir(s) as dictated by the split of the brake hydraulic system.

2. Loosen tension on the parking brake cable at the equalizer, then disconnect the cable from the lever on the caliper.

3. Unscrew the retaining nut, and remove the parking brake lever, return spring, lever seal, and anti-friction washer.

4. Use a C-clamp to bottom the caliper piston in its bore as described earlier.

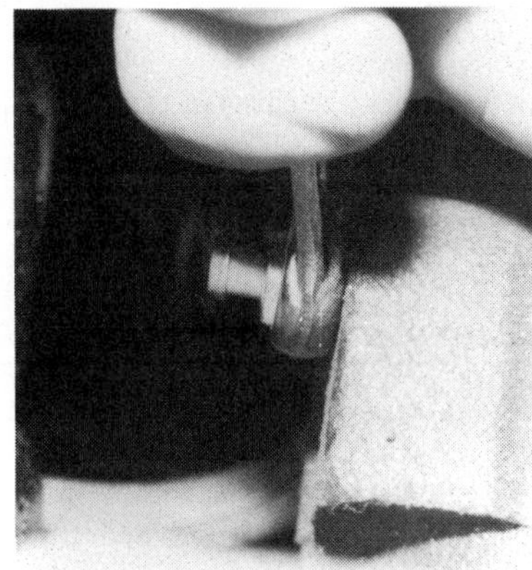
5. If the caliper is being rebuilt, make sure the bleeder screw is free, then disconnect the brake hose from the caliper fluid inlet.

6. Unscrew and remove the caliper mounting bolts.

7. Lift the caliper off of the rotor, and hang it from the suspension on a wire if it is not being rebuilt.

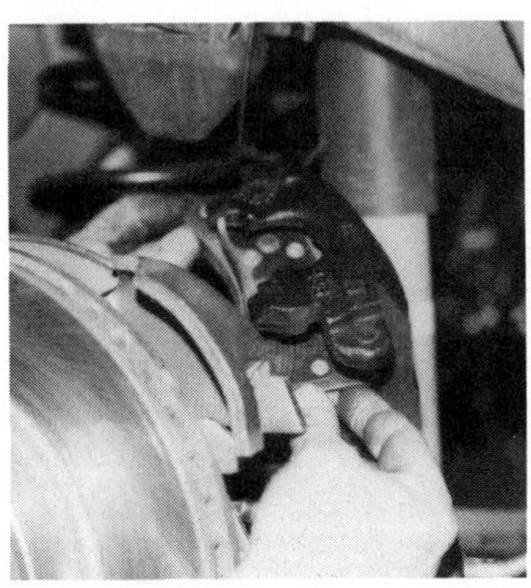
8. Remove the outboard brake pad from the caliper.

9. Remove the inboard brake pad from the caliper.

GENERAL MOTORS SINGLE-PISTON REAR CALIPER SERVICE

10. Use a small screwdriver to pry the two-way check valve out of the piston.

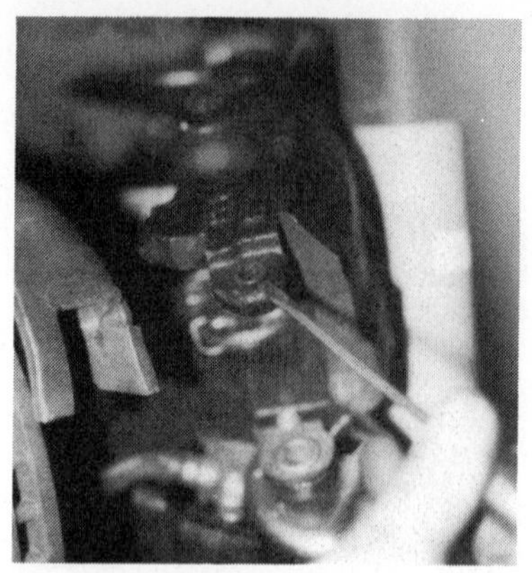

11. Remove the sleeves and bushings from the caliper mounting bolt holes. If the caliper is not being rebuilt, go to Step 30.

12. Remove the brake pad damping spring from the end of the piston.

13. Mount the caliper in a vise, then use a wrench to rotate the actuator screw and force the piston out of its bore.

14. Remove the balance spring from the caliper bore.

15. Remove the actuator screw and thrust washer from the caliper bore. Remove the shaft seal from its groove in the actuator screw.

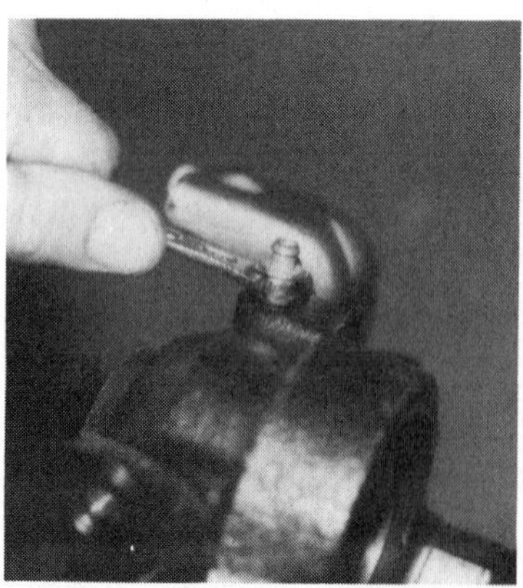

16. Remove the bleeder screw.

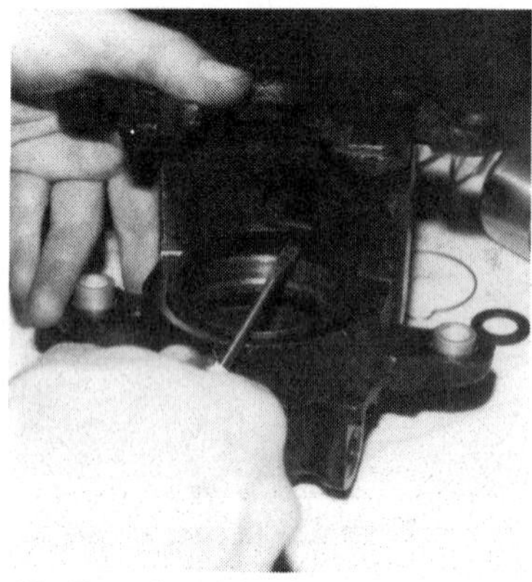

17. Pry the dust boot out of the caliper body with a screwdriver.

18. Remove the piston seal from its groove in the caliper bore.

GENERAL MOTORS SINGLE-PISTON REAR CALIPER SERVICE

19. Clean and inspect the caliper body and other components. Hone the caliper bore if necessary, and measure the piston-to-bore clearance.

20. Install the bleeder screw in the caliper body and tighten it to 80 to 140 in-lb (9 to 16 Nm).

21. Lubricate the caliper bore and the new piston seal. Install the seal into its groove in the caliper bore.

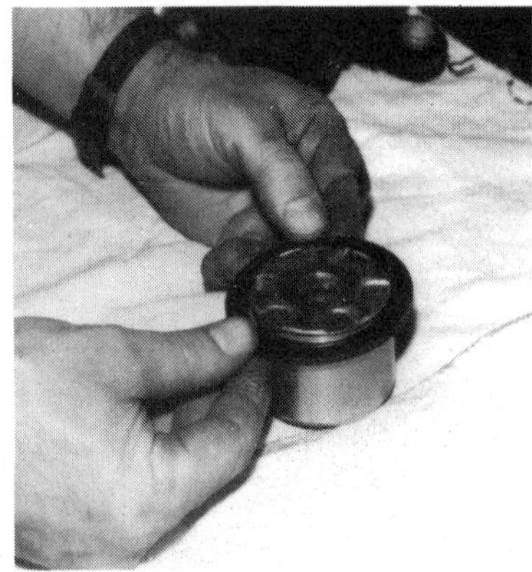

22. Install the new dust boot into its groove on the piston.

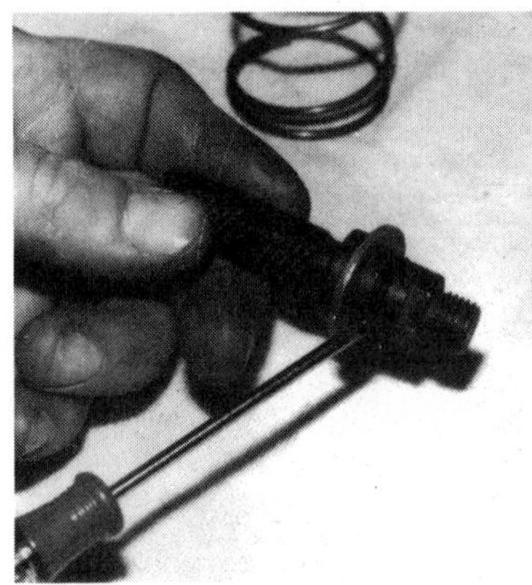

23. Lubricate the new shaft seal and install it into its groove on the actuator screw.

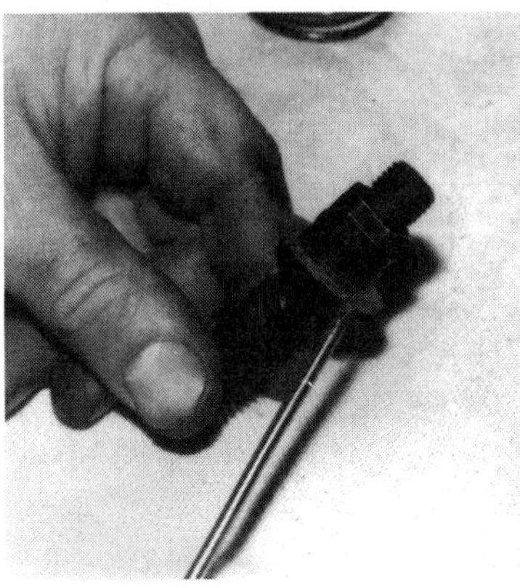

24. Position the thrust washer on the actuator screw so the rounded edge will contact the caliper body when the screw is installed.

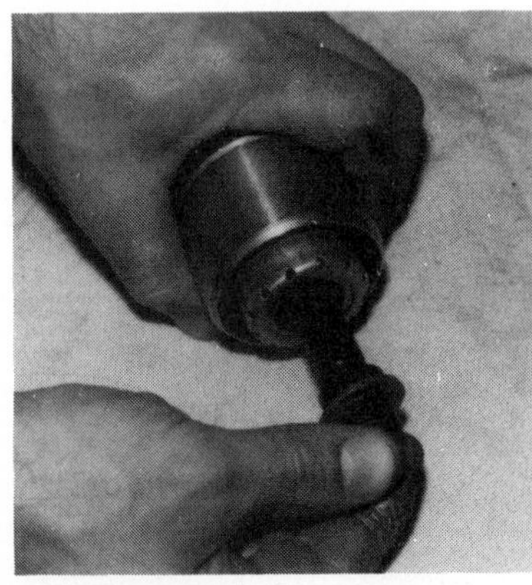

25. Lubricate the actuator screw threads, and install the screw into the piston.

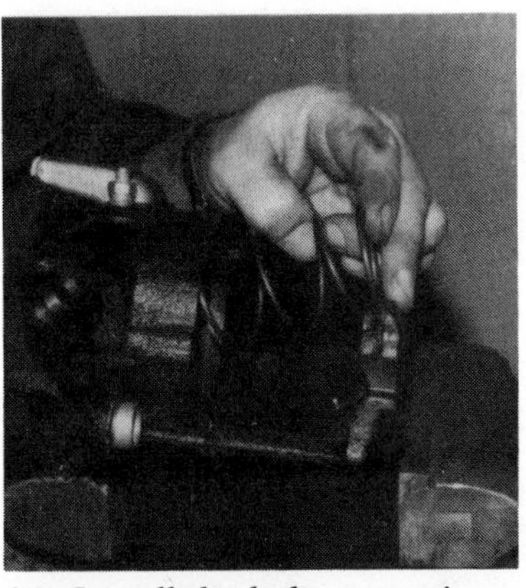

26. Install the balance spring in the caliper bore.

27. Lubricate the piston and install it into the caliper bore until it is bottomed.

GENERAL MOTORS SINGLE-PISTON REAR CALIPER SERVICE

28. Seat the reinforcing ring at the outer edge of the dust boot into the caliper body with a hammer and boot driver.

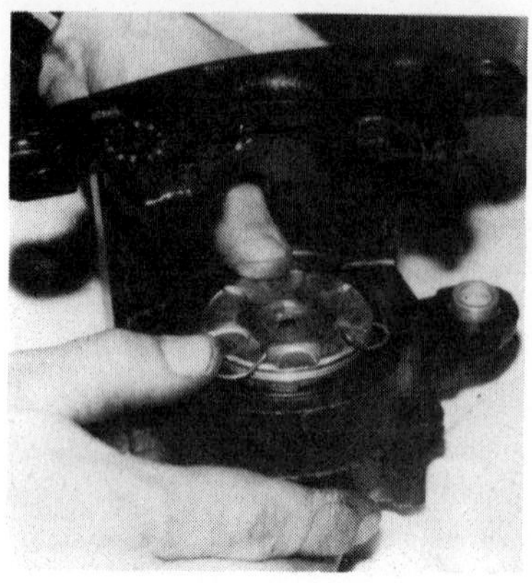
29. Install the brake pad damping spring in the groove at the end of the piston.

30. Lubricate the new two-way check valve with brake fluid and install it into the piston.

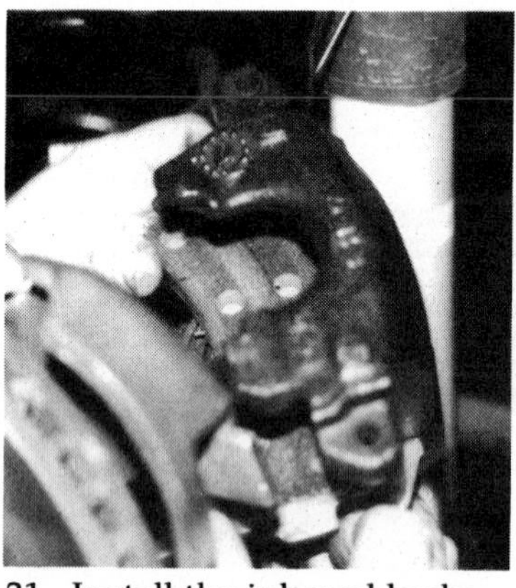
31. Install the inboard brake pad in the caliper as described earlier.

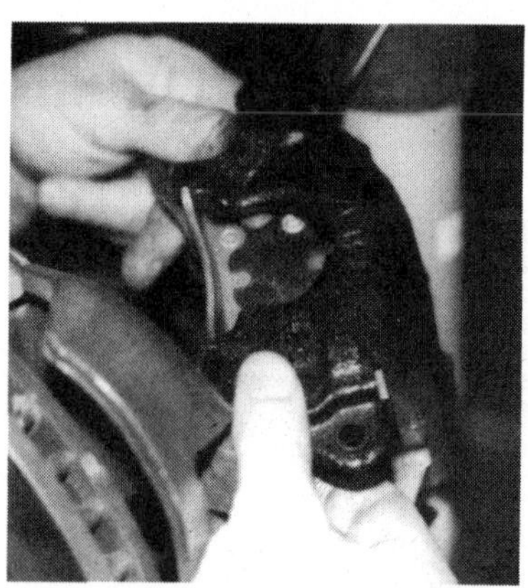
32. Install the outboard pad in the caliper as described earlier.

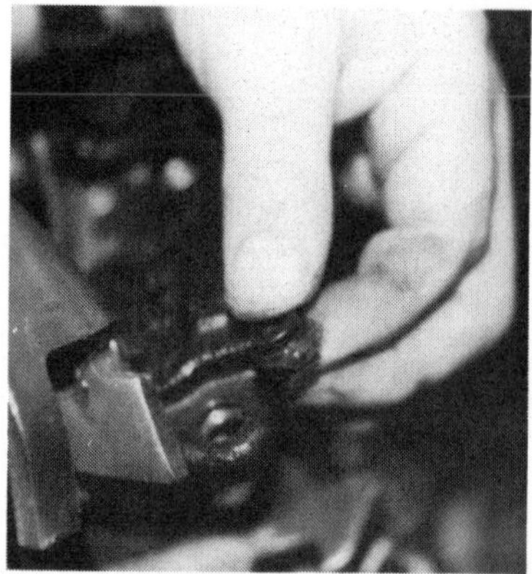
33. Lubricate the new bushings with brake grease and install them into the mounting bolt holes in the caliper body.

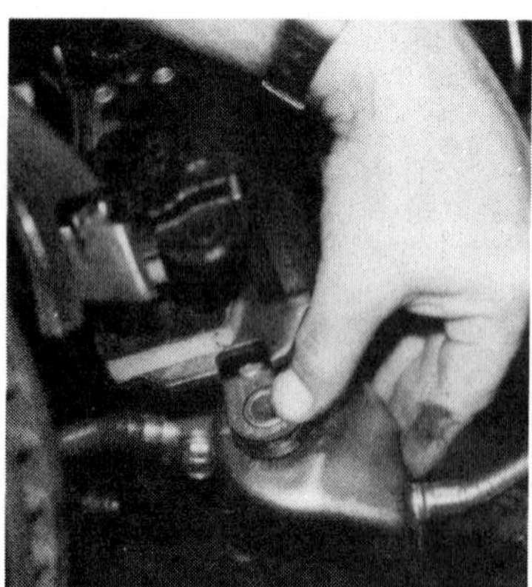
34. Lubricate the outside of the new sleeves with brake grease and install them into the mounting bolt holes.

35. Position the caliper over the rotor, then lubricate and install the caliper mounting bolts. Torque them to 30 to 45 ft-lb (41 to 61 Nm).

36. Grease the new anti-friction washer and lever seal, and install them on the end of the actuator screw.

GENERAL MOTORS SINGLE-PISTON REAR CALIPER SERVICE

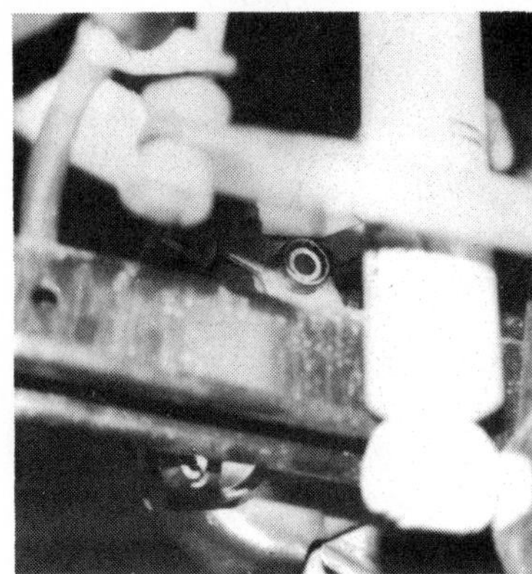

37. Install the parking brake lever, and while holding it away from the stop, tighten the retaining nut to 30 to 40 ft-lb (41 to 54 Nm).

38. Connect the parking brake cable to the lever, and install the return spring. Adjust the cables as described in Chapter 9.

39. If the caliper was rebuilt, connect the brake hose to the caliper fluid inlet. Torque the bolt to 22 to 33 ft-lb (30 to 45 Nm). Bleed the brakes as described in Chapter 3.

Chapter

8

Brake Drum, Rotor, and Shoe Machining

Brake drums and rotors provide the surfaces that the shoes and pads rub against to create the friction that stops the car. Although drums and rotors are not considered disposable parts like the brake linings that help them do their job, they are subjected to great deal of heat and friction. As a result, drums and rotors suffer several types of wear, damage, and distortion. If these problems are minor, a drum or rotor can be machined to restore the friction surface, if the problems are severe, the drum or rotor must be replaced.

In the case of drum brakes, not only is the condition of the drum friction surface important, but the fit of the brake shoes in the drum is a major consideration. The shoe linings must have the proper arc to match the curvature of the drum for the brakes to work at maximum efficiency. If the fit of the brake shoes in the drum is not correct, the linings can be machined, or arced, to fix the problem.

The bulk of this chapter describes the procedures used to inspect and machine brake drums and rotors. Instructions are also provided on how to replace brake drums that are riveted or swaged onto their hub. The final section of the chapter explains how to check the fit of the brake shoes in the drum, and how to arc the shoes to improve their fit.

DRUM AND ROTOR SERVICE

As discussed in the *Classroom Manual*, the size, shape, and finish of drum and rotor friction surfaces are critical to safe and efficient braking. If these factors are not maintained within specific tolerances, brake shoe or pad wear increases, braking force becomes erratic, the brake pedal pulsates, and brake noise and vibration occur. In extreme cases, a drum or rotor can fail entirely.

To identify problems and determine if additional service is required, you should inspect the drums and rotors: whenever you service the brakes, if there is noise and vibration from the brakes, or if a road test reveals a pulsating brake pedal. A brake drum must be removed from the vehicle (see Chapter 6) before it can be inspected, but you can inspect the outboard friction surface of a brake rotor simply by removing the wheel. To thoroughly inspect a rotor, however, you may have to remove the brake caliper (see Chapter 7) to get a clear view of the inboard friction surface.

Some shops bypass the inspection and automatically turn drums and rotors any time they replace the shoes or pads. This is generally a good policy on drum brakes because the drums are already off the car, and drums are subject

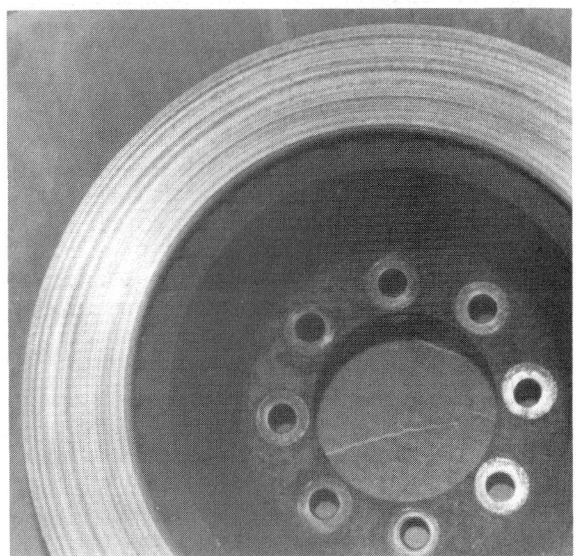

Figure 8-1. Inspect for scores and grooves in the friction surface.

to a number of problems that can be difficult to spot until a drum is machined in a lathe. In addition, today's lightly loaded rear drum brakes are serviced at infrequent intervals, which makes a complete brake job, including turning the drums, a good way to go.

On disc brakes, however, automatically turning the rotors is not a good policy. All of the problems that require rotors to be machined can be identified by an inspection. In addition, the extra work required to remove and reinstall the rotors increases the cost of the job, while any metal removed from a rotor that is in serviceable condition needlessly shortens its service life.

Brake drum and rotor inspections consist of two parts, a visual inspection followed by one or more careful measurements. Once you have completed a thorough inspection, you will know if the drum or rotor is in serviceable condition, must be machined to restore the friction surface, or is beyond saving and must be replaced.

VISUAL DRUM AND ROTOR INSPECTION

The visual inspection checks a drum or rotor for obvious defects such as scoring, cracking, heat checking, and hard spots. Before you start the inspection, wipe the drum or rotor friction surface clean with a shop towel soaked in brake cleaner; this will make it easier to spot any problems. If you are inspecting a rotor that has rust buildup on its friction surfaces as a result of disuse, sand away the rust with medium-grit sandpaper or abrasive cloth. If the rust buildup

Figure 8-2. Inspect the entire surface of a drum or rotor for cracks.

cannot be easily removed by hand, resurface the rotor as described later in the chapter.

During the visual inspection, if you find any problem that requires the drum or rotor to be machined, immediately measure the drum or rotor as described later in the chapter. If the visual inspection reveals no problems, but there are other symptoms of brake trouble such as noise, vibration, or brake pedal pulsation, measure the drum or rotor anyway. If a visual inspection reveals no problems, and there are no other symptoms of brake trouble, the drum or rotor is most likely in serviceable condition. However, you should *always* measure the drum inside diameter, or rotor thickness, to make sure it is within legal limits.

Visual Inspection Procedure

First, inspect the drum or rotor friction surface for scoring and grooves, figure 8-1. If a badly scored or grooved friction surface is not refinished, lining wear will increase, and the brakes may be noisy. To determine the depth of any scores or grooves, use a micrometer with a pointed anvil designed for this purpose. Scoring or grooves up to about .010″ (.25 mm) deep are not usually harmful to brake performance, and the drum or rotor need not be turned. If the damage is worse than this, machine the drum or rotor.

Second, inspect the drum or rotor for cracks, figure 8-2. A cracked drum or rotor is unsafe and may fail completely if left in service. Cracks can occur anywhere, but drums usually crack near the bolt circle or web, or at the open edge of the friction surface. Rotors usually crack at the outer edge of their friction surface. Do not confuse small surface cracks with cracks that reach deeply into the structure of the drum or

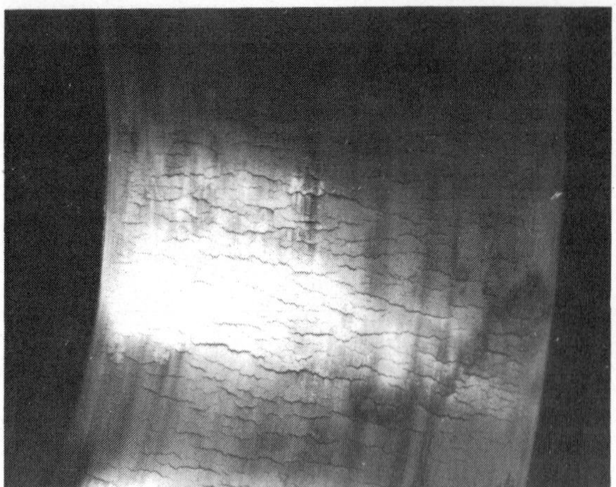
Figure 8-3. Inspect the friction surface for heat checking.

Figure 8-4. Inspect the friction surface for hard spots.

rotor. If any cracks are visible, replace the drum or rotor.

Third, inspect the drum or rotor for heat checking, figure 8-3. Heat checking creates a rough friction surface that increases lining wear and may cause a slight pedal pulsation or brake noise. Heat checking appears as many small, interlaced, cracks on the friction surface. If the heat checking is minor and the drum or rotor checks good in other respects, machine the drum or rotor. If heat checking is widespread, and there are other problems, it may be better to replace the drum or rotor.

Finally, inspect the drum or rotor for hard spots, figure 8-4. These are round, bluish/gold, glassy appearing areas on the friction surface

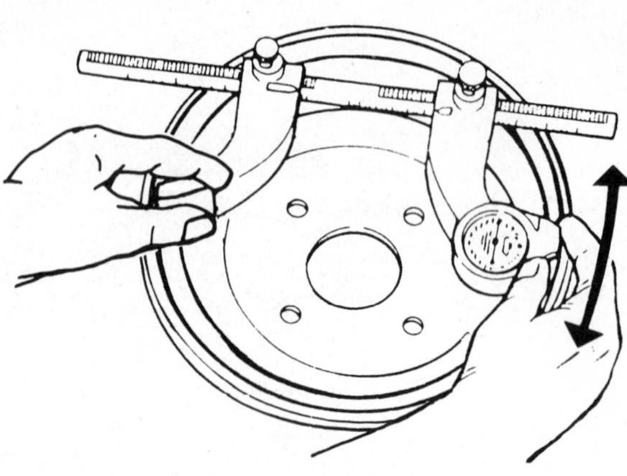
Figure 8-5. Use a drum micrometer to check the inside diameter of the brake drum.

that can increase lining wear, or cause brake chatter and a pulsating pedal. It is possible to machine hard spots flush with the friction surface, however, most hard spots cannot be removed entirely because they penetrate too far into the metal. Machining down hard spots requires special equipment and is time consuming, so most manufacturers today recommend that drums and rotors with hard spots be replaced.

BRAKE DRUM MEASUREMENT

Brake drums are measured to identify wear and distortion that is not visually apparent. When drums wear, they become oversize, tapered, or barrel-shaped. Distorted drums become bell-mouthed, out-of-round, or eccentric. You can measure most of these problems using a drum micrometer or an inside micrometer. However, some forms of drum wear and distortion cannot be identified until the drum is turned in a brake lathe.

Drum Inside Diameter

Anytime a brake drum is removed from the car you should measure its inside diameter to check for wear. To do this, note the discard diameter stamped or cast into the drum, then position the drum so the open side is facing up. On early brake drums that do not have a discard diameter marked on them, consult a shop manual for the proper dimension. Next, adjust a drum micrometer to the nominal drum diameter as described in Chapter 1, and insert the micrometer into the brake drum, figure 8-5. Hold the anvil steady against the friction surface, and slide the dial end of the micrometer back and forth until the highest reading is obtained on the dial scale. Repeat this process at two or

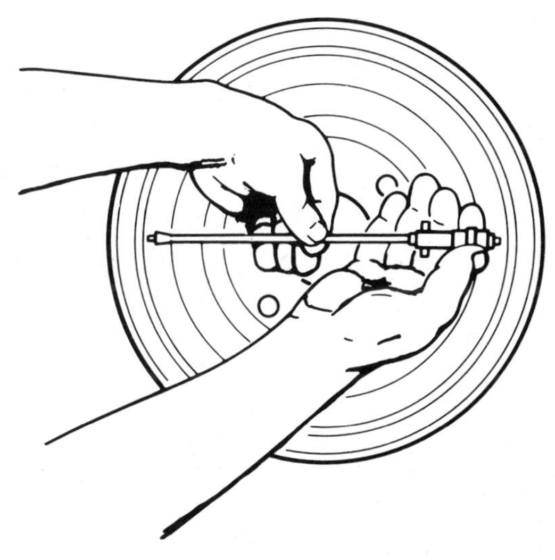

Figure 8-6. Use an inside micrometer to check a drum for taper wear, barrel wear, or bellmouth distortion.

three locations around the drum, then compare the largest reading to the discard diameter.

On most cars, if the inside diameter is not at least .030″ (.75 mm) smaller than the drum discard diameter, replace the drum. The amount of additional metal required to allow for wear in service varies depending on the manufacturer; check the shop manual of the car you are servicing for the exact value. If the drum needs to be turned, there must be sufficient metal remaining so the inside diameter will be at least .030″ (.75 mm) smaller than the discard diameter *after* the drum is turned. Ford Motor Company drums, and those on some other cars, are marked differently. Replace one of these drums whenever its inside diameter exceeds the "maximum diameter" stamped or cast into the drum. If you are unsure of what the measurements on a drum mean, consult a shop manual to be sure.

Drum Taper Wear, Barrel Wear, and Bellmouth Distortion

Taper wear, barrel wear, and bellmouth distortion are problems that cause variations in brake drum diameter between the open and closed edges of the friction surface. A drum with taper wear has a larger diameter at the closed edge than at the open edge. A drum with barrel wear has a larger diameter at the center than at either edge. A drum with bellmouth distortion has a larger diameter at the open edge than at the closed edge. If new brake shoes are installed into a drum with any of these problems, a spongy brake pedal, brake fade caused by poor lining-to-drum contact, and uneven lining wear will result.

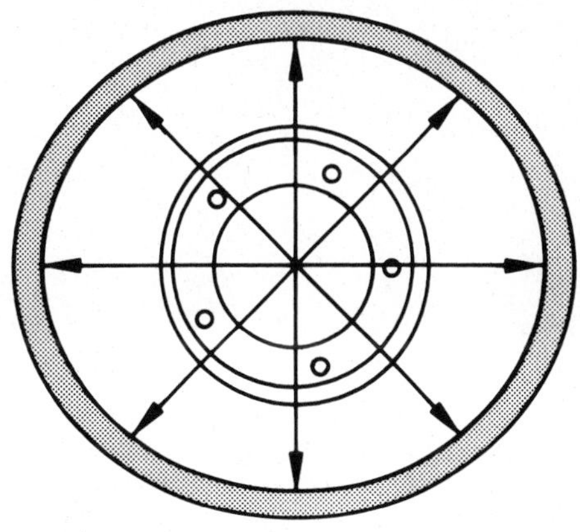

Figure 8-7. Measure the drum inside diameter at four places around the drum to check for out-of-roundness.

Taper wear can sometimes cause a spongy brake pedal, but barrel wear and bellmouth distortion have no symptoms that are obvious to the driver. These problems can sometimes be spotted by ridges or lips worn into the drum friction surface; other times, unusual wear patterns on the brake linings will reveal the problem. Generally, however, these types of wear and distortion are discovered when the drum is turned on a lathe and the tool bit contacts the drum only at one, the other, or both edges.

You can also identify these problems by measuring the drum inside diameter at several points across the friction surface. A drum micrometer cannot reach deeply enough into the drum to make these measurements, so you must use an inside micrometer instead. Position the micrometer as shown in figure 8-6 and take three measurements, one at the open edge of the drum, one at the center of the friction surface, and one at the closed edge of the drum. If the highest and lowest of these measurements vary by more than .006″ (.15 mm), machine the drum.

Out-Of-Round Drum Distortion

The diameter of an out-of-round drum varies when measured at several points around its circumference. This causes a pulsating brake pedal, brake vibration, and sometimes, grabby, erratic braking. To check for an out-of-round drum, use a brake drum micrometer to measure the drum inside diameter at four locations 45 degrees apart from one another, figure 8-7. If the highest and lowest measurements vary by more than .006″ (.15 mm), machine the drum.

Figure 8-8. Measure the thickness, parallelism, and taper variation of a rotor with an outside micrometer.

Eccentric Drum Distortion

Eccentric brake drum distortion exists when the geometric center of the friction surface is different from that of the hub. This makes the drum rotate with a cam-like motion that causes the shoe contact pads on the backing plate to wear, and creates noise whenever the brakes are applied. Eccentric drum distortion cannot be detected visually or with common measuring tools. This condition is usually identified when a drum is turned on a lathe and the tool bit contacts the friction surface on only one side of the drum.

BRAKE ROTOR MEASUREMENT

Like brake drums, brake rotors are measured to identify wear and distortion that is not visually apparent. When rotors wear, they become too thin or have taper variation. Distorted rotors have lateral runout, or the two friction surfaces are no longer parallel. You can measure all types of rotor wear and distortion using an outside micrometer and a dial indicator. If your measurements reveal that a Ford rotor must be turned, you will also need a special depth micrometer to check whether sufficient metal remains on the inner friction surface for the rotor to be machined safely.

Brake rotor dimensions are held to much tighter tolerances than drum dimensions. This makes the proper use of accurate measuring tools a critical part of rotor inspection. If you have any questions on how to use and read a micrometer, see Chapter 1 for instructions.

Rotor Thickness

Anytime you inspect a brake rotor, measure its thickness to check for wear. On some cars, it may be necessary to remove the caliper in order to do this. To check a rotor for wear, note the discard dimension stamped or cast into the rotor. On early brake rotors that do not have a discard dimension marked on them, consult a shop manual for the proper measurement. Next, position an outside micrometer one inch (25 mm) in from the outer edge of the friction surface, and measure the rotor thickness, figure 8-8. Compare the measured thickness to the discard dimension.

If the rotor thickness is not at least .015″ to .030″ (.40 to .75 mm) larger than the rotor discard dimension, replace the rotor. The amount of additional metal required to allow for wear in service varies depending on the manufacturer; check the shop manual of the car you are servicing for the exact value. If the rotor needs to be turned, there must be sufficient metal remaining so the thickness will be at least .015″ to .030″ (.40 to .75 mm) larger than the discard dimension *after* the rotor is turned. Ford Motor Company rotors, and those on some other cars, are marked differently. Replace one of these rotors whenever its thickness is less than the "minimum thickness" stamped or cast into the outside of the rotor. If you are unsure of what the measurements on a rotor mean, consult a shop manual to be sure.

Rotor Taper Variation

A rotor with taper variation has a different thickness at the outer edge of its friction surface than at the inner edge. If the variation is too great, new brake pads will not contact the rotor squarely, and the caliper pistons may bind in their bores.

To check for taper variation, use an outside micrometer with a deep frame to measure the rotor thickness at the outer edge just below the ridge, then at the inner edge of the area swept by the brake pads. If a micrometer with a deep frame is unavailable, use a conventional outside micrometer to make the outer measurement as described above, then measure inward as far as the micrometer can reach. In either case, subtract the smaller measurement from the larger one to obtain the taper variation. Repeat these measurements at four points around the rotor. If the variation is greater than .003″ (.08 mm) at any point, machine the rotor.

Rotor Lateral Runout

Lateral runout is a side-to-side movement of the rotor as it turns. Excessive runout can cause brake pedal pulsations, vibration during braking, and increased brake pedal travel from too much pad knockback. For the best braking performance, the lateral runout should be less than .003″ (.08 mm). However, depending on the

Figure 8-9. Check the lateral runout of a rotor with a dial indicator.

Figure 8-10. Turn the outside of the dial to zero the dial indicator.

vehicle manufacturer, anywhere between .002″ and .008″ (.05 and .20 mm) runout is allowed. It is only necessary to check lateral runout on one side of the rotor; runout never varies significantly between the two sides.

You check for lateral runout while the rotor is mounted on the car and rotating on the wheel bearings. When making this check, it is very important not to mistake bearing play for lateral runout. Adjustable wheel bearings can be tightened to eliminate play as a factor; non-adjustable wheel bearings cannot. As a result, a different procedure is required to check rotor runout with each type of wheel bearing.

To check the lateral runout of a rotor mounted on adjustable wheel bearings:

1. Raise and properly support the car so the wheel with the rotor to be checked hangs free, then remove the wheel.
2. If the car has floating rotors, install two lug nuts to hold the rotor tightly on the hub.

Figure 8-11. When the rotor is rotated, the dial needle shows the lateral runout.

3. Pry the brake pads back so they do not drag against the rotor. Use one of the methods described in Chapter 7.
4. Tighten the wheel bearing adjusting nut with a wrench to a snug fit. There should be no play, but the rotor should still turn without binding. See Chapter 11 for information on adjusting wheel bearings.
5. Mount a dial indicator on the suspension, and position the plunger so it contacts the rotor at a 90 degree angle about one inch (25 mm) from the outer edge, figure 8-9.
6. Rotate the rotor until the lowest reading shows on the indicator dial, then zero the dial, figure 8-10.
7. Rotate the rotor until the highest reading shows on the dial; this is the lateral runout, figure 8-11.

Because the play cannot be adjusted out of non-adjustable wheel bearings, it must be subtracted from the final reading on the dial indicator to determine the true runout. To check the lateral runout of a rotor mounted on non-adjustable wheel bearings:

1. Raise and properly support the car so the wheel with the rotor to be checked hangs free, then remove the wheel.
2. If the car has floating rotors, install two lug nuts to hold the rotor tightly on the hub.
3. Pry the brake pads back so they do not drag against the rotor. Use one of the methods described in Chapter 7.
4. Mount a dial indicator on the suspension, and position the plunger so it contacts the rotor at a 90-degree angle about one inch (25 mm) from the outer edge, figure 8-9.
5. Push inward on the hub with moderate pressure, and rotate the rotor until the lowest reading shows on the indicator dial; set the dial to zero, figure 8-10.

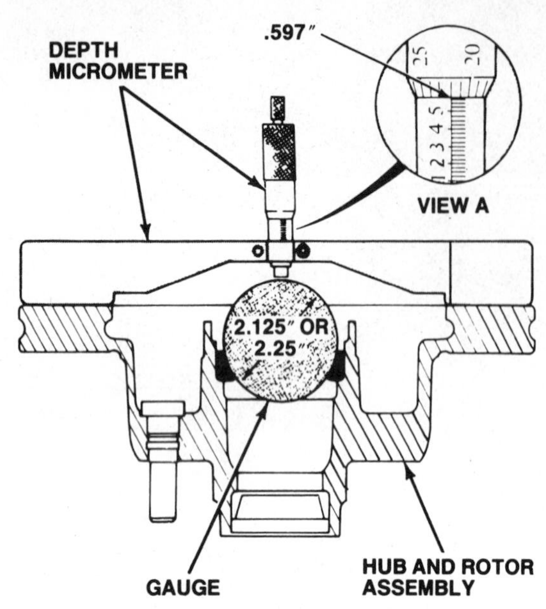

Figure 8-12. A gauge ball is used to measure the inner brake surface on Ford fixed rotors.

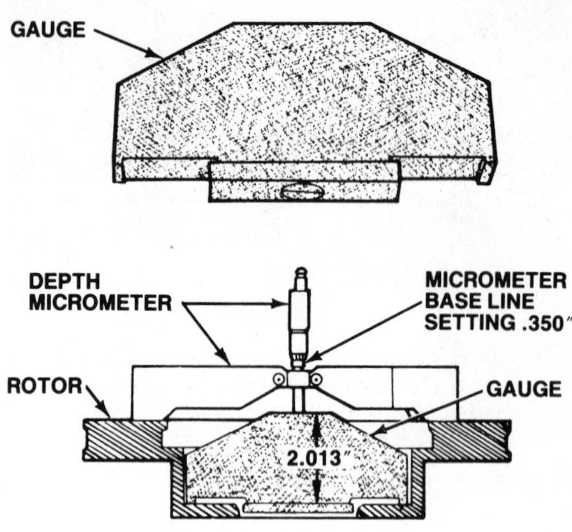

Figure 8-13. A gauge template is used to measure the inner brake surface on Ford floating rotors.

6. Pull outward on the hub with moderate pressure, and record the measurement shown on the indicator dial; this is the bearing end play.
7. Rotate the rotor until the highest reading shows on the indicator dial, then pull outward on the hub with moderate pressure; record the reading shown on the indicator dial.
8. Subtract the bearing end play reading in Step 6 from the total reading in Step 7; the result is the rotor lateral runout.

Rotor Lack of Parallelism

A rotor that lacks parallelism varies in thickness at different places around its friction surfaces. Lack of parallelism is the single biggest cause of brake pedal pulsation, and also causes braking vibration. For these reasons parallelism is a critical form of rotor distortion that requires very careful measurement.

To check for variations in parallelism, use an outside micrometer to measure the rotor thickness at 6 to 12 equally spaced points around the friction surface. Make all of the measurements the same distance in from the outer edge of the rotor so taper variation will not affect the measurements. If the thickness variation between any two points is greater than .0005″ (.013 mm), and there is noticeable brake pedal pulsation, machine the rotor.

Ford Inner Friction Surface Measurement

Once you have removed a Ford brake rotor for machining, you must measure the inner friction surface to determine how much, if any, metal can safely be removed. This is done using a special depth micrometer and gauge available from Ford. The gauge for fixed rotors is a large steel ball, for floating rotors it is a pyramid-shaped template.

To measure the inner friction surface on a Ford brake rotor:

1. Make sure the rotor inner friction surface is free of rust and dirt. On fixed rotors, remove the seal and inner wheel bearing, then clean the bearing race of grease. On floating rotors, make sure the inside of the rotor hat is free of rust and dirt.
2. Carefully place the gauge ball, figure 8-12, or template, figure 8-13, into position in the rotor. Do not drop the ball into place or the bearing race may be damaged.
3. Set the depth micrometer to the baseline setting given in the factory shop manual, and place it on the inner friction surface of the rotor so it is centered over the gauge.
4. Turn the micrometer down until it contacts the top of the gauge. Subtract the reading on the micrometer from the baseline setting; the result is the maximum amount of metal that can be removed from the inner friction surface.
5. If the micrometer touches the gauge at the baseline setting, or must be retracted to allow both legs of the micrometer to rest flat on the rotor friction surface, replace the rotor.

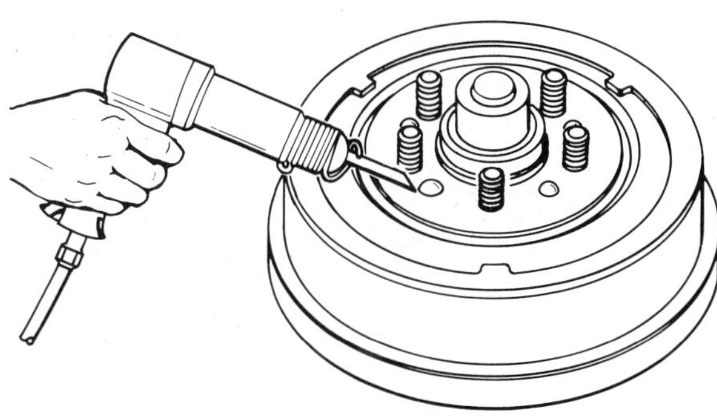

Figure 8-14. Using an air chisel helps remove rivets that hold a drum and hub together.

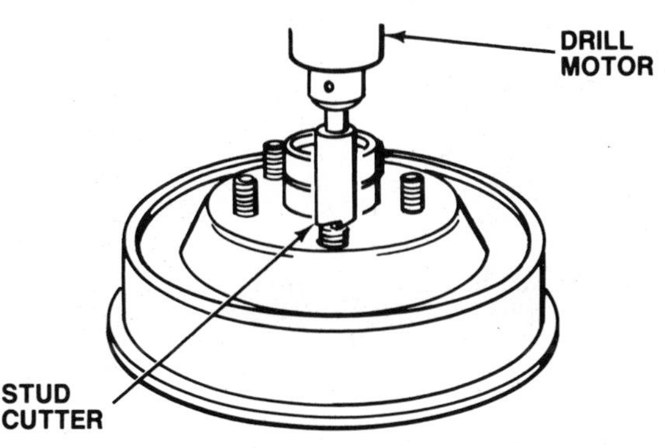

Figure 8-15. Using a cutter to remove the swaged portion of a wheel stud.

DRUM REPLACEMENT

A few floating brake drums are attached to their hubs with rivets or swaged studs. This assembly is generally machined in the same manner as a fixed drum. However, if the drum is badly damaged, it can be removed from the hub and replaced separately. Once the old drum and hub are separated, the new drum generally does not have to be secured in place. Instead, it is machined and installed like a floating drum.

The exact removal procedure varies depending on whether the drum is riveted or swaged in place. To replace a riveted drum, cut off the heads of the rivets with an air chisel, figure 8-14. Drill a small hole through the remaining portion of each rivet, then drive the rivets out with a hammer and punch to separate the drum and hub.

To replace a drum that is held in place by swaged studs you will need a special cutter designed to machine away the swaged portion of the stud shoulders. Chuck the cutter in a drill motor, position it over the stud, start the drill motor, and press the cutter down firmly for about 10 seconds, figure 8-15. Remove the cutter and determine if all of the swaged metal has been removed. Repeat this process as needed until the swaged portion on all of the studs has been cut away, then separate the drum and hub.

DRUM AND ROTOR REFINISHING

As discussed in the *Classroom Manual*, there are three methods used to machine or refinish brake drums and rotors. *Turning* removes metal from drum and rotor friction surfaces with a steel tool bit to repair most forms of wear, damage, and distortion. *Grinding* removes metal from drum friction surfaces with an abrasive stone wheel to machine hard spots level with the rest of the friction surface. *Resurfacing* removes very small amounts of metal from rotor friction surfaces with a spinning abrasive disc, and is commonly done to create a non-directional finish that speeds the bedding-in of new brake pads. Resurfacing can also be done to remove rust, brake lining deposits, and minor rotor damage. Except when a rotor is turned on the car, all of these operations are performed with the drum or rotor off the car and mounted in a brake lathe.

Metal Removal Considerations

Regardless of the procedure you use to machine a drum or rotor, you should only remove the minimum amount of metal necessary to restore the friction surface. This helps ensure the longest possible service life for the drum or rotor. On rotors used with sliding or floating calipers, you can machine different amounts of metal from each friction surface; however, you must machine rotors used with fixed calipers equally on both sides.

Never machine a drum or rotor unless you also machine the drum or rotor on the other end of the same axle. This keeps braking force and fade resistance equal from side to side, and prevents brake pull. This is especially important with drum brakes where the inside diameters of drums on the same axle should be kept within .010″ to .020″ (.25 to .51 mm) of each other. To help keep the drum diameters equal, machine the more badly worn drum first, then machine the other drum to match. In the same manner, machine a new drum to match the diameter of an old drum on the same axle.

Machining New Drums and Rotors

Because they become distorted through improper storage and rough handling, you should

Figure 8-16. Wipe the arbor shaft clean before mounting a drum or rotor on a brake lathe.

Figure 8-17. Dirt and metal chips on the arbor shaft and adapters can greatly affect the accuracy of a lathe.

always turn a new drum or rotor before you install it. A light cut is usually all that is needed to ensure that the friction surfaces are true and smooth. Before you machine a new drum or rotor, wipe the friction surfaces with a shop towel soaked in brake cleaner to remove any anti-rust coating applied by the manufacturer. If the coating is not removed, it can be driven into the friction surface by the brake lathe tool bit. If this happens, the first time the drum or rotor gets hot in service, the coating will come out of the pores in the metal and contaminate the brake linings.

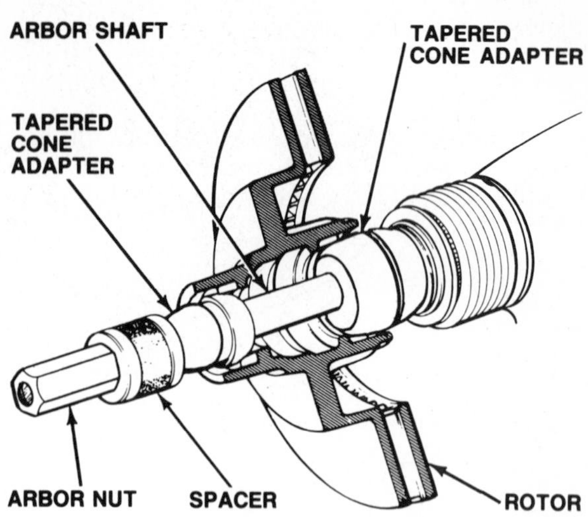

Figure 8-18. A typical mounting for a fixed drum or rotor.

BRAKE LATHE OPERATION

Once you inspect a brake drum or rotor and determine it needs to be machined, the next step is to mount and center the drum or rotor in a brake lathe. If you are servicing drum brakes, you removed the drums from the car for inspection. If you are servicing disc brakes, remove the rotors at this time unless you are using an on-car lathe to machine them.

To remove a brake rotor, first remove the brake caliper as described in Chapter 7. Then, you can remove a floating rotor by removing any attaching hardware, and sliding it off the hub. Remove a fixed rotor in the same manner as a fixed drum, and remember that whenever you machine a fixed drum or rotor you must also service the wheel bearings. Details on fixed drum and rotor removal, as well as wheel bearing service, are contained in Chapter 11.

Lathe Care

Before you mount a drum or rotor on a brake lathe, make sure the arbor shaft and adapters are wiped clean, figure 8-16. Any dirt, metal chips, or other contaminants, figure 8-17, will affect the alignment of the drum or rotor on the lathe, and the quality of the final cut. Always handle lathe parts carefully — scratches, nicks, and dents on these parts will also adversely affect the accuracy of the machining operation.

To ensure a good friction surface finish, always use sharp tool bits when turning a drum or rotor. Also, make sure the bits are rounded and not pointed. This is especially important on drum brakes where a pointed bit can cut a "thread" into the friction surface. This can

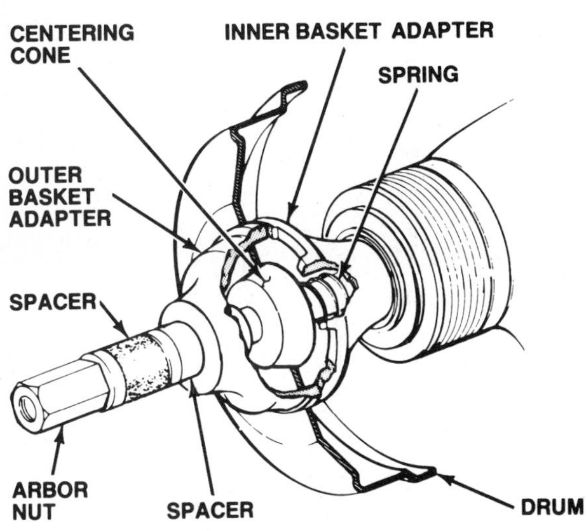

Figure 8-19. A typical mounting for a floating drum or rotor.

Figure 8-20. Check the bearing races before you install a drum or rotor on the lathe.

cause the linings to "thread" outward when the brakes are applied. The result will be shoe misalignment with the drum, less effective braking, increased lining wear, and a clicking noise as the shoes snap back into place when the brakes are released.

Mounting Drums and Rotors

Fixed drums and rotors are mounted on the brake lathe arbor shaft differently than floating drums and rotors. Fixed drums and rotors mount using tapered or radiused cone adapters, figure 8-18, that fit into the wheel bearing races in the drum or rotor hub. Floating drums and rotors mount using two basket adapters, figure 8-19, that press against the inner and outer sides of the drum web or rotor hat; the drum or rotor is centered by a spring-loaded cone, that fits into the center hole.

Figure 8-21. Cone adapters used to mount fixed drums and rotors on a brake lathe.

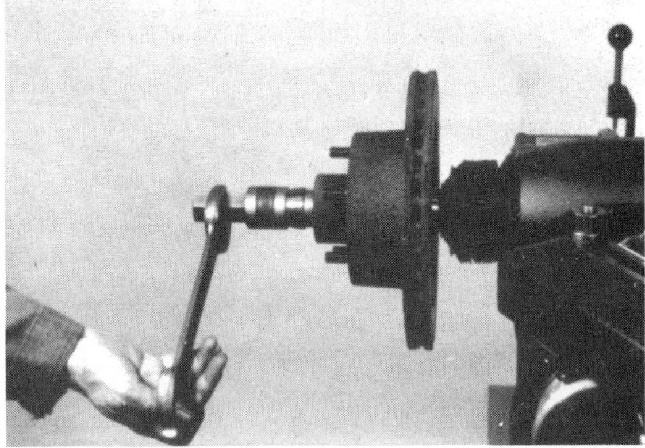

Figure 8-22. Tighten the arbor nut to hold the drum or rotor onto the arbor shaft.

To mount a fixed drum or rotor on a brake lathe:

1. Use a shop towel to wipe the bearing races clean of all dirt and grease.
2. Inspect the races for fit and damage:
   a. If you can turn any race with your fingers, figure 8-20, replace the drum or rotor along with the bearings.
   b. Replace any damaged or worn races as described in Chapter 11.
3. Select the cone adapters that fit into the bearing races, figure 8-21. Generally, a small cone fits the outer race and a large cone fits the inner race.
4. Slide the cone adapter for the inner bearing race onto the arbor shaft so the tapered side faces out.
5. Slide the drum or rotor onto the arbor shaft so that the inner bearing race fits over the adapter.
6. Slide the cone adapter for the outer bearing race onto the arbor shaft, tapered side in, and install it into the outer bearing race.
7. Place spacers between the outer cone adapter and the arbor nut as needed, figure 8-22, then securely tighten the nut so the drum or rotor is held solidly in position.

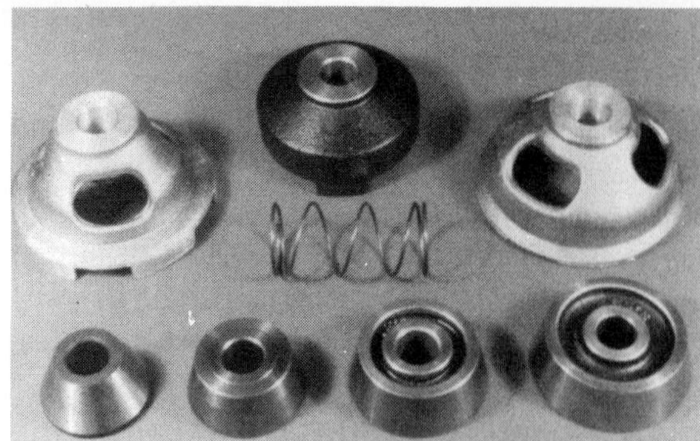

Figure 8-23. Basket adapters and centering cones used to mount floating drums and rotors on a brake lathe.

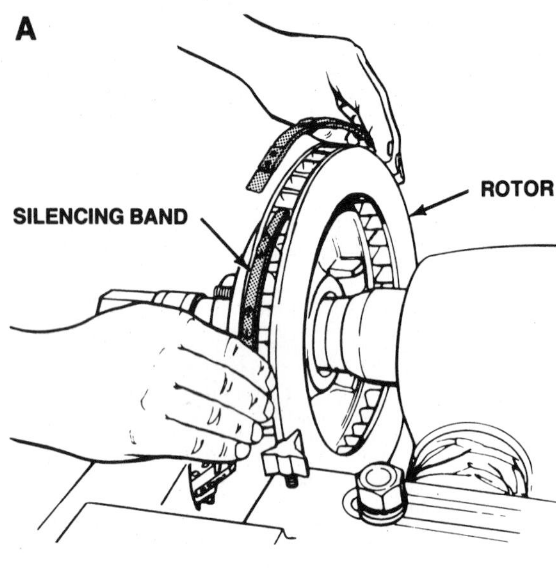

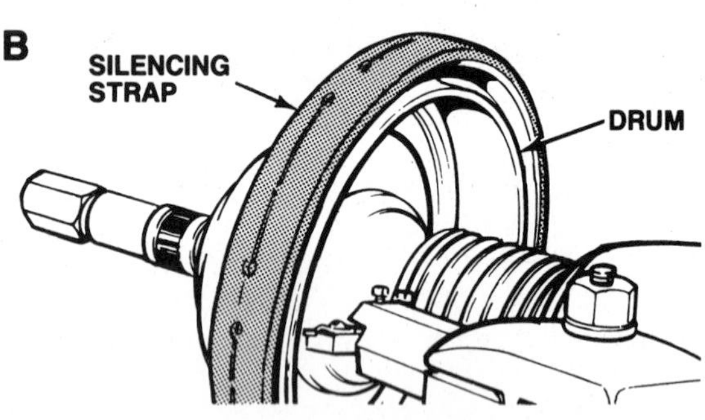

Figure 8-24. A silencing band or strap improves the quality of the machined surface.

To mount a floating drum or rotor on a brake lathe:

1. Clean the inside, outside, and center hole of the drum web or rotor hat with a wire brush so it is free of rust and dirt.
2. Scrape the inside, and lightly file the outside, of the drum web or rotor hat so both surfaces are clean and free of high spots.

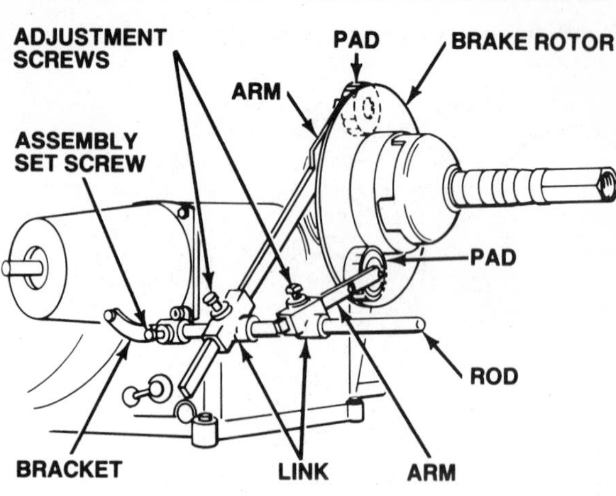

Figure 8-25. Use a damper to quell vibrations on solid rotors.

3. Select the centering cone that fits the center hole in the drum or rotor, figure 8-23.
4. Slide one of the basket adapters onto the arbor shaft with the open side facing out.
5. Slide the spring onto the arbor shaft.
6. Slide the centering cone onto the arbor shaft with the tapered side facing out.
7. Install the drum or rotor onto the arbor shaft with the outside of the drum web or rotor hat facing out.
8. Install the other basket adapter onto the arbor shaft with the open side pressing against the drum or rotor.
9. Place spacers between the outer basket adapter and the arbor nut as needed, figure 8-22, then securely tighten the nut so the drum or rotor is held solidly in position.

Silencing bands and straps

After the brake drum or rotor is mounted on the lathe, install a silencing band or strap around the outer edge. The band or strap prevents vibrations during machining; without it, the cutting tool will chatter on the friction surface, making it impossible to produce a smooth finish. Brake rotors usually use a rubber silencing band, figure 8-24A, while brake drums use a wider, rubber strap, figure 8-24B. To install the band or strap, stretch it around the rotor or drum, and secure it so it remains tightly in place.

Solid brake rotors, or vented rotors less than ½ inch (13 mm) thick, may be too thin for a silencing band to remain in place. In these cases, a special damper like that shown figure 8-25 can be used. To install the damper, bolt it to the lathe and position the adjustable arms so the pads contact the rotor on opposite sides, about 180 degrees apart.

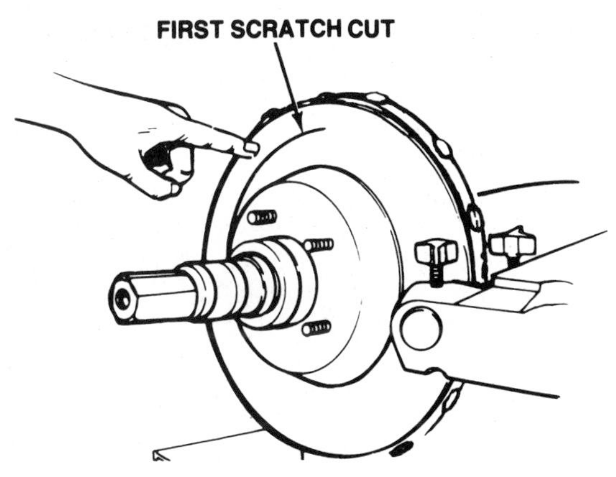

Figure 8-26. A scratch cut in only part of the friction surface may indicate a mounting problem or lathe damage.

Centering Drums and Rotors

Once the silencing band or strap is in position, the next step is to make sure the drum or rotor is properly centered. If a drum or rotor is not centered accurately, you can easily cut more runout into it than it had to begin with! Use the procedure below to check the centering of a drum or rotor. When you check a rotor, it is only necessary to make scratch cuts on one of the friction surfaces. Always wear safety glasses when operating a brake lathe to prevent eye injury from flying metal chips.

To check that the drum or rotor is centered:

1. Back the tool bit or bits away from the drum or rotor, and rotate the drum or rotor by hand through at least one full turn to make sure everything clears.
2. Start the lathe, and advance the tool bit until it just touches the friction surface at about its mid-point. This scratch cut should be not more than .001″ (.025 mm) deep.
3. Back the tool bit away from the friction surface and stop the lathe.
   a. If the scratch cut appears all around the friction surface, the drum or rotor is properly mounted and centered, and you can proceed with machining.
   b. If the scratch cut appears only part of the way around the friction surface, figure 8-26, go to step 4.

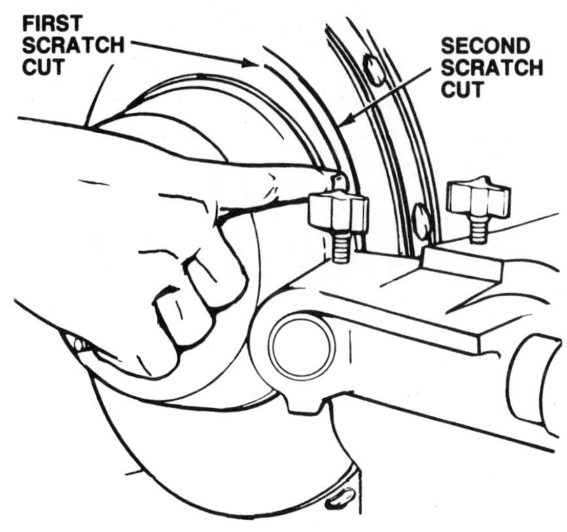

Figure 8-27. If the second scratch cut is next to the first, the drum or rotor is mounted properly and the lathe is okay.

4. Loosen the arbor nut, rotate the drum or rotor 180 degrees on the arbor shaft, and retighten the arbor nut.
5. Start the lathe and advance the tool bit so it again just touches the friction surface, this time at a point next to the first scratch cut. As with the first cut, the second scratch cut should be not more than .001″ (.025 mm) deep.
6. Back the tool bit away from the friction surface and stop the lathe.
   a. If the second scratch cut appears on the friction surface at the same location as the first cut, figure 8-27, the drum is out of round or the rotor has runout. The drum or rotor is properly centered, and you can proceed with machining.
   b. If the second scratch cut appears in the friction surface opposite the first cut, figure 8-28, go to step 7.
7. Remove the drum or rotor from the lathe, and make sure the mounting adapters fit properly. Clean the adapters and arbor shaft, and inspect them for dirt, metal chips, rust, nicks, dents, and burrs.

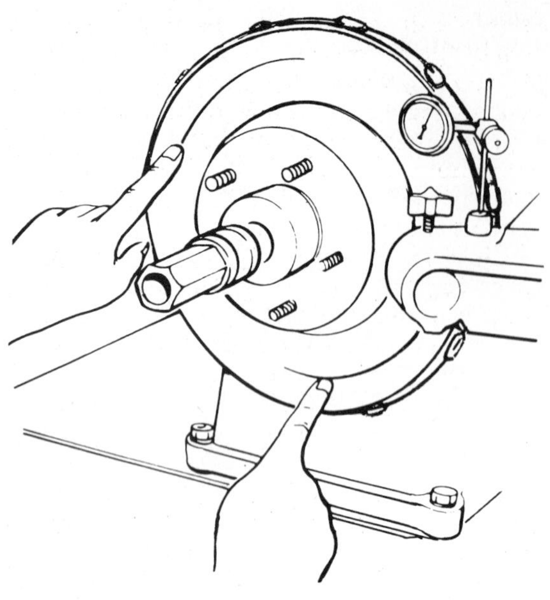

Figure 8-28. If the second scratch cut is opposite the first, check the mounting of the drum or rotor.

8. Remount the drum or rotor, and repeat Steps 2 through 6. If the two scratch cuts are still on opposite sides of the friction surface, have your tool representative check the lathe and arbor shaft for runout or damage.

Lathe Settings

Once you have mounted and centered the brake drum or rotor on the lathe, it is ready to be machined. However, in order to obtain the proper friction surface finish, you must adjust three settings: rpm, crossfeed, and depth of cut. The rpm is the speed at which the drum or rotor rotates on the lathe. The crossfeed is the distance the tool bit moves across the friction surface for each revolution of the drum or rotor. The depth of cut is the amount of metal the tool bit removes from the overall brake drum diameter, or one friction surface of the rotor. There are many brands of brake lathes on the market, but they all have some means of adjusting these three settings.

The drum or rotor rpm is usually constant throughout the machining operation; however, the crossfeed and depth of cut vary depending on whether you are making a rough cut or a finish cut. One or more rough cuts are used to remove major damage, and quickly get the drum or rotor close to finished size. The final cut is a finish cut that produces a smooth finish on the friction surface. Rough cuts are made with a faster crossfeed and a deeper depth of cut than those used for finish cuts. As a general rule, faster rpm, slower crossfeed, and shallower depth of cut, all contribute to a smoother friction surface finish.

Lathe setting recommendations

To recap, the rpm is the speed at which the drum or rotor rotates on the lathe. If you do not want to constantly readjust the lathe speed, you can turn both drums and rotors at 150 rpm; however, rotors end up with a better friction surface finish if they are turned at 200 rpm. The rpm should be the same for both the rough cuts and the finish cut.

The crossfeed is the distance the tool bit moves across the friction surface for each revolution of the drum or rotor. When you turn brake drums, use a crossfeed of .020″ (.50 mm) per revolution for the rough cut, and a much slower .005″ (.15 mm) per revolution for the finish cut. When you turn brake rotors, use a crossfeed of .006″ to .010″ (.15 to .25 mm) per revolution for the rough cut, and a much slower .002″ (.05 mm) per revolution for the finish cut.

The depth of cut is the amount of metal the tool bit removes from the overall brake drum diameter, or one friction surface of the rotor. When you turn drums, use up to a .015″ (.40 mm) depth of cut for the rough cut, and a much shallower .005″ (.15 mm) depth of cut for the finish cut. When you turn rotors that will run against organic or synthetic brake pads, use a .006″ (.15 mm) depth of cut for both the rough cut and the finish cut. When you turn rotors that will run against semi-metallic brake pads, use a .006″ (.15 mm) depth of cut for the rough cut, and a much shallower .002″ (.05 mm) depth of cut for the finish cut. Some technicians prefer to take *only* .002″ (.05 mm) cuts on a rotor that will be used with semi-metallic pads.

TURNING A BRAKE DRUM

Once you have mounted and centered the drum on the lathe, it is ready to be turned. Because every brake lathe operates differently, read the instructions for the particular lathe you are using, and familiarize yourself with the controls. The following procedure is a general guide for machining a drum. The photo sequence shows the steps required to turn a drum on one specific type of brake lathe. Always wear safety glasses when operating a brake lathe to prevent eye injury from flying metal chips.

First, adjust the lathe rpm to the proper setting. Advance the tool bit to the open edge of the drum and machine away the ridge of rust and metal that forms there. Next, move the tool bit to the closed edge of the drum and remove the ridge of rust and metal there as well. As you remove these ridges, note the point on the friction surface where the drum diameter is smallest. Position the tool bit in this location and zero the micrometer scale on the handwheel controlling depth of cut. This is your reference point for setting the depth of cut.

Move the tool bit to the closed edge of the drum (if it is not already there) and adjust the depth of cut for a rough cut as described earlier in the chapter. To a large extent, the depth of cut is determined by the condition of the drum, the greater the damage, the deeper the cut you can take, up to a maximum of .015″ (.40 mm). All drum lathe handwheels that control depth of cut are calibrated to show the amount of metal removed from the overall drum diameter. For example, if you set the handwheel on .010″ (.25 mm), the lathe really cuts .005″ (.13 mm) deep into the friction surface.

Adjust the crossfeed for a rough cut as described earlier in the chapter, and engage the mechanism. The tool bit will automatically move from the closed edge of the drum out to the open edge of the drum. Turn the drum with rough cuts as needed until all defects are removed or nearly removed. Complete the drum turning operation with a finish cut that removes the last traces of defects from the drum and provides a smooth friction surface. After the finish cut, some sources recommend you lightly sand across the friction surface with sandpaper or abrasive cloth. This removes any trace of "thread" left by the tool bit that could increase lining wear and cause a brake noise.

Once you are completely finished machining the drum, wipe the friction surface with a shop towel soaked in brake cleaner. This removes loose metal shavings and other particles that can become embedded in the brake linings and score the drum friction surface. Finally, remeasure the drum inside diameter as described earlier to make sure it is within legal limits.

TURNING A BRAKE DRUM

1. Loosen the boring bar lock nut.

2. Pull the boring bar back to prevent damage to the bar or tool bit when mounting the drum.

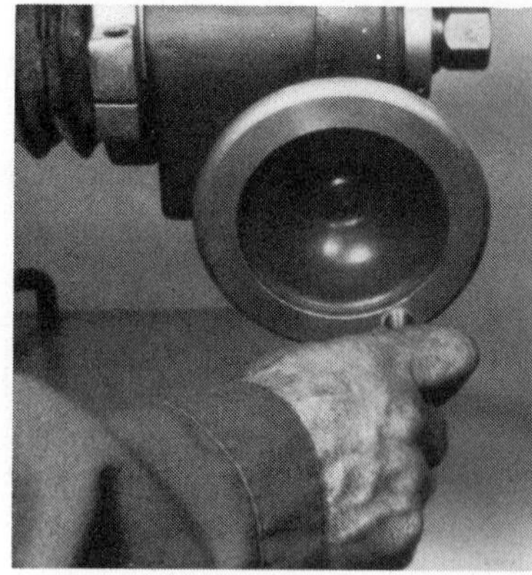

3. Crank the spindle handwheel back fully, then forward approximately five turns.

4. Crank the depth-of-cut handwheel in fully, then back out approximately two turns.

5. On fixed drums, check the bearing races for looseness in the hub. Replace the drum if a race is loose.

6. Install a fixed drum on the arbor shaft using the appropriate adapters as described earlier.

7. Install a floating drum on the arbor shaft using the appropriate adapters as described earlier.

8. Tighten the arbor shaft nut, but do not overtighten it.

9. Install the rubber silencer strap on the drum.

TURNING A BRAKE DRUM

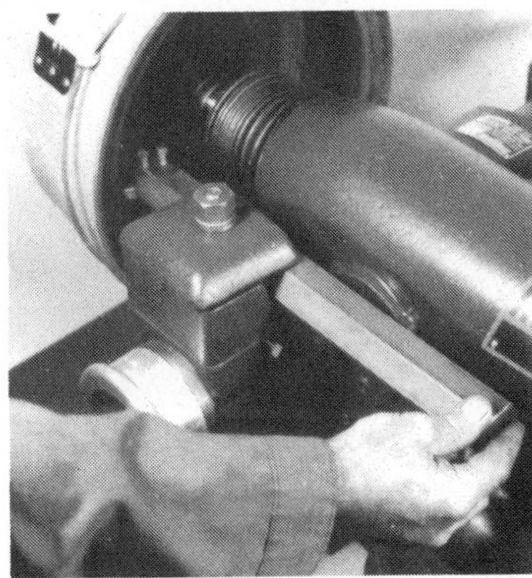

10. Slide the boring bar and pivot the bar holder until the tool bit contacts the friction surface just beyond the rust ridge.

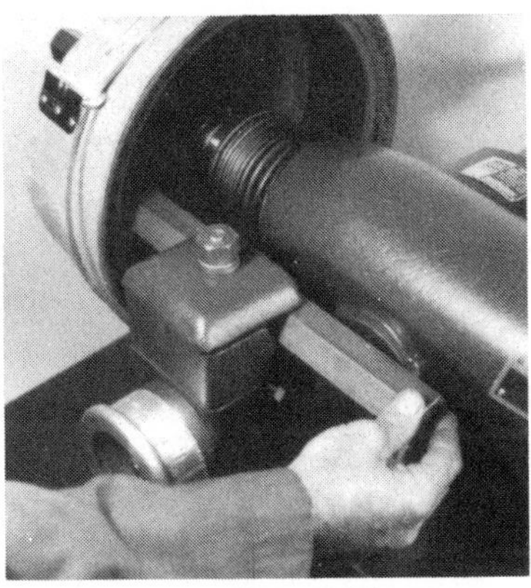

11. Keeping the tool bit in contact with the friction surface, bottom the boring bar in the drum.

12. Tighten the boring bar lock nut.

13. Crank the depth-of-cut handwheel one turn inward to move the tool bit away from the friction surface.

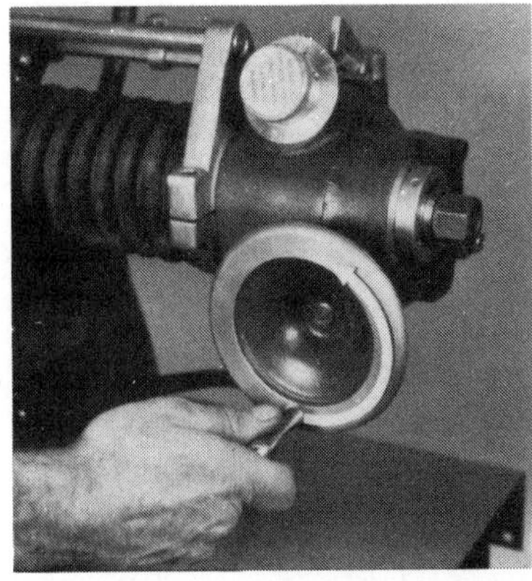

14. Crank the spindle (drum) outward . . .

15. . . . until the tool bit is ⅛ inch (3 mm) away from the open edge of the drum.

16. Set the crossfeed shutoff bushing against its left stop and tighten the knurled lock knob.

17. Make sure the drive belt is in the proper pulley for the correct turning rpm.

18. Start the lathe, and make a pair of scratch cuts to check for drum runout and centering as described earlier.

TURNING A BRAKE DRUM

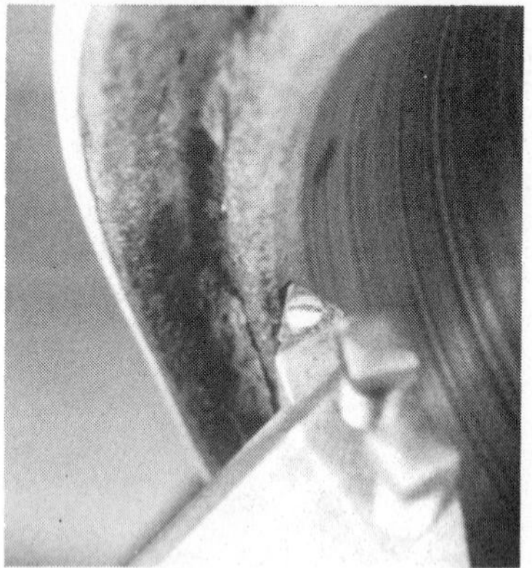

19. Position the spindle so the tool bit is centered over the rust ridge.

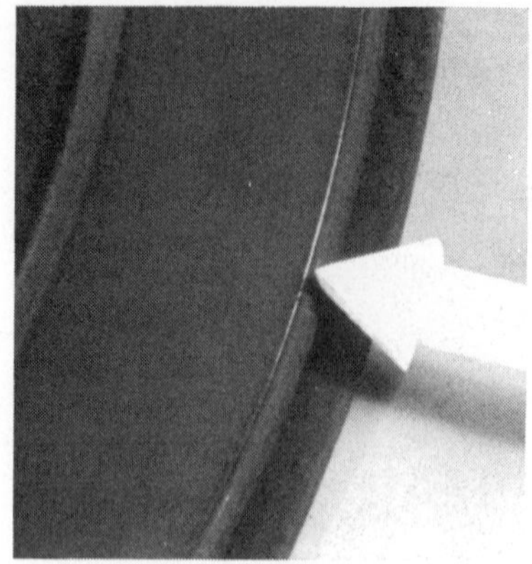

20. Advance the tool bit toward the rust ridge until very light contact is made.

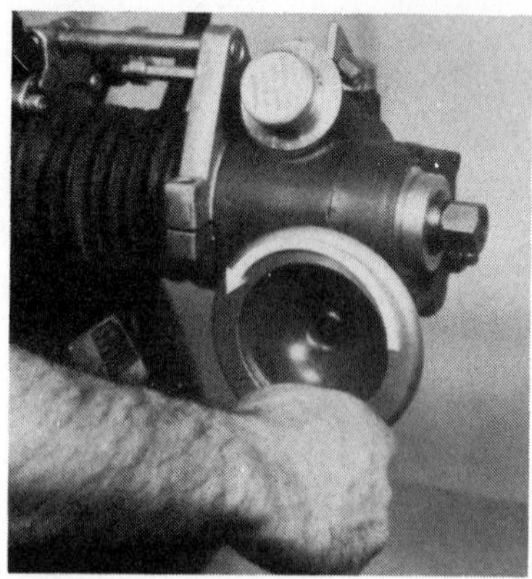

21. Crank the spindle handwheel until the edge of the drum clears the tool bit.

22. Turn the depth-of-cut handwheel counterclockwise about .005″ (2½ graduations).

23. Hold the depth-of-cut handwheel stationary, and crank the spindle handwheel to advance the drum . . .

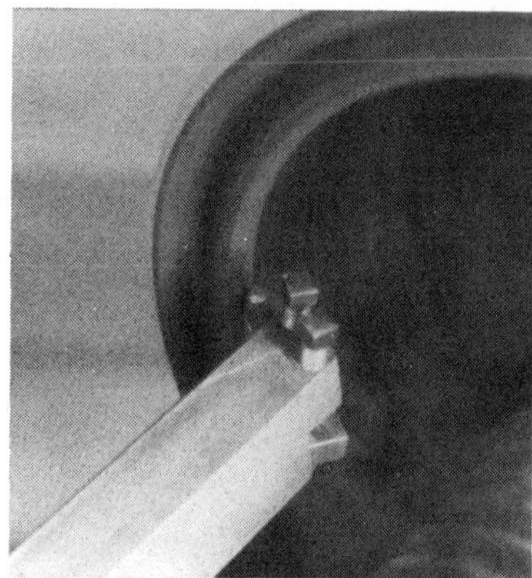

24. . . . and machine away a portion of the rust ridge.

25. Repeat Steps 20 through 23 until the ridge is removed and the tool bit contacts the worn portion of the friction surface.

26. Advance the tool bit to the unworn shoulder at the closed edge of the drum . . .

27. . . . adjusting the depth-of-cut handwheel as needed to maintain light contact with the friction surface.

TURNING A BRAKE DRUM

28. Advance the tool bit slowly by hand to machine away the shoulder of unworn metal at the closed edge of the drum.

29. The bit will make a scraping noise when it contacts the bottom of the drum.

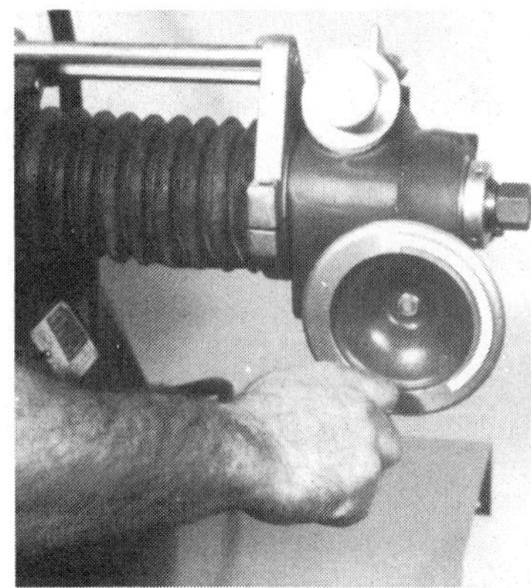

30. Crank the spindle handwheel counterclockwise until the scraping sound stops.

31. Zero the micrometer dial of the depth-of-cut handwheel then tighten the dial lockscrew.

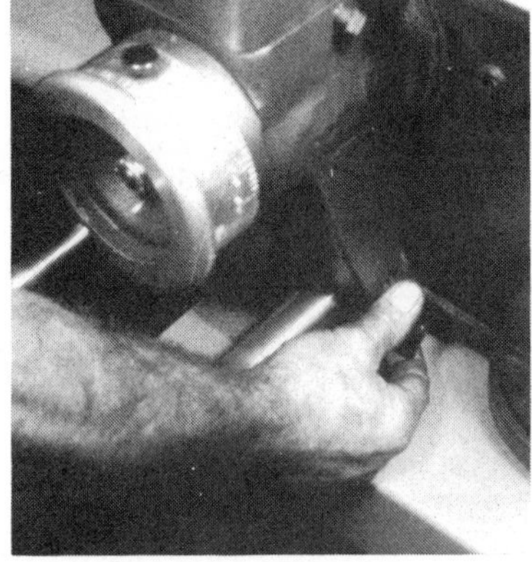

32. Rotate the depth-of-cut handwheel to advance the tool bit into the drum the desired amount; tighten the handwheel lock.

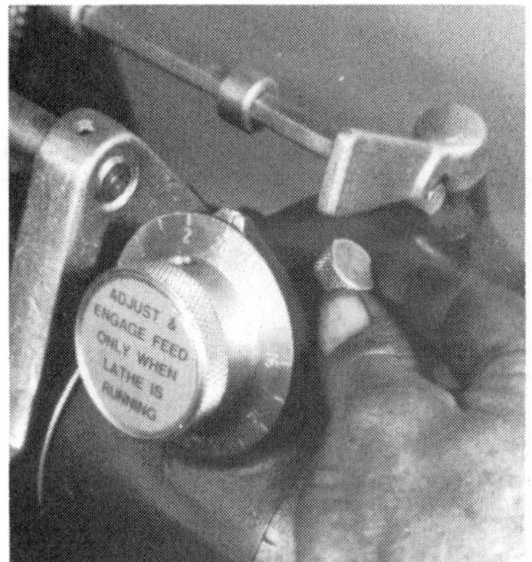

33. Loosen the crossfeed lockscrew.

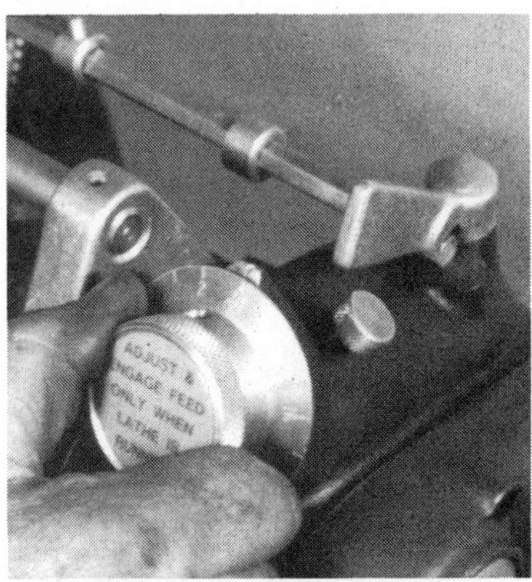

34. Set the crossfeed control to the desired speed, and tighten the lockscrew.

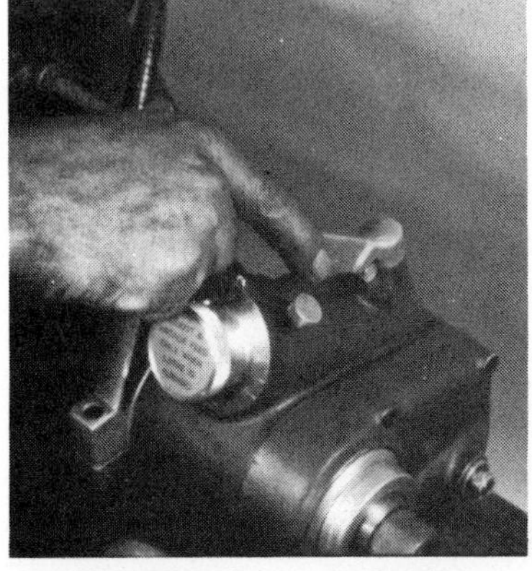

35. Engage the crossfeed mechanism. The shutoff bushing will turn off the lathe when the tool bit clears the drum.

36. Inspect the drum; repeat steps 29 through 31, making rough and finish cuts, until the friction surface is smooth.

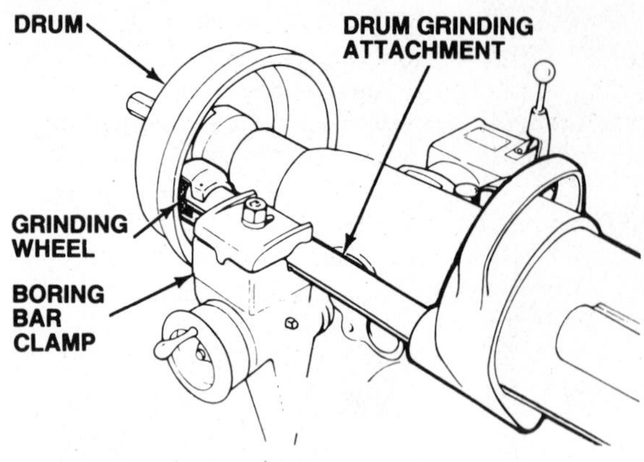

Figure 8-29. A typical drum grinding attachment.

GRINDING A BRAKE DRUM

If turning a drum reveals hard spots and you want to grind them flush with the rest of the friction surface, you should do so immediately after the turning operation while the drum is still mounted and centered on the lathe. This ensures that the cut made by the abrasive stone of the grinding attachment will be parallel to the drum friction surface. Only one pass across the friction surface is needed to machine down the hard spots; any other drum defects will have been corrected by turning.

Before you can use the grinding attachment to smooth out the hard spots, you must first true the face of the abrasive stone so it is parallel to the drum friction surface. Provided the boring bar clamp position is not changed, and the stone does not load up with metal, the abrasive stone needs to be trued only once to grind a complete set of drums.

To grind a brake drum:

1. Replace the brake lathe boring bar with a grinding attachment, figure 8-29.
2. Mount the truing-stone assembly on the brake drum as shown in figure 8-30; you may have to remove the silencing strap to do this.
3. Following the instructions of the brake lathe manufacturer, move the diamond tip of the truing-stone assembly across the face of the spinning grinding wheel to cut it parallel to the drum friction surface.
4. Remove the truing-stone assembly from the drum, and install the silencing strap if it was removed in Step 2.
5. Set the lathe rotation speed to between 65 and 100 rpm, and start the lathe.
6. Move the spinning abrasive stone to the closed edge of the drum, and set the depth of cut at .002″ to .005″ (.05 to .13 mm). Bearing deflection in the grinding attachment can make it difficult to accurately set the depth of cut, so it is often easier to simply move the stone slowly into contact with the friction surface until sparks trail one-quarter of the way around the drum.
7. Set the crossfeed at .002″ (.05 mm), and engage the mechanism. After one pass with the abrasive stone, stop the lathe and inspect for hard spots.
8. If any hard spots remain, repeat Steps 6 and 7.

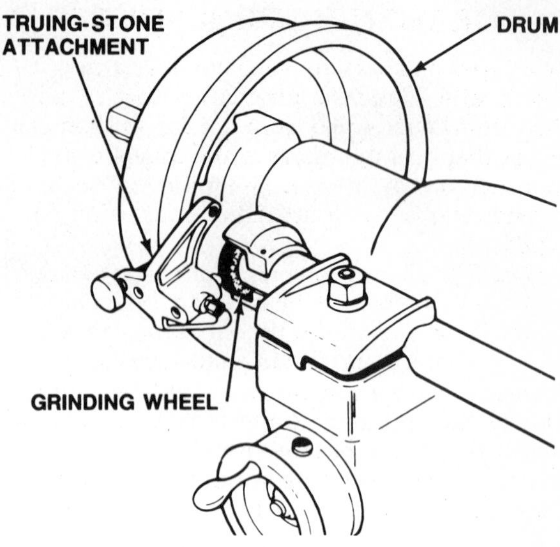

Figure 8-30. True the grinding wheel before you grind a drum.

ROTOR TURNING PROCEDURES

Two types of lathes are available for turning rotors, off-car free standing lathes that require the rotors be removed from the car, and on-car lathes that turn the rotors while they are still mounted on the vehicle. The following sections provide step-by-step procedures for using both types.

Virtually all modern brake lathes have two tool bits that straddle the rotor and refinish both friction surfaces at the same time. This makes the job go faster, and produces a better friction surface finish because distortion is reduced when pressure is applied equally to both sides of the rotor. Because two-bit lathes (no pun intended) are the most common type, they are the only kind dealt with below.

TURNING A BRAKE ROTOR OFF THE CAR

Once you have mounted and centered the rotor on the brake lathe as described earlier, the rotor is ready to be turned. Because every lathe operates differently, read the instructions for the particular unit you are using, and familiarize yourself with the controls. The following procedure is a general guide for turning a rotor. The photo sequence shows the steps required to turn a rotor using one specific type of brake lathe. Always wear safety glasses when operating a brake lathe to prevent eye injury from flying metal chips.

First, adjust the lathe rpm to the proper setting. Advance the tool bit to the outer edge of the rotor and machine away the ridge of rust and metal that forms there, then move the tool bit to the inner edge of the friction surface and remove the ridge of rust and metal there as well.

Position the tool bits at approximately the center of the rotor, and advance each bit until it lightly contacts the rotor surface. Zero the micrometer scales on the handwheels controlling depth of cut. These are your reference points for setting the depth of cut. Move the tool bits to the inner edge of the rotor, and turn the handwheels to set the depth of cut on both bits for a rough cut as described earlier in the chapter. To a large extent, the depth of cut is determined by the condition of the rotor, the greater the damage, the deeper the cuts you can take, up to a maximum of .006″ (.015 mm) on each side. All rotor lathe handwheels that control depth of cut are calibrated to show the amount of metal removed from the friction surface. If you set each handwheel at .005″ (.13 mm), the lathe will cut a total of .010″ (.25 mm) from the thickness of the rotor.

Adjust the crossfeed for a rough cut as described earlier in the chapter, and engage the mechanism. The tool bit will automatically move from the inner edge of the rotor to the outer edge. Turn the rotor with rough cuts as needed until all defects are removed or nearly removed. Finish turning the rotor with a finish cut that removes the last traces of defects from the rotor, and provides a friction surface finish compatible with the type of brake lining material used. Many manufacturers recommend that after the finish cut you should apply a non-directional finish to the friction surfaces with a resurfacing attachment as described later in the chapter.

Once you are completely finished machining a rotor, wipe the friction surfaces clean with a shop towel soaked in brake cleaner. This removes loose metal shavings that could become embedded in the brake linings and score the rotor friction surface. Remeasure the rotor thickness as described earlier in the chapter to make sure it is within legal limits.

TURNING A BRAKE ROTOR OFF THE CAR

1. If the lathe is designed to machine both drums and rotors, install the saddle mount cutting tool assembly.

2. Install the washers over the boring bar clamp stud with the convex and concave sides facing one another.

3. Install the lock nut on the boring bar clamp stud, but do not tighten it completely at this time.

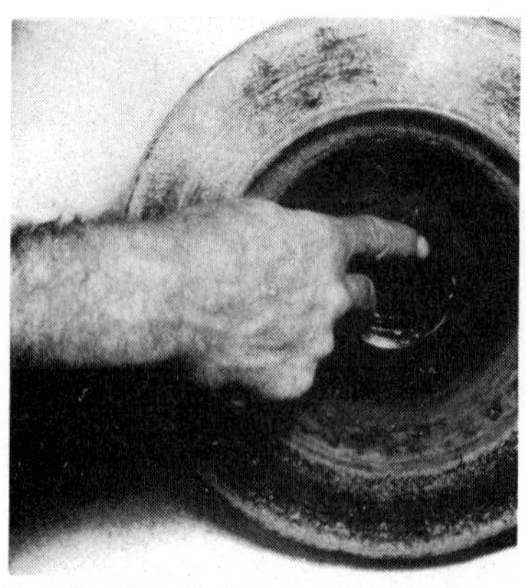

4. On fixed rotors, check the bearing races for looseness in the hub. Replace the rotor if a race is loose.

5. Install a fixed drum on the arbor shaft using the appropriate adapters as described earlier.

6. Install a floating drum on the arbor shaft using the appropriate adapters as described earlier.

7. Tighten the arbor shaft nut, but do not overtighten it.

8. Install a rubber silencing band on vented rotors.

9. Install a damper assembly on rotors where a rubber silencing band will not work.

TURNING A BRAKE ROTOR OFF THE CAR

10. Swing the cutting tool assembly into position.

11. Position the tool bits ½ inch in from the outer edge of the rotor, and equal distances from the friction surfaces.

12. Tighten the lock nut on the boring bar clamp stud.

13. Make sure the drive belt is in the proper pulley for the correct turning rpm.

14. Install the safety shield to stop flying metal chips.

15. Start the lathe, and make a pair of scratch cuts to check for rotor runout and centering as described earlier.

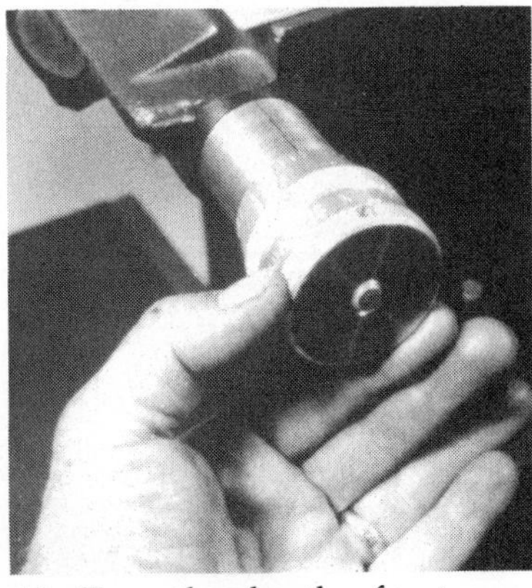
16. Turn the depth-of-cut handwheels until each tool bit lightly contacts the rotor friction surface.

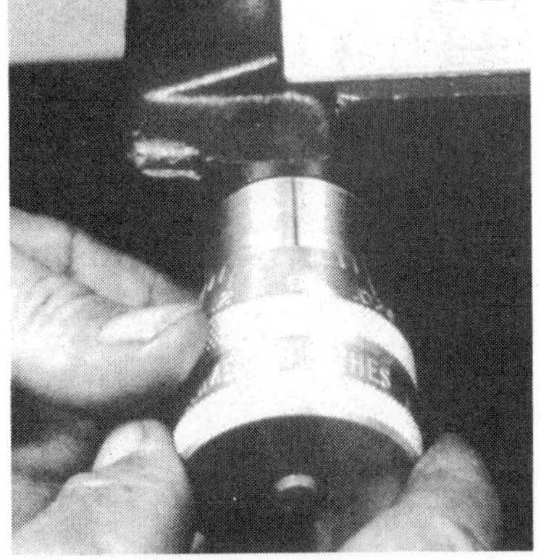
17. Hold the outer knurled portions of the handwheels stationary, and zero the depth-of-cut collars.

18. Turn the crossfeed handwheel until the tool bit on the outer face of the rotor is at the inside edge of the friction surface.

TURNING A BRAKE ROTOR OFF THE CAR

19. The tool bit on the inner friction surface will be off the friction surface entirely.

20. Rotate the depth-of-cut handwheels to advance the tool bits into the rotor the desired amount; tighten the handwheel locks.

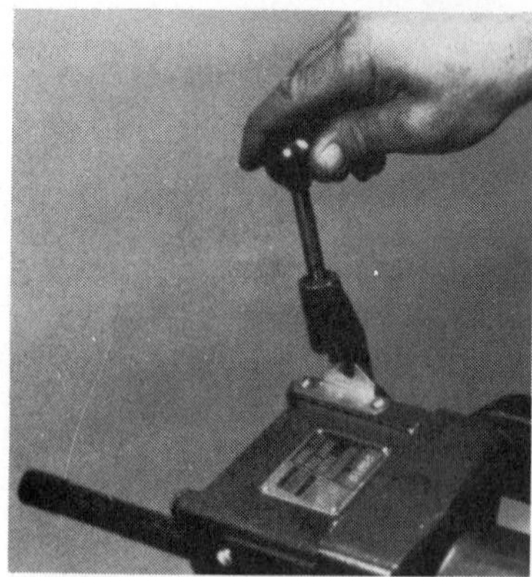
21. Engage the crossfeed mechanism by moving the gearbox control to the "slow" position.

22. Once the crossfeed is engaged, the tool bits will advance to the outer edge of the rotor.

23. When the tool bits clear the rotor, disengage the crossfeed by moving the gearbox lever to the "off" position.

24. Stop the lathe and inspect both the outer . . .

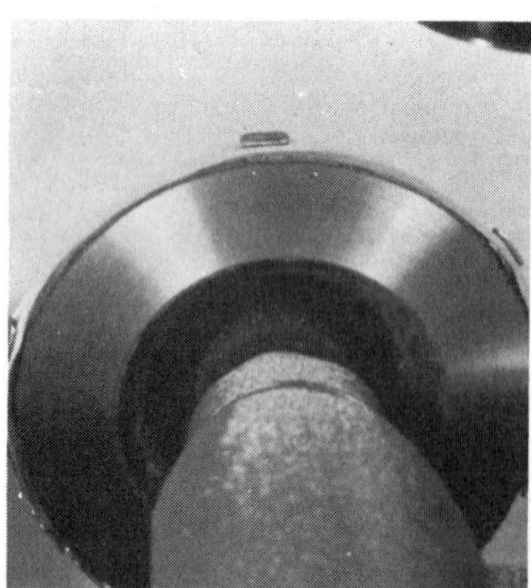
25. . . . and the inner friction surfaces of the rotor.

26. Repeat steps 18 through 25, making rough and finish cuts, until both friction surfaces are smooth.

27. Measure the rotor thickness to make sure it is within legal limits.

TURNING A BRAKE ROTOR ON THE CAR

The on-car lathe is a relatively new tool that machines rotors while they are still on the car. The lathe bolts in place of the brake caliper, and uses the vehicle powertrain or a separate electric motor to turn the rotor. Because the spindle and wheel bearings serve as the arbor shaft for the lathe, it is very important that the wheel bearings be in good condition and properly adjusted (where possible).

The photo sequence details the procedure for turning a rotor using one type of on-car lathe. Every lathe operates differently, however, so read the instructions for the unit you are using, and familiarize yourself with its controls.

If you are using engine power to turn the rotor, lock the opposite wheel to keep it from turning unless the car has a limited slip differential. During machining, the rotor must turn into the tool bits. If you are using the vehicle powertrain to turn the rotor, run the car at idle in either first or reverse gear depending on the mounting position of the lathe, and the side of the car on which it is located.

The crossfeed of an on-car lathe may be either automatic or operated by hand. With a hand feed, advance the tool bits slowly and steadily across the rotor. Set an automatic crossfeed at .003" (.08 mm) per revolution, and use a slightly slower crossfeed for heavily grooved rotors. With either type of feed, make only light cuts with a depth of .002" to .005" (.05 to .13 mm); deeper cuts cause a poor rotor finish.

TURNING A BRAKE ROTOR ON THE CAR

1. Raise and properly support the vehicle so the wheels with the rotors to be turned hang free.

2. Remove the wheel from the rotor to be turned. If engine power is being used to turn the rotor, lock the opposite wheel to keep it from turning.

3. On a floating rotor, install the lug nuts to hold the rotor tightly on the hub.

4. Remove the brake caliper and hang it by a wire from the suspension.

5. Remove any dirt and rust from the caliper mounting bracket.

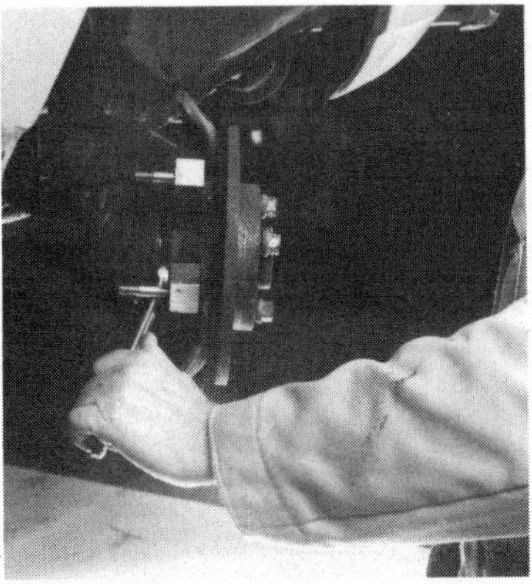

6. Install the lathe mounting links on the caliper mounting bracket. Snug the bolts so the links can still be rotated.

TURNING A BRAKE ROTOR ON THE CAR

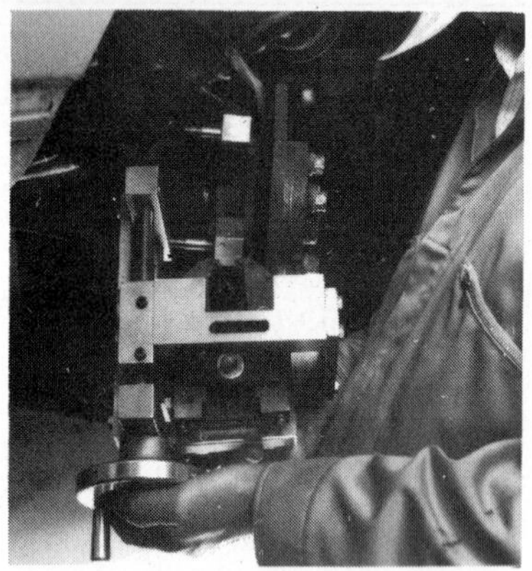

7. Mount the lathe on the links, then install the retaining washers and nuts on the studs, but do not tighten them.

8. Install the aligning bar, and position the lathe so the bar end is centered on the hub. Tighten all mounting hardware.

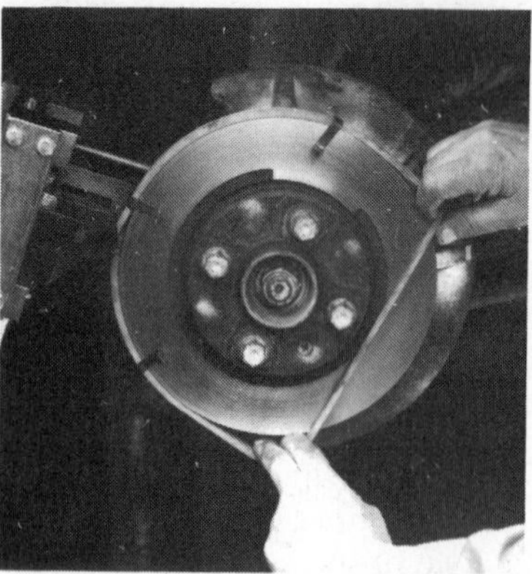

9. Install the silencing band. The clips are used only to aid installation; remove them before machining the rotor.

10. Install the protective band over the studs and lug nuts to prevent clothes from catching on the spinning rotor.

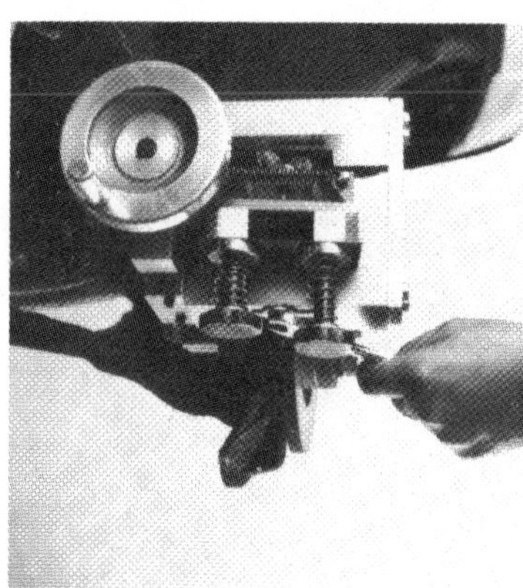

11. Loosen the lock bolt, center the tool bit holder over the rotor, then tighten the lock bolt securely.

12. Center the tool bits on the friction surface. Turn the rotor by hand and adjust each bit to lightly contact the friction surface.

13. Position the tool bits off the friction surface inner edge. Advance each bit one graduation (.002") on the depth-of-cut indicator ring.

14. Lock the tool bit holder in place to retain the depth-of-cut settings.

15. Start the engine to turn the rotor. Crank the crossfeed handwheel to move the tool bits slowly across the rotor surface.

Figure 8-31. A brake rotor resurfacing attachment.

RESURFACING A BRAKE ROTOR

If the brake lathe tool bits are sharp, and you follow the recommended rpm, crossfeed, and depth-of-cut settings when turning a rotor, you will generally get an acceptable surface finish. However, many manufacturers recommend that the rotor friction surface also be resurfaced after turning to give it a non-directional finish. This is especially important when semi-metallic brake pads are fitted. Brake rotors are resurfaced with a lathe attachment that applies a spinning abrasive disc against the friction surfaces. You should resurface a brake rotor immediately after it has been turned, while it is still mounted and centered on the brake lathe.

To resurface a rotor, mount the special attachment onto the brake lathe, figure 8-31, and install an abrasive disc of the proper grit on the disc holder. For organic or synthetic brake pads, use 50- to 80-grit abrasive paper. For semi-metallic brake pads, 120-grit paper will provide the smoother surface needed for proper break-in.

Start the lathe, and advance the spindle until the abrasive disc contacts the rotor with moderate pressure. The disc holder is spring loaded to provide the proper pressure; do not bottom the spring. Hold the abrasive disc against the rotor for 15 seconds, then stop the lathe and inspect the rotor friction surface for a good crosshatch pattern. Repeat this process as necessary, than reposition the attachment and resurface the other side of the rotor.

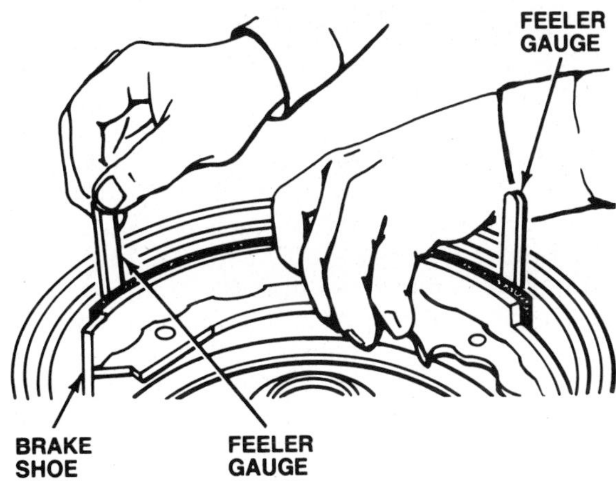

Figure 8-32. Checking the end clearance of a brake shoe lining.

ARC GRINDING A BRAKE SHOE

Brake shoe arcing is the process of machining the linings so they properly fit the curvature of the drum. Arcing eliminates high spots on the linings, shortens the break-in period, improves lining-to-drum contact, gives better brake pedal feel, and allows maximum lining life. Most modern brake shoes are arced at the factory for their intended application, but before you install new or relined brake shoes, always check their fit in the drum they will run against. If the shoes are properly arced, the linings will have a small amount of clearance at each end that keeps them from binding against the drum and causing the brakes to grab or squeal.

To check the fit of a brake shoe, place it into the drum with the center of the lining against the friction surface. The shoe should rock slightly, indicating that there is some end clearance. If the shoe does not rock, arc the linings. Next, measure the amount of end clearance. You should be able to place two feeler gauges of the same thickness between the ends of the lining and the drum to remove all clearance, figure 8-32. On non-servo brakes, use .005″ to .007″ (.13 to .18 mm) feeler gauges. On dual-servo brakes, use .010″ to .012″ (.25 to .30 mm) feeler gauges. If there is more or less clearance than these gauges can compensate for, arc the linings.

Arc Grinding Precautions

Brake shoe arc grinding produces asbestos dust that is a grave health hazard. Only operate an arc grinder that is equipped with an approved dust collection system. Before arcing a set of shoes, be sure the dust collector is installed correctly, and the disposal bag is not full. Do not remove more than .005" to .010" (.13 to .25 mm) of friction material from the lining on each pass or you will overstress the motor, produce more asbestos dust than the collection system can easily deal with, and cause a rough finish on the lining surface. On each pass, smoothly move the lining past the grinding wheel at a moderate rate; do not stop in the middle of a pass, or move the lining too quickly. Stop arcing the shoe as soon as the entire surface of the lining is cleaned up; continued arcing wastes time and lining material.

Arc Grinding Procedure

Undersize brake shoe arcing for either standard or fixed anchors is the only kind done at the shop level. The procedures for both jobs are essentially the same; they differ only in how the shoe is clamped to the grinder. Follow the equipment manufacturer's instructions for mounting the shoe on the grinder.

Some older arc grinders require the operator to dial in the amount the shoe is to be ground undersize; however, most newer grinders automatically arc the shoes to a diameter .030" (.75 mm) smaller than the drum diameter. Again, consult the instructions to determine how the grinder you use is calibrated. The following photo sequence provides a general set of instructions on how to machine a brake shoe using a typical arc grinder.

ARC GRINDING A BRAKE SHOE

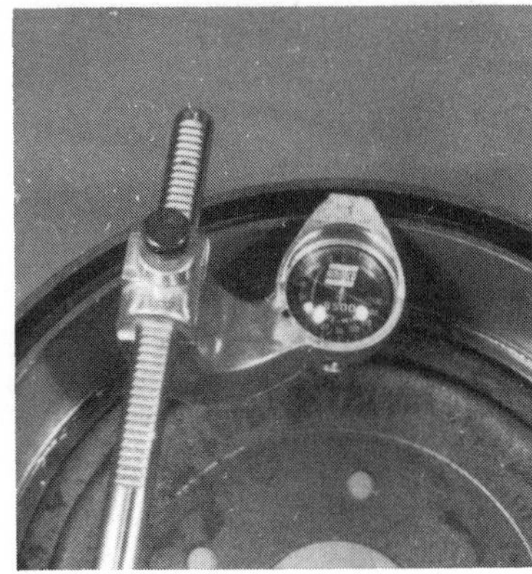

1. Measure the actual drum diameter with a drum micrometer.

2. Unlock the grinder table by loosening the locking lever.

3. Set the micrometer dial to zero.

4. Pull out on the locking knob, and set the grinder diameter scale to the drum "whole inch" size. For example, on a 10½-inch drum, set the pointer to 10.

5. Set the micrometer dial to the decimal measurement of the drum "partial inch" size. For the 10½-inch drum, set the dial at .50 inches.

6. Lock the grinder table by tightening the locking lever.

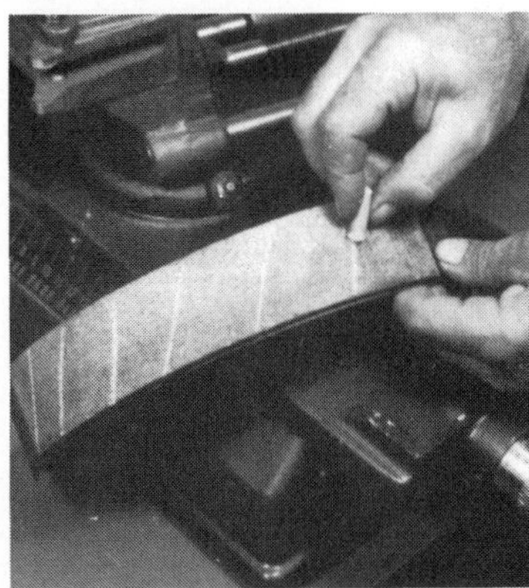

7. Mark across the face of the lining with chalk.

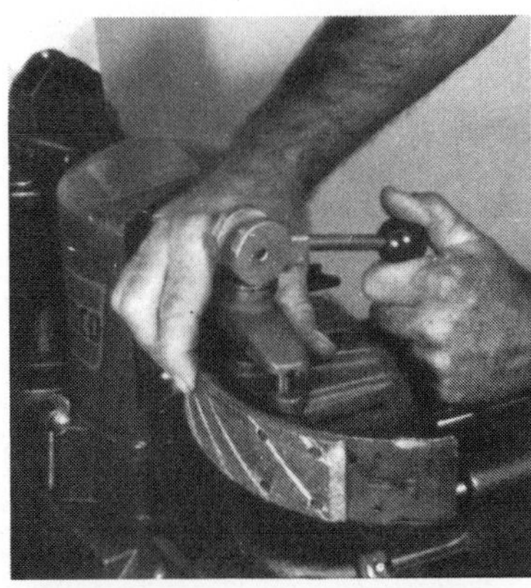

8. Set-up the shoe in the arc grinder as instructed by the manufacturer.

9. With the grinder OFF, lift the operating handle and push the clamp and shoe forward until the lining contacts the grinding wheel.

ARC GRINDING A BRAKE SHOE

10. Lower the operating handle and rotate it counterclockwise until the lining just clears the grinding wheel.

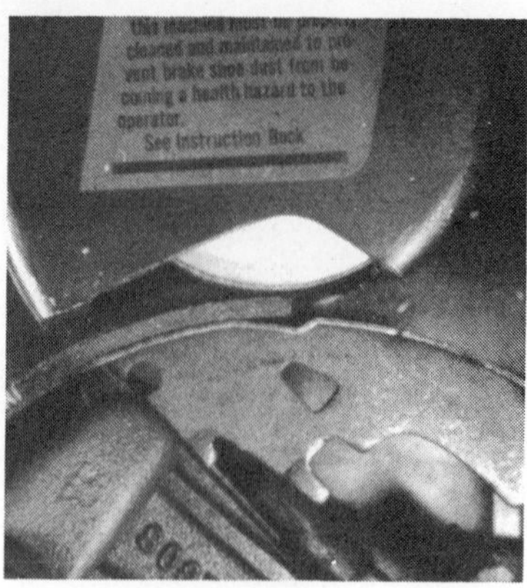

11. Swing the operating handle back and forth to make sure the lining clears the grinding wheel at all points.

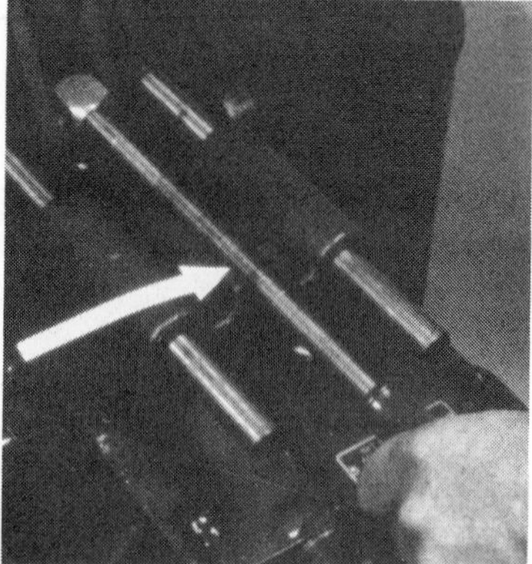

12. Turn the grinder ON, and rotate the operating handle clockwise until the lining lightly contacts the grinding wheel.

13. Smoothly swing the handle back and forth at slow speed to make two full-length sweeps across the lining — one in each direction.

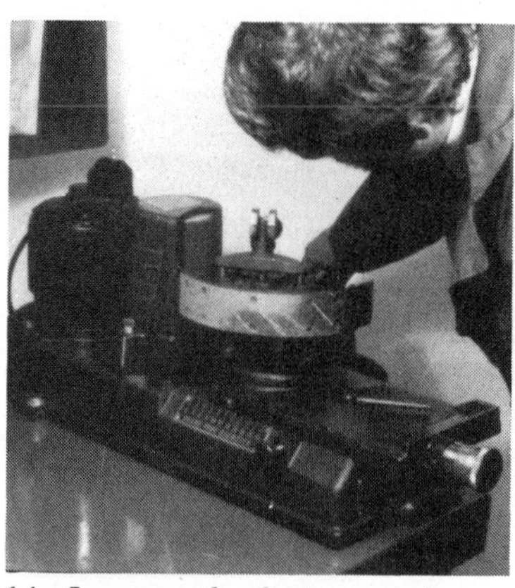

14. Inspect the lining surface to see if additional grinding is necessary.

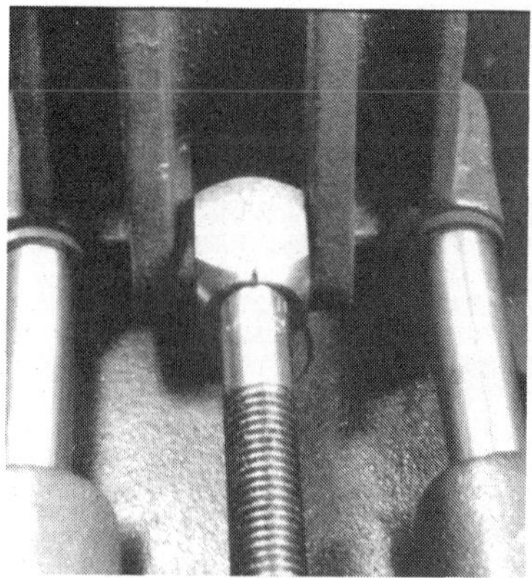

15. Rotate the operating lever clockwise until up to .005" (.15 mm) of shoe movement of is indicated on the grinding scale.

16. Repeat Steps 13 through 15 until the chalk marks are gone and the surface of the lining is ground smooth.

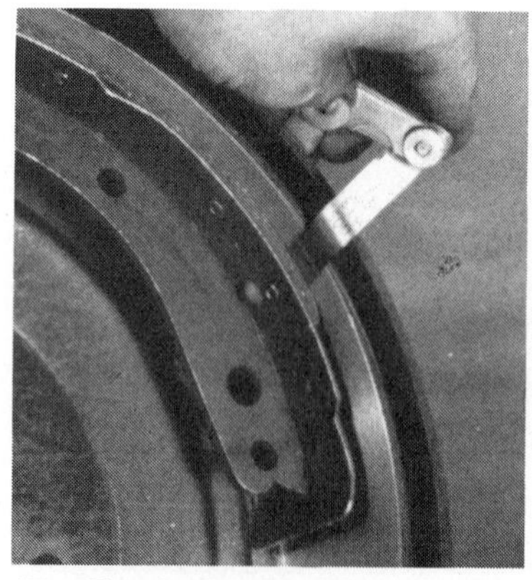

17. Check the fit of the shoe in the drum to make sure it is within specifications.

18. Chamfer the heel and toe of the lining on the grinding wheel to reduce the chances of brake noise.

PART FOUR

Brake Subsystem Testing and Service

Chapter

Parking Brake Service

The parking brakes keep the vehicle from rolling when it is parked, and provide a limited amount of stopping power if there is a failure in the service brake hydraulic system. Three main types of parking brakes are currently used on automobiles and light trucks: integral drum parking brakes that apply the shoes of the rear drum service brakes, auxiliary drum parking brakes that apply a set of small brake shoes against drums inside the rear brake rotors, and caliper-actuated disc parking brakes that use a mechanical linkage to apply the service brake pads of the brake calipers.

Parking brake service consists of three basic operations: testing the parking brake system, adjusting the parking brake linkage, and replacing the parking brake cables. In addition to these basic jobs, auxiliary drum parking brakes require periodic replacement of the parking brake shoes, and the vacuum parking brake release mechanism on some cars may require diagnosis and repair.

PARKING BRAKE TESTING

Parking brake problems, like almost all automotive problems, are easiest to diagnose if you approach them in an organized way. You can isolate a problem in the parking brake system using a four-step test procedure. First, perform a brake pedal travel test to confirm that the parking brake is the source of the problem. Second, perform a parking brake control test to determine if the parking brake adjustment is correct. Third, perform a release test to make sure the parking brake is not causing brake drag. And finally, do a performance test to find out if the parking brake has sufficient holding power.

Brake Pedal Travel Test — Drum Brakes

The integral drum parking brakes used on most cars apply the shoes of the rear drum service brakes. As a result, the adjustment of the service brakes affects the performance of the parking brake as well. If there is excessive lining-to-drum clearance, the parking brake linkage may not have sufficient travel to take up the slack and still apply the brake with enough force to hold the car in place.

To test the adjustment of the service brakes, apply the brake pedal, figure 9-1, and note the amount of travel. If the travel is excessive, you must adjust or repair the service brakes before you adjust the parking brakes. Refer to Chapter 6 for complete drum brake diagnosis and service procedures.

Figure 9-1. You must have a high and firm service brake pedal before you adjust the parking brakes.

Brake Pedal Travel Test — Disc Brakes

A low brake pedal on a car with a caliper-actuated disc parking brake also indicates a problem in the service brake system. However, if the low pedal is accompanied by excessive travel of the parking brake control, the problem is most likely caused by a defective adjuster in one or both of the brake calipers that function as the parking brake.

In order to apply the caliper for parking brake service, the rear disc brakes on General Motors and Ford cars have mechanical adjusters built into their caliper pistons. If these adjusters fail to operate properly, the piston will retract too far into the caliper bore, causing the symptoms described above. To determine if this is the problem, follow the procedures described in the next two sections. In rare cases, you may be able to free the adjuster and re-establish the proper lining-to-rotor clearance; however, most of the time you will have to rebuild the caliper and replace the piston/adjuster assembly.

General Motors rear disc parking brake

The rear calipers on General Motors cars with disc parking brakes are known for their problems with frozen adjusters. On these vehicles it is extremely important that the car owner apply the parking brake regularly. This keeps the actuator screw threads clean so the automatic adjusting mechanism in the piston will work properly. If both the service brake pedal and parking brake control have excessive travel on one of these cars, attempt to adjust the caliper pistons.

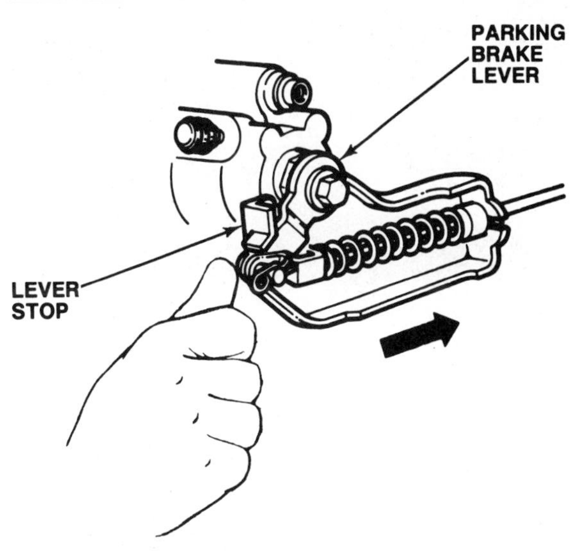

Figure 9-2. Push the parking brake lever forward to check the piston adjustment in a Ford rear brake caliper.

While driving the vehicle at 15 to 20 mph (25 to 30 kph), pull and hold the parking brake pedal release handle to prevent the pedal mechanism from locking. At the same time, repeatedly apply the parking brake pedal. If the travel of both the service brake pedal and the parking brake control does not decrease, the adjuster is frozen and you must rebuild the caliper and replace the piston/adjuster assembly as described in Chapter 6.

Ford rear disc parking brake

The rear calipers on Ford Motor Company cars with disc parking brakes are less subject to adjuster problems than the General Motors design, but they do have occasional failures. To determine if the piston in a Ford rear caliper is properly adjusted, push the parking brake lever in the apply direction under firm hand pressure, figure 9-2. If the lever moves more than 20 degrees, attempt to adjust the caliper pistons.

To adjust the pistons with the engine idling, apply the brake pedal 40 times with approximately 15 lb (67 N) of force; allow one second between each pedal application. With the engine not running, apply the pedal 30 times with approximately 90 lbs (400 N) of force; allow one second between each pedal application. If the travel of both the service brake pedal and the parking brake control does not decrease, the adjuster is frozen and you must rebuild the caliper and replace the piston/adjuster assembly as described in Chapter 7.

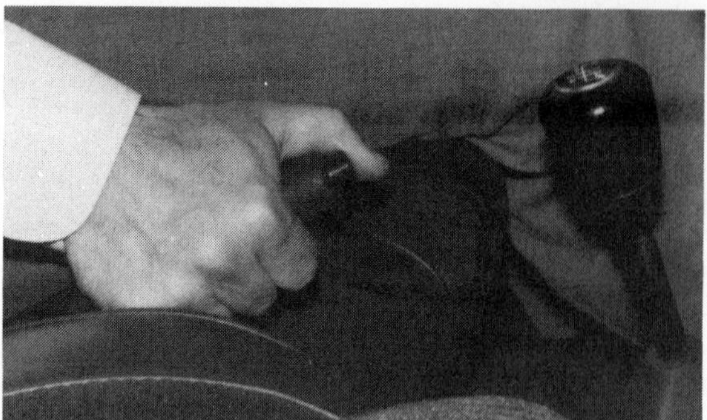

Figure 9-3. Apply the parking brake control to check the linkage adjustment.

Parking Brake Control Test

If the service brake pedal is high and firm, the next step is to apply the parking brake control, figure 9-3, and check its travel. The parking brake should be fully applied when the control has moved through one- to two-thirds of its available travel. Some manufacturers specify that the brake should be fully applied when the parking brake control ratchet has made a certain number of clicks, usually five to seven.

If the control travels more than the specified amount, the parking brake adjustment is too loose. A loose adjustment can leave the parking brake linkage with insufficient travel to take up the slack in the linkage and still apply the brake with enough force to hold the car in place.
If the control travels less than the specified amount, the parking brake adjustment is too tight. A tight adjustment may cause the brake shoes or pads to drag against the drum or rotor. This creates heat that can lead to brake fade, rapid wear of the brake linings, and distortion of the drum or rotor. Adjust the parking brakes if the control requires more or less movement than specified to apply the brakes.

Release Test

If both the service brake pedal and parking brake control have acceptable travel, raise and support the car so the wheels with the parking brakes hang free. With the parking brake control fully released, rotate the wheels by hand and check for drag, figure 9-4. If either brake drags, check for sticking or frozen parking brake cables, or a problem in the service brake friction assembly.

Inspect parking brake cables for broken strands along the sections of cable that are visible. Then, have an assistant operate the parking brake control while you observe that the cables move freely in and out of their housings.

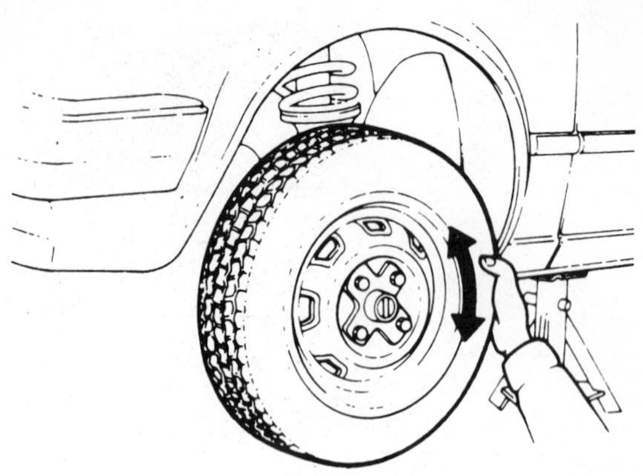

Figure 9-4. There should be no drag at the wheels with the parking brake control released.

If any cable has broken strands or does not move freely, replace it as described later in the chapter.

On cars with dual-servo brakes, a slightly overtightened parking brake will sometimes cause the brakes to drag in one direction but not the other. This occurs because the overtightened adjustment prevents the shoe with the weaker return spring from fully returning against the anchor when the service brakes are released.

As with any disc brake, a small amount of drag is considered normal in a rear brake caliper. However, if there is excessive drag on cars with rear disc parking brakes, make sure the parking brake levers on both rear calipers are fully returned. If the calipers are fitted with external lever stops, the levers should contact the stops when the parking brake control is released. If the levers are not fully returned or against their stops, adjust the parking brake. If adjusting the parking brake does not eliminate the drag, the caliper has a sticking piston, or the caliper floating or sliding surfaces are in need of lubrication. If there is excessive drag in a rear brake caliper, locate and repair the problem as described in Chapter 7.

Performance Test

If both the service brake pedal and parking brake control have acceptable travel, and the wheels with the parking brakes turn freely, the final test is to check the parking brake performance. Stop the vehicle facing uphill on a grade of approximately 30 percent, then firmly apply the parking brake and release the service brakes. The vehicle should hold its position and not creep or roll. Repeat the test with the vehicle facing downhill.

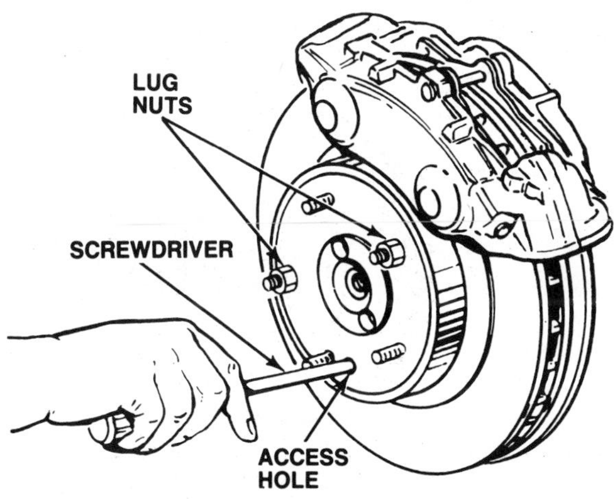

Figure 9-5. Auxiliary drum parking brake shoes are adjusted through an access hole in the rotor/drum.

If the vehicle fails the performance test, it usually indicates a problem in the service brake friction assembly. A worn out service brake, or one with contaminated linings, cannot generate enough friction for the parking brake to work properly. However, some parking brakes, particularly caliper-actuated disc designs, have marginal holding power unless a great deal of force is applied to the parking brake control. Some drivers who complain of parking brake problems may simply be unable to apply the parking brake control hard enough. To a certain extent, you must depend on experience to help identify when a parking brake has sufficient braking power, and when it does not.

PARKING BRAKE ADJUSTMENT

Most vehicle manufacturers recommend that the parking brake be tested and adjusted (if necessary) at regular intervals, generally twice a year or every 7,500 miles (12 000 km). More frequent adjustment may be required if the vehicle is used in severe service that involves extensive use of the parking brake. Naturally, the parking brake must also be adjusted whenever a cable is replaced, or the wheel brakes that contain the parking brakes are serviced.

Parking brake adjustment involves two basic procedures. The first, parking brake shoe adjustment, is required only on cars that have auxiliary drum parking brakes. The second procedure, cable adjustment, is required on all parking brake systems, both disc and drum.

PARKING BRAKE SHOE ADJUSTMENT

As with any drum brake, the shoes in auxiliary drum parking brakes must be adjusted periodically to maintain the proper lining-to-drum clearance. On a few imported cars, this adjustment is made by adjusting the parking brake cables. However, most of these brakes use manual starwheel adjusters to set the lining-to-drum clearance independent of the cable adjustment.

On most cars, you reach the adjuster through an access hole in the outer face of the rotor/drum, figure 9-5. However, on some imported cars, one of the wheel lug bolt holes serves as the access hole; this allows adjustment with the wheel installed. Because access to the adjusters on auxiliary drum parking brakes is relatively easy, a standard screwdriver is often used in place of a brake spoon to rotate the starwheel.

All auxiliary drum parking brake shoes are adjusted the same way, and the procedure is essentially the same as for a drum service brake with a manual starwheel adjuster. To adjust an auxiliary drum parking brake:

1. Shift the transmission into neutral, and release the parking brake control.
2. Raise and properly support the vehicle so the wheels to be adjusted hang free.
3. If the brake is adjusted through an access hole in the outer face of the rotor/drum, remove the wheel and reinstall two of the lug nuts to retain the rotor/drum securely in place. If the brake is adjusted through a lug bolt hole, remove one of the lug bolts.
4. Loosen the parking brake cable adjustment until there is slack in the cables.
5. Turn the rotor/drum or wheel until the access hole aligns with the starwheel inside the friction assembly.
6. Insert a standard screwdriver through the access hole and use a lever action to rotate the starwheel and reduce the lining-to-drum clearance. The direction of rotation used to tighten or loosen the adjustment varies from car to car, and usually from one side of the vehicle to the other. As you rotate the starwheel, the spring holding it in position should snap from one tooth to the next and cause a clicking sound.
7. Tighten the adjuster until there is heavy drag at the wheel or the brake locks (it will be necessary to remove the screwdriver to check this), then lever the starwheel in the opposite direction to increase the lining-to-drum clearance. Loosen the adjustment the number of clicks specified by the vehicle manufacturer, typically five to seven, or until the wheel just turns free.

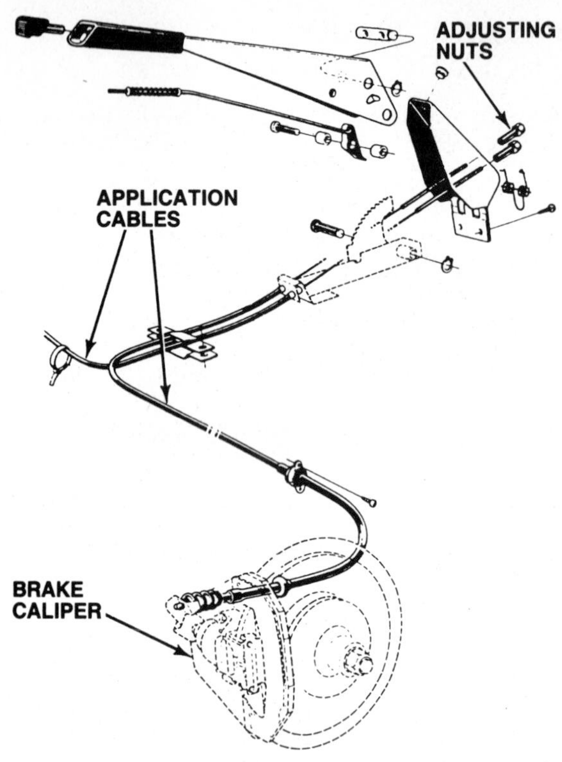

Figure 9-6. Parking brakes with a separate cable for each wheel require that the service brakes be properly adjusted.

PARKING BRAKE CABLE ADJUSTMENT

Whenever the parking brake is applied, there is constant tension on the cables that connect the parking brake control to the friction assemblies. Regular use of the parking brake stretches the cables, and increases the amount of parking brake control travel required to apply the parking brake. When the amount of travel becomes too great, the cable adjustment is tightened to shorten the working length of the cables and reduce parking brake control travel to the proper amount.

The service brakes must be in good operating condition before you can adjust the parking brake cables. If the brake pedal travel tests described earlier revealed a low brake pedal, adjust or repair the drum service brakes as described in Chapter 6, or the disc service brakes as described in Chapter 7, before you adjust the parking brake cables.

In addition to the reasons already given, you must adjust the service brakes before the parking brakes because certain cars have a parking brake linkage with a separate cable for each friction assembly, figure 9-6. No equalizer is used

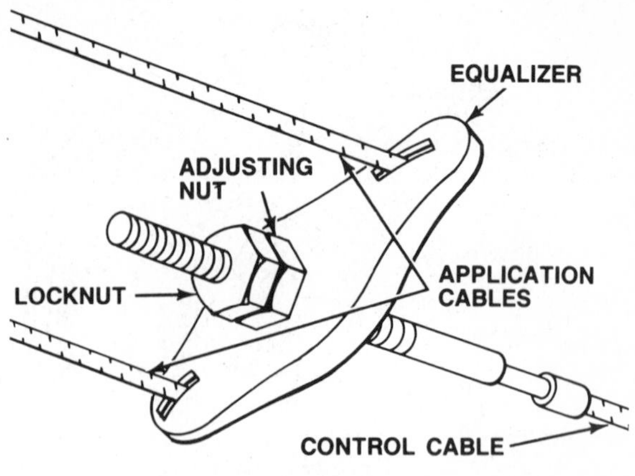

Figure 9-7. Many parking brake cables are adjusted under the car at an equalizer.

in this system, and the parking brake control moves each cable the same amount. If the wheel brakes do not have similar lining-to-drum clearance, the more loosely adjusted brake will not grab as hard when the parking brake is applied, and the tighter brake may drag when the parking brake is released.

Cable Adjusters

When the parking brake is controlled by a foot pedal or underdash handle, the cable adjuster is generally located under the car at an intermediate lever or equalizer, figure 9-7. If the parking brake is controlled by a floor-mounted lever, the cable adjuster is usually inside the car on the lever assembly, figure 9-8. When the adjuster is mounted inside the car, you often have to remove a rubber boot or plastic cover to reach it.

Most cable adjusters use two jam nuts — an adjusting nut and a lock nut — to set the cable length. A few cables have a single self-locking adjusting nut. If the nuts are rusted, corroded, or seized to the threaded adjuster rod, soak the assembly with penetrating oil before you attempt to make the adjustment; this will help prevent the nuts and rod from being stripped.

To adjust the cable, hold the adjusting nut in place with an open-end wrench, then loosen the lock nut with a second wrench, figure 9-9. Rotate the adjusting nut to draw the end of the cable through the lever or equalizer to shorten the working length of the cable. Once the adjustment is complete, hold the adjusting nut in place with an open-end wrench, and tighten the lock nut against it with a second wrench. On systems with a separate adjustment for each cable, tighten the cables equally.

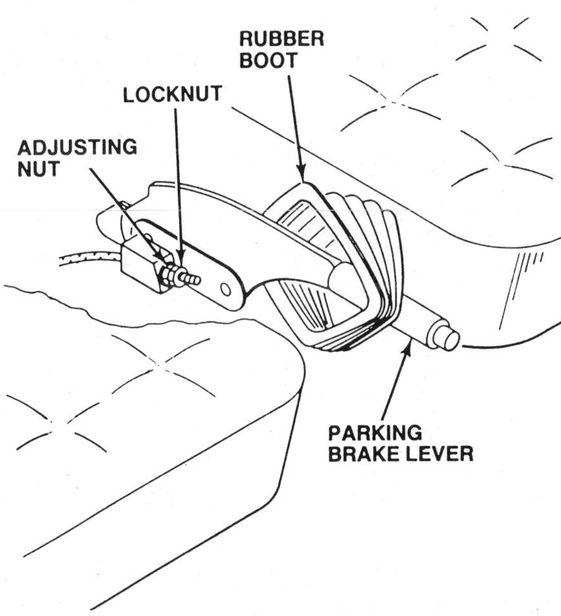

Figure 9-8. Some parking brake cables are adjusted inside the car at the parking brake lever.

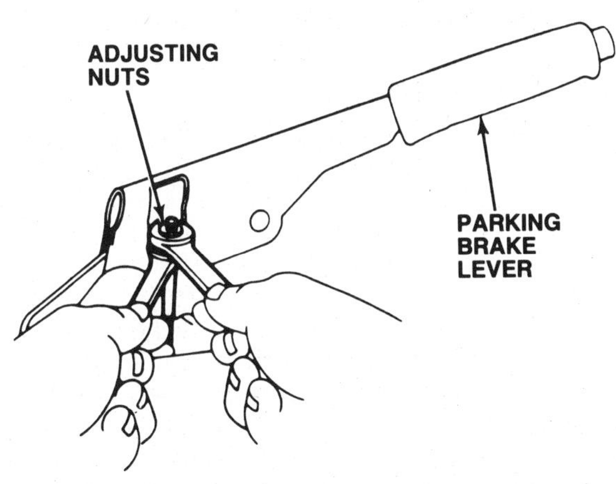

Figure 9-9. Two open-end wrenches are normally used to adjust parking brake cables.

It is very important not to twist the cables when making an adjustment. This places the cables under additional stress, and leads to premature failure. If necessary, use a pair of locking pliers on an unthreaded section of the threaded rod to hold the rod stationary while you tighten the adjusting nut. Some adjusters have a slot in the end of the rod, figure 9-10, so you can insert a screwdriver to stop the cable from twisting.

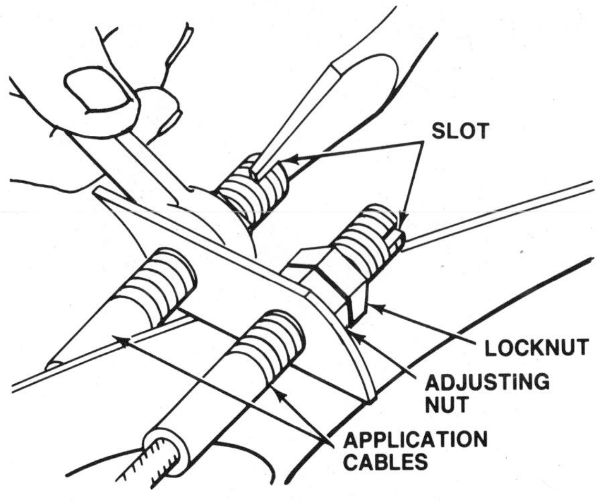

Figure 9-10. Always hold the parking brake cable stationary when you turn the adjusting nut.

Cable Adjustment — Drum Brakes

The cable adjustment for any kind of drum parking brake is similar whether it is an integral drum parking brake or a rear-disc auxiliary drum parking brake. To adjust the parking brake cable:

1. Shift the transmission into neutral, and fully release the parking brake control.
2. Raise and properly support the vehicle so the wheels with the parking brakes hang free.
3. Tighten the cable adjustment until the brakes begin to drag as you rotate them by hand.
4. Loosen the cable adjustment until the brakes just turn free.
5. Apply the parking brake and make sure it is fully engaged when the parking brake control has moved through one- to two-thirds of its available travel.
6. Apply and release the parking brake several times, then check that the wheels still turn free when the parking brake control is fully released. Loosen the adjustment slightly if necessary.

Some vehicle manufacturers recommend an alternative parking brake adjusting procedure similar to the one above, except that the parking brake control is partially applied during the adjustment. This helps prevent overadjustment by ensuring that there will be sufficient slack in the cables when the parking brake is released. To adjust the parking brake in this manner:

1. Shift the transmission into neutral, and apply the parking brake control as specified by the vehicle manufacturer. Most recommend that the control be engaged from one to seven clicks.

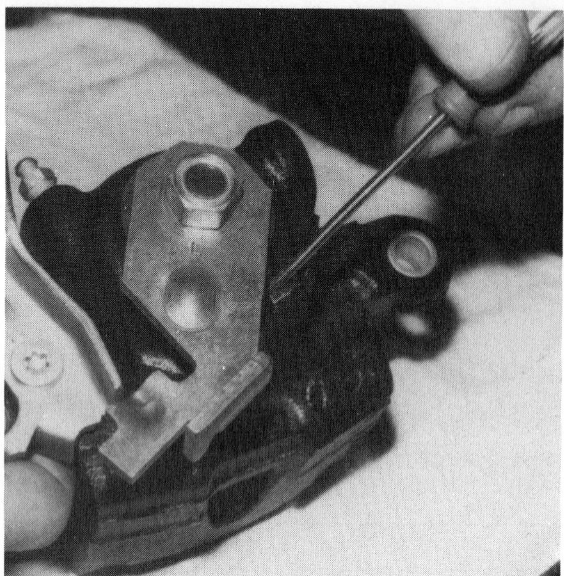

Figure 9-11. The parking brake levers on disc brake calipers must be fully returned when the brake is released.

2. Raise and properly support the vehicle so the wheels with the parking brakes hang free.
3. Tighten the cable adjustment until the brakes begin to drag as you rotate them by hand.
4. Fully release the parking brake control, and check that the wheels spin freely. Loosen the adjustment slightly if necessary.

Cable Adjustment — Disc Brakes

The cables that operate both General Motors and Ford caliper-actuated rear disc parking brakes can be adjusted using the following common procedure:

1. Shift the transmission into neutral, and fully release the parking brake control.
2. Raise and properly support the vehicle so the wheels with the parking brakes hang free.
3. Loosen the cable adjustment until there is slack in the cables, then make sure the parking brake levers on both calipers are fully returned. If the calipers are fitted with external lever stops, the levers should contact the stops when the parking brake control is released.
4. Tighten the cable adjustment until all of the slack is removed and either parking brake lever just starts to move.
5. Loosen the cable adjustment until both levers are again fully returned, figure 9-11.
6. Apply and release the parking brake several times, then make sure the parking brake levers still return all the way when the parking brake control is fully released. Loosen the adjustment slightly if necessary.

If the parking brake control travel is still excessive after you have adjusted the cables, the automatic adjuster in one or both caliper pistons is not operating properly. You can attempt to adjust the piston as described earlier in the chapter, but you will probably have to overhaul the caliper and replace the piston/adjuster assembly as described in Chapter 7.

PARKING BRAKE CABLE REPLACEMENT

In dry climates, parking brake cables often last the life of the car. However, in areas with lots of wet and cold weather, cables are subjected to rain, snow, and road salt that greatly increase rust and corrosion. This damage results in two distinct problems. First, the inner cable may stick or seize in its housing. And second, damage to exposed sections of cable can cause individual strands to snap until the cable is frayed to the point where it breaks entirely.

Various attempts have been made to help prevent parking brake cable damage. A few older cables have a grease fitting for lubrication, and some automakers recommend that exposed sections of inner cable be lightly greased. However, for the most part, modern parking brake cables are sealed and do not require regular maintenance. If a parking brake cable is frayed or broken, you must replace it. Generally, sticking and seized cables should also be replaced because the time and effort required to free them can seldom be justified.

Freeing Sticking and Seized Cables

If a replacement cable is unavailable and you must free up a sticking or seized parking brake cable, use penetrating oil to soak the inner cable where it enters the housing. Most cable damage is confined to within a few inches of where the cable enters the housing. Allow the oil to soak in for several minutes, then work the inner cable back and forth until it loosens up. If used with extreme care, heat applied near the ends of the housing can sometimes help free a sticking or seized cable.

Once the inner cable begins to move, pull it out of the housing as far as possible and clean the contamination from its surface with solvents and abrasive cloth. Continue to apply penetrating oil, move the cable in and out of the housing, and clean the cable as needed. When the cable is completely free, grease the exposed sections with brake grease to help prevent future problems.

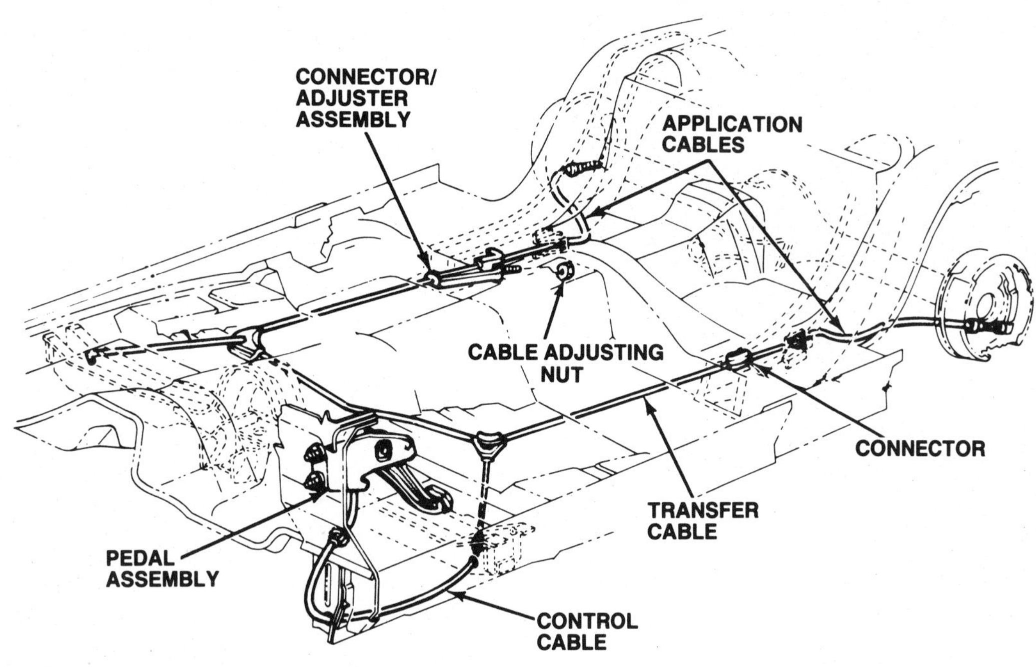

Figure 9-12. A typical four-cable parking brake linkage.

Cable Replacement Procedures

Parking brake linkages have from one to four cables that can be arranged in a wide variety of configurations. A typical domestic linkage, figure 9-12, has a control cable that runs from the parking brake control to the equalizer or adjuster, a transfer cable that runs from the equalizer or adjuster to near the rear wheels, and two application cables that run from connectors at the ends of the transfer cable to the parking brake levers at the wheels. The sections below describe typical procedures used to remove and install these three types of cables. Most other cables are replaced in a manner similar to one of these cables, or using some combination of the procedures described.

Whenever you replace a parking brake cable, check to see if you will need any additional parts before you order the new cable. Cable mounting hardware is often badly corroded, and will break apart when you remove the old cable. If the mounting hardware is in good condition, transfer it to the new cable. After you have installed the new cable, adjust it as described earlier, then apply the parking brake hard three or four times to pre-stretch the cable; readjust the cable if necessary.

Control cable replacement

The control cable is often attached to the parking brake pedal assembly under the dash. Typically, the cable will be concealed in the underdash wiring where it will be difficult to reach the cable end and disconnect it from the pedal linkage. In these situations, you may want to disconnect the battery ground cable to eliminate any chance of creating an electrical short circuit while working around the wiring. However, keep in mind that doing this on a modern car with computer controls will erase the computer memories for such convenience items as the radio, climate control, and power seats.

Installing a new parking brake control cable can also be difficult if you have to thread the cable through a hard-to-reach opening in the firewall or floor pan. To make the job easier, connect a wire about 50 inches (130 cm) long to the end of the cable that is removed last. Pull the old cable through the firewall or floor pan, then disconnect the wire from the cable, leaving the wire in the place normally occupied by the cable. Connect the wire to the end of the new cable that is installed first, then use the wire to guide and pull the new cable into position through the firewall or floor pan.

To replace a parking brake control cable:

1. Fully release the parking brake control.
2. Raise and properly support the vehicle.
3. Disconnect the lower end of the cable from the equalizer or adjuster.
4. Disconnect the brackets, clips, and screws that hold the cable to the frame or body.
5. Disconnect the upper end of the cable from the parking brake control.

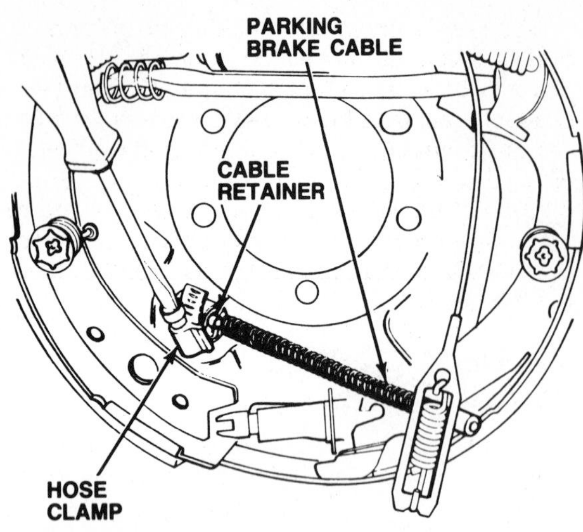

Figure 9-13. Compress the parking brake cable retainer with pliers or a hose clamp.

6. Remove the cable by pulling the upper end down through the hole in the firewall or floor pan. On some cars, the lower end of the cable is pulled up through the inside of the vehicle.
7. Insert the new cable through the firewall or floor pan as described above.
8. Connect the upper end of the cable to the parking brake control.
9. Attach the mounting hardware to the new cable and screw or clip the cable to the frame or body.
10. Connect the lower end of the cable to the equalizer or adjuster.

Transfer cable replacement

Parking brake transfer cables are the easiest to replace because the job is done entirely from under the car; there is no need to work inside the vehicle, or disassemble a friction assembly as is required when you replace an application cable.

To replace a parking brake transfer cable:

1. Fully release the parking brake control.
2. Raise and properly support the vehicle.
3. Loosen the parking brake adjustment, then disconnect the transfer cable from the adjuster or equalizer, and the application cable connectors.
4. Disconnect the brackets, clips, and screws that hold the cable to the frame or body.
5. Connect the new transfer cable to the adjuster or equalizer.
6. Attach the mounting hardware to the new cable and screw or clip the cable to the frame or body.
7. Connect the cable to the application cable connectors.

Application cable replacement

Parking brake application cables are easy to install on cars that have caliper-actuated disc parking brakes, however, they are somewhat more difficult to install when they actuate an integral drum parking brake. In these cases, you must disassemble the drum brake friction assembly in order to disconnect the cable end from the parking brake lever on the brake shoe, and release the cable housing from the brake backing plate. Details on drum brake disassembly and assembly can be found in Chapter 6.

To replace a parking brake application cable on a car with either an integral drum or caliper-actuated disc parking brake:

1. Fully release the parking brake control.
2. Raise and properly support the vehicle.
3. Loosen the parking brake adjustment, and disconnect the application cable from the equalizer or adjuster, or the transfer cable connector.
4. Disconnect the brackets, clips, and screws that hold the cable to the frame or body:
   a. With caliper-actuated disc parking brakes, go to Step 5.
   b. With integral drum parking brakes, go to Step 7.
5. Disconnect the cable from the parking brake lever on the caliper.
6. Connect the new cable to the parking brake lever on the caliper. Go to Step 11.
7. Disassemble the service brake as necessary, then disconnect the end of the cable from the parking brake lever on the brake shoe.
8. Use a pair of pliers or a hose clamp to compress the fingers of the cable retaining clip, figure 9-13, then remove the cable from the backing plate.
9. Install the new cable through the backing plate, and make sure the retaining clip locks into place.
10. Connect the end of the cable to the parking brake lever on the brake shoe, then reassemble the service brake.
11. Attach mounting hardware to the new cable, and screw or clip the cable to the frame or body.
12. Connect the end of the cable to the equalizer or adjuster, or the transfer cable connector.

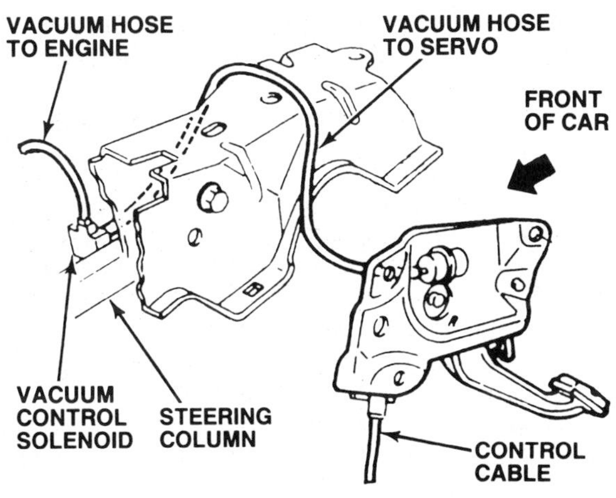

Figure 9-14. A vacuum parking brake release mechanism.

PARKING BRAKE SHOE AND PAD SERVICE

Integral drum and caliper-actuated disc parking brakes use the service brake shoes and pads as the parking brake friction material. In these applications, you simply change the brake shoes or pads as needed in the course of normal service. The inspection and replacement procedures for shoes and pads are covered in Chapter 6 for drum brakes, and Chapter 7 for disc brakes.

The linings of the small shoes used in auxiliary drum parking brakes are made of soft friction materials that provide a high coefficient of friction for better holding power. These shoes usually last the life of the car, or require replacement only at high-mileage intervals, because the parking brake only holds the vehicle in place and is not normally called upon to slow or stop the car. However, because the linings are softer than those in service brake systems, they will wear very quickly if the parking brake is overadjusted and drags, or if the driver forgets and drives even a short distance with the parking brake engaged.

For the same reasons described above, the friction surface of small parking brake drum inside the rotor is not normally subject to significant wear. This is important because, unlike a service-brake drum, the parking brake drum cannot be turned to refinish the friction surface. To do so would excessively weaken the rotor structure, which could lead to a brake failure.

Shoe replacement on auxiliary drum parking brakes is very similar to the same job on service drum brakes. If anything, the job is easier because the parking brake friction assemblies do not have wheel cylinders or automatic adjusters to deal with. To replace the auxiliary drum parking brake shoes, remove the caliper and rotor as described in Chapter 7. Then, refer to the drum brake inspection and service procedures in Chapter 6.

VACUUM RELEASE PARKING BRAKE

Some full-size cars have a vacuum-actuated automatic release mechanism on their parking brake pedal, figure 9-14. All of these systems operate in basically the same way. When the automatic transmission is placed in Drive or Reverse, a switch closes and completes an electrical circuit that activates a vacuum control solenoid. The solenoid then supplies vacuum to a servo on the parking brake pedal assembly, and the servo pulls a lever to release the parking brake pedal. On all of these systems, the lever can also be operated manually to override the vacuum release system in the event of a failure.

Most automatic parking brake release systems require a minimum of 10 in. Hg (35 kPa) of vacuum for proper operation. Before you proceed to more involved diagnosis, make sure the engine has good manifold vacuum at idle, and that all the vacuum hoses in the system are in good condition and tightly attached. To check the operation of the automatic parking brake release system:

1. Apply the parking brake, then pull the manual release lever, figure 9-15:
   a. If the parking brake releases, go to Step 2.
   b. If the parking brake does not release, locate and repair the problem in the pedal assembly.
2. With the service brakes applied, start the engine and allow it to idle.
3. Place the transmission shift lever in Neutral, then apply the parking brake.
4. Move the shift lever to Drive or Reverse:
   a. If the parking brake pedal releases, the system is operating properly and no further service is required.
   b. If the parking brake does not release, go to Step 5.

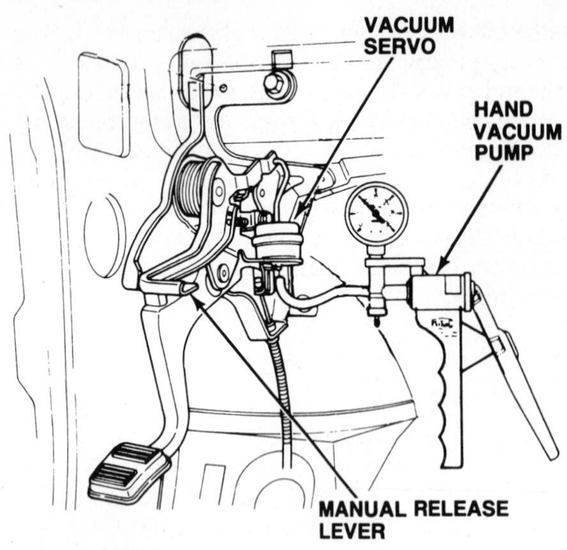

Figure 9-15. Using a vacuum pump to test the parking brake vacuum servo.

5. Remove the vacuum hose from the servo unit and connect a vacuum gauge to the hose.

6. Have an assistant apply the service brakes and move the shift lever to Drive or Reverse:
   a. If intake manifold vacuum is shown on the gauge, go to Step 7.
   b. If low or no vacuum is shown on the gauge, go to Step 8.

7. Use a hand pump to apply a minimum of 10 in. Hg (35 kPa) of vacuum to the servo, figure 9-15:
   a. If the servo does not hold a vacuum, replace the servo.
   b. If the servo holds a vacuum, check the servo link connection to the release lever, or locate and repair the problem with the pedal assembly.

8. Remove the vacuum supply hose from the vacuum control solenoid, and connect a vacuum gauge to the hose:
   a. If manifold vacuum is shown on the gauge, locate and repair the electrical problem with the solenoid, switch, or wiring of the system.
   b. If no or low vacuum is shown on the gauge, replace the obstructed vacuum supply hose.

Chapter

10

Power Brake and Anti-Lock Brake Service

Passenger car power brake systems use a vacuum or hydraulic power booster to increase brake application force and reduce the amount of foot pressure required on the brake pedal. Late-model anti-lock brake systems use computer-controlled valves to modulate brake system hydraulic pressure and prevent the brakes from locking. A failure in either a power brake or anti-lock system will not prevent the brake system from stopping the car, however braking does require greater effort without power brakes, and greater skill without anti-lock brakes.

As discussed in the *Classroom Manual,* three types of power boosters are used on cars today. The most common is the vacuum booster, which is used by all automakers and comes in both single- and dual-diaphragm forms. The second most common design is the Bendix Hydro-Boost hydraulic booster that is powered by the power steering pump; Hydro-Boost is used primarily by General Motors and the Ford Motor Company. The third, and newest, design is the General Motors electro-hydraulic Powermaster booster that combines the master cylinder, an electro-hydraulic pump, and a hydraulic booster into one compact unit. Although a variety of anti-lock brake systems are used on European cars, the two most common systems on cars sold in the U.S. are the Bosch ABS II and Teves designs.

Power brake service consists primarily of testing and replacing the three types of power boosters. Although it is possible to rebuild Hydro-Boost units and some vacuum boosters, most shops today simply install a factory-rebuilt assembly. When a vacuum booster is rebuilt or replaced, its output pushrod may need adjustment. And, when a hydraulic or electro-hydraulic booster is replaced, the power brake system must be bled of air in order to function properly. Because anti-lock brake systems are complex, service and diagnosis is limited to a few basic tests and inspections unless you have the factory shop manual and special test equipment for the system being repaired.

VACUUM BOOSTER TESTING

Vacuum boosters are generally trouble free, and do not require regular service other than a function test during routine brake system inspection. A problem with the vacuum power booster is unusual in a vehicle less than five years old, and many vacuum boosters last the life of the car. When a vacuum booster does fail, the service brake pedal height remains normal, but pedal feel becomes much harder, and significantly greater pedal force is required to

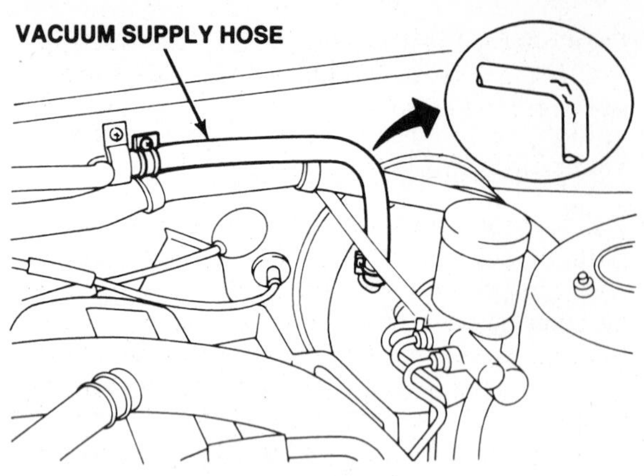

Figure 10-1. Vacuum supply hoses often crack where they bend.

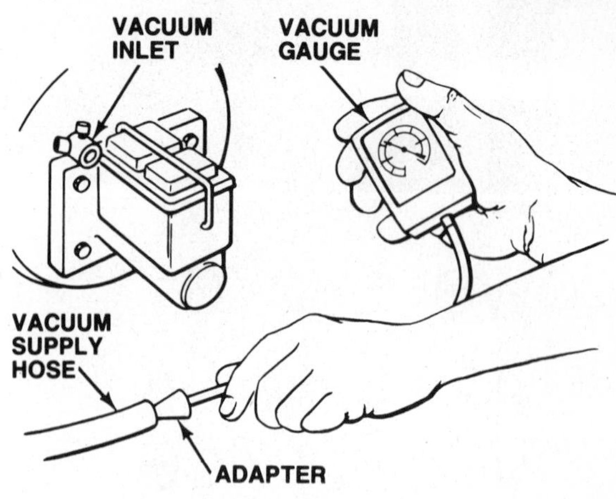

Figure 10-2. Use a vacuum gauge to check the vacuum supply to the booster.

slow and stop the car. Usually, it is impossible to apply enough force at the pedal to lock the brakes.

Because other problems in the brake system can also increase the force required to apply the brakes, the first step in vacuum booster diagnosis is to perform a function test to determine if the booster is the source of the problem. Depending on the function test results, you may also perform vacuum supply and booster leak tests described in this section to further identify the problem.

Vacuum Booster Function Test

The booster function test determines if the booster is receiving adequate vacuum and creating brake application force. With the ignition OFF, apply the brake pedal repeatedly with medium pressure until the booster reserve is depleted. There should be a power-assisted feel for at least two brake applications before the pedal becomes hard. If the pedal feels hard immediately, or after only one brake application, there may be a problem with the booster vacuum supply, or the booster may have a vacuum leak.

Once the reserve is depleted, hold medium pressure on the brake pedal and start the engine. If the booster is working properly, the pedal will drop slightly toward the floor as the engine begins to run, and less force will be required to hold the pedal in place. If the booster passes the function test, proceed to the booster leak test. If there is no noticeable change in the pedal position or feel, the booster is not operating properly, and you should perform the vacuum supply test.

Vacuum Supply Test

The vacuum supply test consists of several checks and inspections that make sure the power booster is receiving enough vacuum to function properly. There are two reasons the booster may not receive sufficient vacuum: either there is an obstruction or leak in the vacuum supply hose, or the amount of vacuum generated by the engine and/or vacuum pump is below specifications.

First, visually inspect the vacuum supply hose to the booster. Look for kinks or other indications that the hose may be blocked. If the car has a vacuum hose filter, remove the filter and blow through it to make sure it is clear. Next, inspect the hose for holes or cracks, figure 10-1. Make sure the fittings are tight where the hose connects to the booster check valve and vacuum source. Start the engine and allow it to idle, then listen along the hose for hissing noises that indicate a vacuum leak.

If the vacuum supply hose is clear and free from leaks, check the level of vacuum supplied by the engine or vacuum pump. With the ignition OFF, pump the brake pedal to deplete the booster reserve. Disconnect the vacuum supply hose from the booster, and connect a vacuum gauge to the hose using a cone-shaped adapter, figure 10-2. On all cars except those with electric vacuum pumps, start the engine and allow it to idle, then observe the vacuum reading on the gauge. The proper amount of vacuum will vary with the application, but the reading at idle should typically be between 15 and 20 in. Hg (50 and 70 kPa).

Sometimes, a vacuum supply hose that is restricted but not completely plugged, will allow a satisfactory reading on a vacuum gauge, but

Figure 10-3. Testing a vacuum booster inlet check valve.

prevent a sufficient "volume" of vacuum from reaching the booster. To quickly check for this problem, disconnect the vacuum supply hose from the booster while the engine is idling. If the engine does not stall almost immediately from the extra air drawn in through the supply hose, check for a restricted hose or filter. If the vacuum reading is within specifications, and the vacuum supply hose is not restricted, perform the booster leak test.

If the vacuum reading is less than 15 in. Hg (50 kPa), repair the problem at the vacuum source. If the vacuum source is a mechanically driven vacuum pump, rebuild or replace the pump as necessary. If the vacuum source is the engine, you will have to locate and repair the cause of the low manifold vacuum. This may be something as simple as improperly adjusted ignition timing, or as major as valvetrain and piston ring wear. After the vacuum source is repaired, repeat the booster function test. If the booster still fails the test, perform the booster leak test.

On cars with an electric vacuum pump, connect a vacuum gauge to the booster vacuum supply hose as described above, then turn the ignition ON, but do not start the engine. The vacuum pump should begin to run, and the reading on the vacuum gauge should increase. When vacuum level reaches the point specified in the factory shop manual, the vacuum pump should stop running. If the pump does not run at all, or continues to run after the proper vacuum level is reached, consult the shop manual for instructions on the diagnosis and repair of the pump electrical circuit. After the vacuum source is repaired, repeat the booster function test. If the booster still fails the test, perform the booster leak test.

Vacuum Booster Leak Test

The vacuum booster leak test consists of two parts that determine if the booster can maintain a vacuum reserve when the brakes are released, and contain vacuum within the booster when the brakes are applied. To check the ability of the booster to hold a vacuum when the brakes are released, run the engine at a fast idle for 30 seconds, then release the throttle and turn the ignition OFF. Wait two minutes, then apply the brake pedal repeatedly with medium force. There should be a power-assisted feel for at least two brake applications before the pedal feel becomes hard. If the booster passes this test, perform the brakes applied leak test below.

If you feel no power assist, either the booster vacuum check valve is leaking, or the booster has an unapplied vacuum leak. To test the vacuum check valve, disconnect the vacuum supply hose from the intake manifold or vacuum pump, and blow into the hose, figure 10-3. If air passes through the valve into the booster, replace the check valve. If air does not pass through the check valve, rebuild or replace the booster.

To check the ability of the booster to hold a vacuum while the brakes are applied, start the engine and allow it to idle. Close the car doors and windows, then listen carefully as you apply the brakes. If the engine begins to run rough, or there is a continuous hissing noise from the area where the brake pedal pushrod enters the booster, the booster has an internal vacuum leak and must be rebuilt or replaced.

VACUUM BOOSTER OUTPUT PUSHROD ADJUSTMENT

Like the input pushrod from the brake pedal, the vacuum booster output pushrod must be adjusted to the proper length. A pushrod that is too long holds the master cylinder in a partially applied position that prevents the brakes from releasing completely. This causes brake drag, premature wear of the pad and shoe linings, and possibly brake fade. A pushrod that is too short does not usually cause significant brake problems, although it does increase brake pedal travel slightly, and can cause a clunk or groaning noise from the booster in some applications.

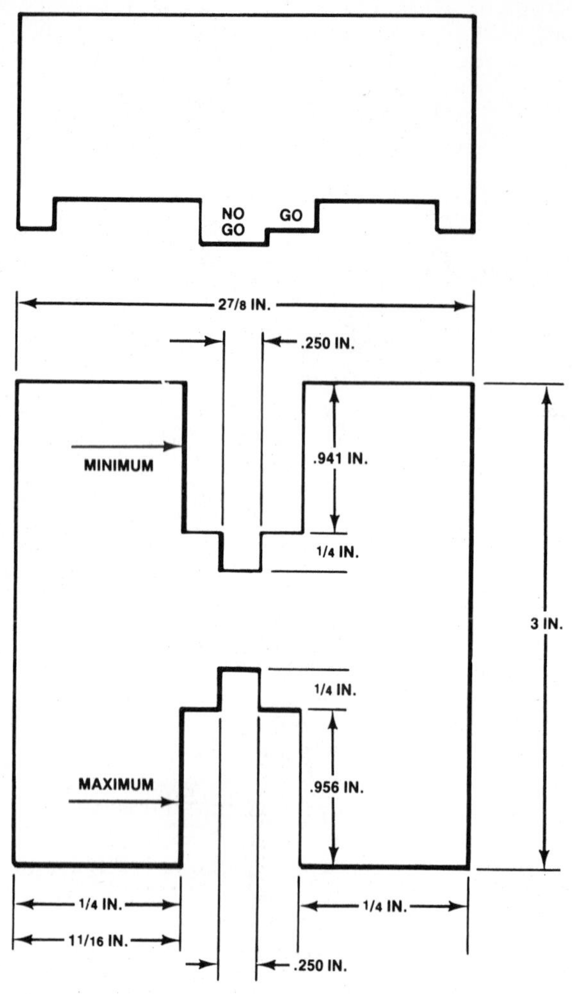

Figure 10-4. Typical vacuum booster output pushrod adjusting gauges.

Pushrod Adjustment Test

To determine if the booster output pushrod is too long, remove the cover or caps from the master cylinder fluid reservoir(s), and have an assistant *gently* apply the brakes. If fluid spurts from the compensating ports as the brakes are applied, the pushrod adjustment is satisfactory. Because brake fluid can spray when performing this test, wear safety glasses to prevent eye injuries, and use a fender cover to protect the vehicle finish.

If there are no spurts of fluid, check and adjust the brake pedal freeplay as described in Chapter 5, then repeat the test. If there are still no spurts, loosen the master cylinder retaining nuts and pull the cylinder away from the booster approximately ⅛ inch (3 mm). Repeat the test. If fluid spurts now appear, adjust the booster output pushrod. If spurts do not appear, the master cylinder should be disassembled to determine why the compensating ports are obstructed.

Pushrod Adjusting Methods

On most cars, you should adjust the booster output pushrod whenever you install a new or rebuilt vacuum power booster, or if the system fails the pushrod adjustment test above. However, some manufacturers do not recommend, or provide specifications for, booster pushrod adjustment. In these applications, you simply install the new booster and hope that the pushrod was properly adjusted at the factory.

To adjust the pushrod on a power booster that is already installed on the car, you must separate the cylinder from the booster. To do this, remove the master cylinder retaining nuts, and pull the cylinder forward and away from the booster. If you do this carefully, you generally do not have to disconnect the brake lines from the cylinder. Once the master cylinder is free from the booster, support it so there is no stress on the brake lines.

Where pushrod adjustment *is* recommended, a go/no-go gauge is commonly used to check the pushrod position. Most gauges are flat pieces of metal or plastic with calibrated notches cut in them, figure 10-4, although some import cars use a variable gauge. The procedure used to adjust the output pushrod is basically the same for all vacuum boosters; the most common variations are described below.

Ford booster pushrod adjustment

Ford power boosters are adjusted when they are installed on the car and the engine is idling. To adjust the output pushrod, place the "minimum" side of the go/no-go gauge over the pushrod. In this position, a 5-lb (22-N) force against the pushrod should seat the legs of the gauge flat against the booster housing, figure 10-5. If necessary, hold the pushrod with a pair of pliers and turn the self-locking adjusting nut with a wrench, figure 10-6, until there is the proper preload when the gauge contacts the pushrod. To double check the adjustment, place the "maximum" side of the go/no-go gauge over the pushrod so the legs of the gauge seat flat against the booster housing. In this position, the end of the pushrod should not contact the gauge.

General Motors pushrod adjustment

General Motors power boosters are adjusted off the car, or on the car with the engine not running. To adjust the output pushrod, place the go/no-go gauge over the pushrod so the legs of

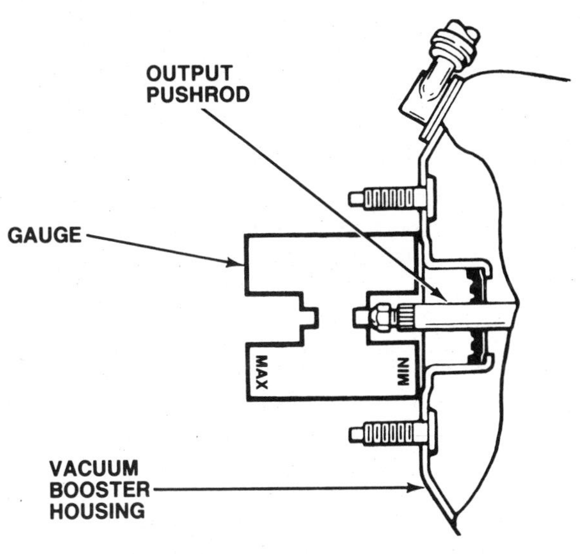

Figure 10-5. Adjusting the output pushrod on a Ford vacuum booster.

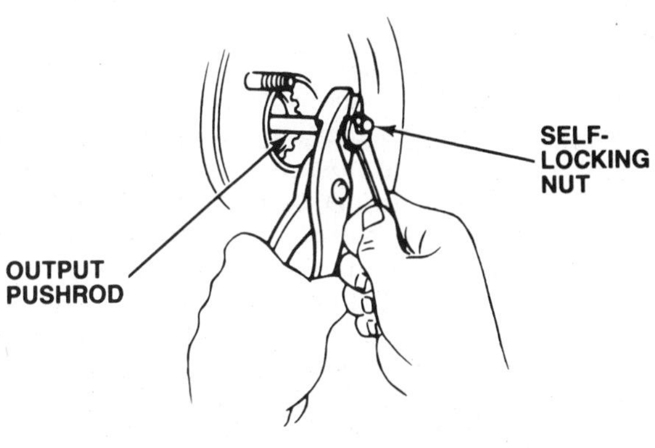

Figure 10-6. Use a wrench and a pair of pliers to adjust the booster output pushrod length.

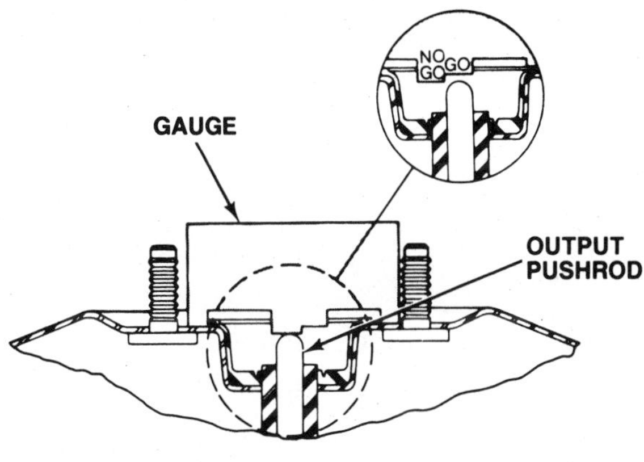

Figure 10-7. A properly adjusted output pushrod on a GM vacuum booster.

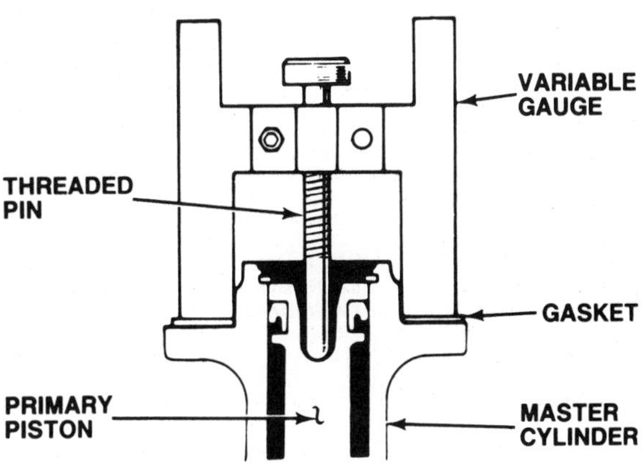

Figure 10-8. Initial setting of a variable gauge used to adjust the booster output pushrod.

the gauge seat flat against the booster housing, figure 10-7. Slide the gauge from side to side to check the pushrod length. The pushrod tip should *always* contact the lower "no-go" section of the gauge, but *never* contact the upper "go" section of the gauge. If the pushrod length is not within these limits, obtain an adjustable pushrod from a General Motors dealer. To set the proper length, hold the pushrod with a pair of pliers and turn the adjusting nut with a wrench, figure 10-6.

Variable gauge pushrod adjustment

Some imported cars use a variable gauge to transfer an adjustment measurement from the master cylinder to the booster output pushrod. To adjust an output pushrod using a variable gauge, figure 10-8, place the gauge on the master cylinder and screw the threaded pin downward until it lightly contacts the master cylinder primary piston. Then, turn the gauge over and place it on the power booster as shown in figure 10-9. Adjust the booster output pushrod until it lightly contacts the head of the threaded pin on the gauge.

VACUUM BOOSTER REPLACEMENT

All vacuum boosters are replaced in basically the same way. The generic procedure below can be used to replace the vacuum power brake booster on most cars.

1. Remove or reposition any wires, brackets, hoses, or other components that obstruct access to the booster.
2. Disconnect the vacuum supply hose from the booster.

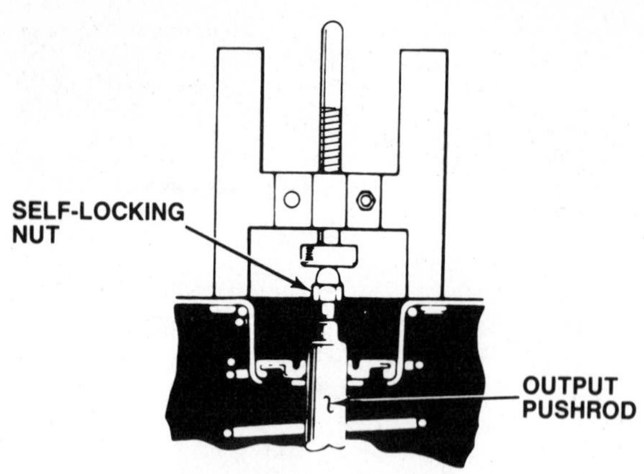

Figure 10-9. Using a variable gauge to check the vacuum booster output pushrod length.

3. Remove the master cylinder retaining nuts, and pull the cylinder away from the booster. In most cases, you do not have to disconnect the brake lines if you do this carefully. Be sure to support the cylinder so there is no stress on the brake lines.
4. Disconnect the brake pedal pushrod from the pedal arm. To do this, it may be necessary to remove the stoplight switch.
5. Remove the retaining nuts that hold the booster to the firewall, figure 10-10, then withdraw the booster from the vehicle.
6. Check and adjust the output pushrod of the replacement booster as described above.
7. Position the new booster on the firewall, then install and tighten the retaining nuts.
8. Connect the brake pedal pushrod to the pedal arm.
9. Install and adjust the stoplight switch if you removed it in Step 4. See Chapter 4 for information on switch adjustment.
10. Position the master cylinder on the booster, then install and tighten the retaining nuts.
11. If you removed the brake lines from the master cylinder in Step 3, reconnect them at this time and bleed the brakes as described in Chapter 3.
12. Connect the vacuum supply hose to the booster.
13. Install any wires, brackets, hoses, or other components removed in Step 1.
14. Start the engine to build vacuum in the booster, then perform a booster function test.

HYDRO-BOOST TESTING

Just as with a vacuum power booster, a failure in a Bendix Hydro-Boost system causes a high, hard brake pedal that makes it difficult for the driver to slow and stop the car. Because the

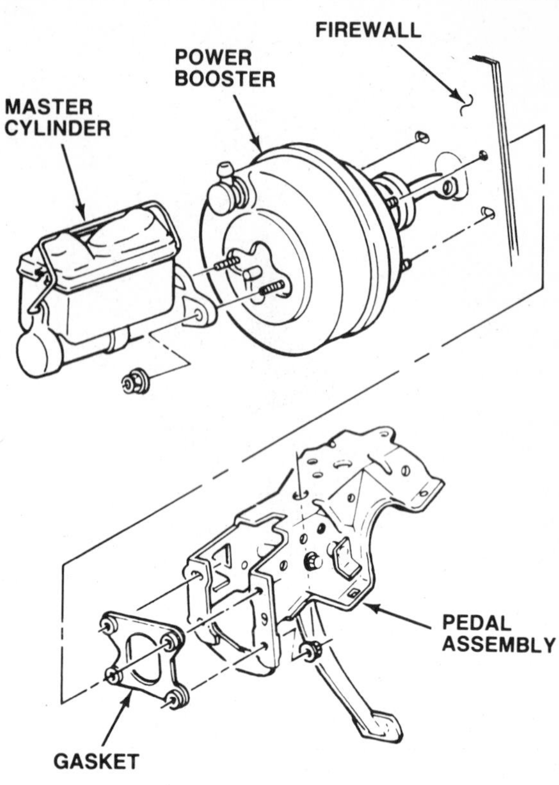

Figure 10-10. A typical vacuum booster installation.

Hydro-Boost system depends on the power steering pump for its boost pressure, a power brake problem may be accompanied by erratic steering feel. And, if air enters the power steering system and causes cavitation, there may be pulsations in the brake pedal as well.

To diagnose a Hydro-Boost malfunction, perform a pre-test inspection to eliminate any outside problems, then perform a function test to determine if the Hydro-Boost unit is the source of the problem. Depending on the function test results, you may also perform one or more of the other tests in this section to help further identify the problem. All of the tests apply to both Hydro-Boost I and Hydro-Boost II systems unless otherwise noted.

Hydro-Boost Pre-Test Inspection

Because the Hydro-Boost system is powered by the power steering system, the first step in any diagnosis is to check for power-steering-related problems. Start by checking the fluid level in the power steering pump reservoir. During parking and other low speed maneuvers, a low fluid level can cause a moan or low-frequency hum, accompanied by a vibration in the brake pedal or steering column.

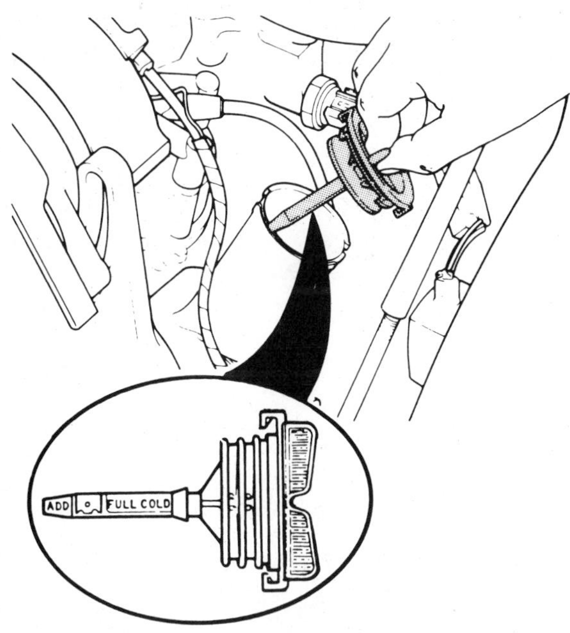

Figure 10-11. The power steering fluid level must be correct for the Hydro-Boost unit to operate properly.

Figure 10-12. Belt problems usually occur on the underside of the belt that runs against the pulleys.

To check the fluid level, remove the cap/dipstick from the power steering reservoir and check the fluid level on the dipstick, figure 10-11. Some manufacturers recommend you check the level with the fluid at operating temperature, others say to check the level when the fluid is cold. Some dipsticks have markings for both hot and cold fluid levels. If the level is

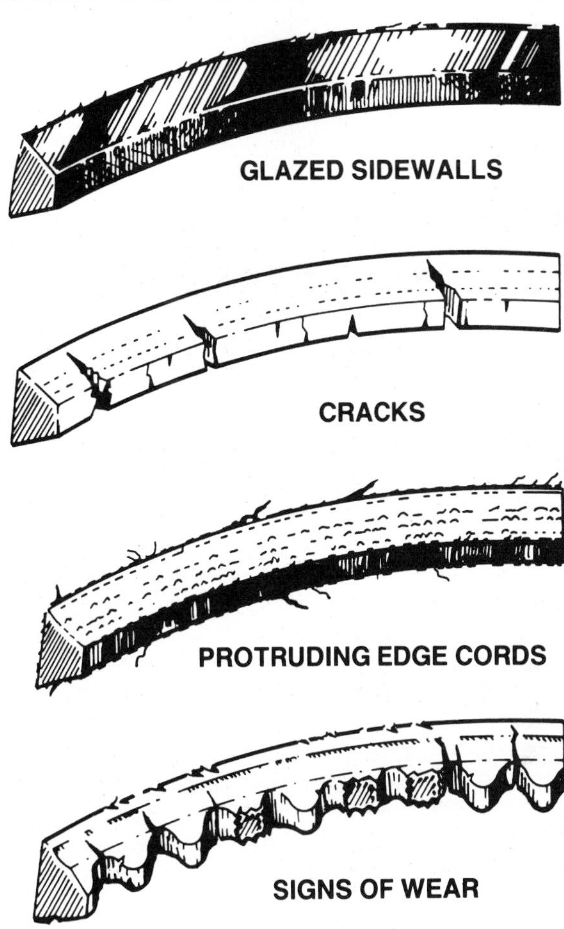

Figure 10-13. Inspect the power steering belt for problems.

low, top up the reservoir with the type of fluid recommended by the vehicle manufacturer. Using the wrong fluid can damage rubber seals and hoses in the power steering and Hydro-Boost systems.

Next, inspect the power steering pump drive belt, figure 10-12. Check the portions of the belt that run against the pulleys for cracks, fraying, separation, brittleness, grease and oil contamination, glazing, and excessive wear, figure 10-13. Replace the belt if any of these conditions are present.

Once you have made sure the pump drive belt is in good condition, or installed a new belt, check and adjust the belt using a strand tension gauge, figure 10-14. If the pump is driven by a flat serpentine belt, use a gauge designed for that type of belt construction. A belt that is too tight quickly wears out. A loose belt slips under load and allows the power steering pump to slow. This reduces the output of the pump and leads to erratic power assist to the

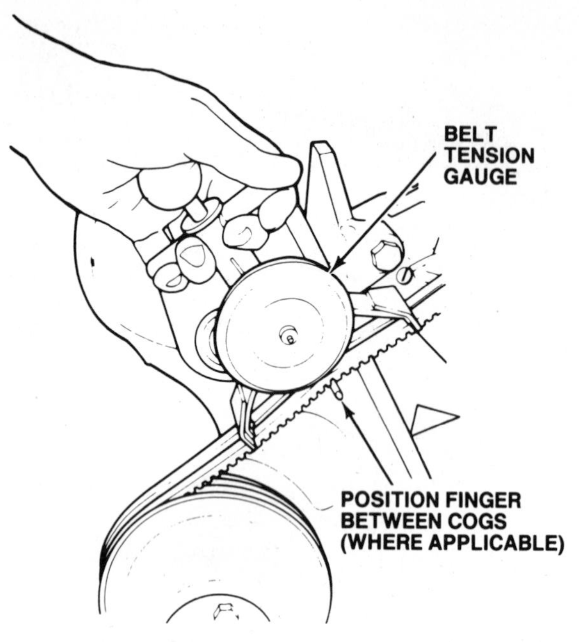

Figure 10-14. Always use a tension gauge to check and adjust belt tightness.

brake and steering systems. Belt tensions differ for new and used belts, and the correct specifications for both can be found in the vehicle owner's and shop manuals. A used belt is any belt that has been tensioned and run for more than 10 or 15 minutes.

Next, inspect all of the power steering and Hydro-Boost lines, hoses, and connections, figure 10-15, for leaks, kinks, and wear. Tighten or replace parts as necessary. To confirm a leak, observe the suspected leak location while an assistant runs the engine at a fast idle and turns the steering wheel to full lock in either direction. This greatly increases pressure in the hydraulic system, and will force fluid out of even a small leak. To prevent damage to the system, do not hold the steering at full lock for more than five seconds. And, for personal safety, wear safety glasses to protect your eyes from high pressure fluid spray.

Next, check the master cylinder fluid level as described in Chapter 3. If the fluid level has dropped low enough for air to enter the system and give the pedal a spongy feel, it will be hard to judge the condition of the Hydro-Boost unit in the tests below. Top up the master cylinder and bleed the brake system as required before proceeding.

Finally, make sure the engine idle speed is set to the proper rpm. If the idle speed is too low, the power steering pump will not turn fast enough to produce the pressure required for the Hydro-Boost unit to operate properly. If this happens, the results of the tests below may not be valid.

Hydro-Boost Function Test

The function test checks the ability of the Hydro-Boost unit to provide power assist. With the engine not running, apply the brake pedal five or more times with medium force to discharge the accumulator. The pedal feel will harden noticeably when the accumulator is discharged. Next, apply the brake pedal with medium force, and start the engine. If the booster is working properly, the pedal will drop toward the floor, then push back upward slightly. If the booster passes this test, perform the accumulator test below. If there is no change in the pedal position or feel, the booster is not working, and you should perform the power steering pump test to determine whether the problem is in the pump or the booster.

Power Steering Pump Test

The power steering pump test checks whether the pump produces enough fluid pressure and flow to enable the Hydro-Boost unit to provide full power brake assist. To perform the power steering system test, drive the vehicle until the power steering fluid is at operating temperature, then shut off the engine. Disconnect the pressure line to the Hydro-Boost unit at the power steering pump end, and install a special pressure/flow analyzer (Kent-Moore No. J 25323 or equivalent) in series between the pump and hose, figure 10-16.

Start the engine, and follow the instructions that come with the analyzer to check the fluid pressure and flow provided by the power steering pump. Typical readings at idle are between 80 and 150 psi (550 and 1035 kPa) of pressure, with a flow rate between 1.25 and 1.75 gallons per minute. The exact specifications vary with the application, so check the factory shop manual for the car you are servicing. If the pump pressure and flow are within specifications, but the booster still fails a function test, rebuild or replace the Hydro-Boost unit. If the readings are below the normal ranges, rebuild or replace the power steering pump, then repeat the function test.

Hydro-Boost Accumulator Test

The accumulator test checks the operation of the booster reserve system by making sure the accumulator can hold a charge of hydraulic pressure. The accumulator test consists of two

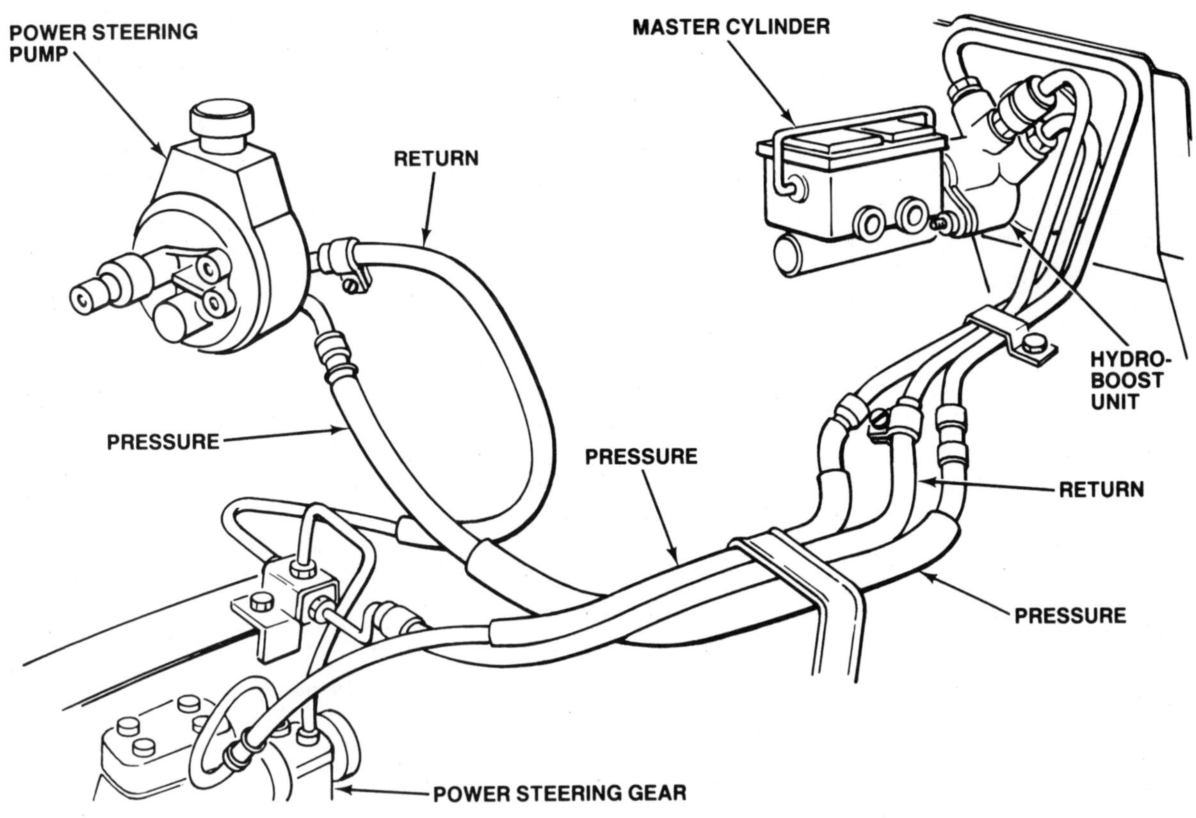

Figure 10-15. The extensive plumbing of the Hydro-Boost system can lead to fluid leaks.

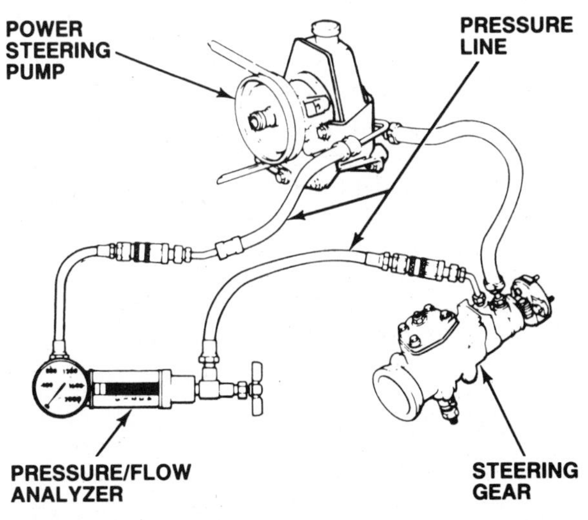

Figure 10-16. Pressure/flow analyzer installation to check the power steering pump output.

parts that check the ability of the reserve system to hold a short-term charge and a long-term charge.

To test the ability of the system to store a short-term charge, start the engine and allow it to idle. Charge the accumulator by turning the steering wheel slowly one time from lock to lock; do not hold the steering at full lock for more than five seconds. Shut off the engine and release the steering wheel, then repeatedly apply the brake pedal with medium force. If the accumulator can hold a charge, a Hydro-Boost I unit will provide two or three power assisted applications; a Hydro-Boost II unit will provide one or two.

To test the ability of the system to store a long-term charge, start the engine and recharge the accumulator as described above. As the accumulator charges on a Hydro-Boost I system, you should hear a slight hissing sound as fluid rushes through the accumulator charging orifice. Once the accumulator is charged, shut off the engine and do not apply the brake pedal for one hour. At the end of the hour, repeatedly apply the brake pedal with medium force. Once again, a Hydro-Boost I unit should provide two or three power assisted applications; a Hydro-Boost II unit should provide one or two.

If the Hydro-Boost unit fails these tests, it usually means the accumulator of a Hydro-Boost I unit, or the accumulator/power-piston assembly of a Hydro-Boost II unit, is leaking and the booster must be rebuilt or replaced. However, if a Hydro-Boost I system fails the test but does not make the hissing sound that indicates the accumulator is charging, the fluid in the system is probably contaminated, and you may be able to solve the problem by flushing the Hydro-Boost system.

Flushing the Hydro-Boost I system

If you suspect fluid contamination in a Hydro-Boost I system, flush the power booster, power steering pump, and fluid lines as follows:

1. Raise and properly support the front of the car so the wheels can be steered freely.
2. Disconnect the fluid pressure line at the steering gear, and place the end of the line in a drain pan.
3. Disable the vehicle ignition system to prevent the engine from starting. On diesel engines, disconnect the power supply wire to the fuel injection pump.
4. Have an assistant crank the engine, and at the same time, pump the brake pedal and slowly turn the steering wheel from lock to lock. To prevent starter damage, do not crank the engine for longer than 30 seconds at a time; allow the starter to cool for two minutes between cranking periods.
5. As the engine is cranking, fluid will be pumped from the system into the drain pan. As this occurs, continuously add new fluid to the power steering reservoir to maintain a minimum fluid level.
6. Once you have added approximately three quarts of fluid, have your assistant stop cranking the engine, then reconnect the pressure line to the power steering gear.
7. Lower the car, check and adjust the power steering fluid level, then start the engine and turn the steering wheel slowly from lock to lock several times.
8. Shut the engine off, then check and adjust the power steering fluid level.
9. Repeat the accumulator test. If the accumulator still will not hold a charge, rebuild or replace the Hydro-Boost unit.

HYDRO-BOOST REPLACEMENT

The accumulator of a Hydro-Boost unit holds over 1,000 psi (6,900 kPa) of pressure. As a result, it is *extremely important* to discharge the accumulator before you disconnect any of the lines or hoses attached to the booster. To discharge the accumulator, apply the brake pedal five or more times with medium force while the engine is not running. The pedal feel will harden noticeably when the accumulator is discharged.

To replace a Hydro-Boost power booster:

1. Remove or reposition any wires, brackets, hoses, or other components that obstruct access to the booster.
2. Remove the master cylinder retaining nuts, and pull the cylinder away from the booster. In most cases, you do not have to disconnect the brake lines if you do this carefully. Be sure to support the cylinder so there is no stress on the brake lines.
3. Remove the three hydraulic lines from the power booster. Cap the lines to avoid fluid loss and prevent contaminants from entering the system.
4. Disconnect the brake pedal pushrod from the pedal arm. To do this, it may be necessary to remove the stoplight switch.
5. Remove the retaining nuts that hold the booster to the firewall, then withdraw the booster from the vehicle.
6. Position the new booster on the firewall, then install and tighten the retaining nuts.
7. Connect the brake pedal pushrod to the pedal arm.
8. Install and adjust the stoplight switch if you removed it in Step 4. See Chapter 4 for information on switch adjustment.
9. Connect the three hydraulic lines to the power booster.
10. Position the master cylinder on the booster, then install and tighten the retaining nuts.
11. If you removed the brake lines from the master cylinder in Step 2, reconnect them at this time and bleed the brakes as described in Chapter 3.
13. Install any wires, brackets, hoses, or other components removed in Step 1.
14. Bleed the Hydro-Boost system as described in the next section, then perform a function test.

HYDRO-BOOST BLEEDING

Hydro-Boost power brake boosters are basically self-bleeding unless large quantities of air get into the power steering system, such as when it is opened for service. Air in the booster hydraulic circuits can cause noises and vibrations in the brake pedal and steering column during parking and other low speed maneuvers. To bleed the Hydro-Boost system, follow the procedure below. Once most of the air has been bled from the system in this manner, normal braking and steering will purge any small air pockets that remain in the system.

After you have finished bleeding a Hydro-Boost I system, there may be a gulping noise when the brake pedal is applied. The noise is caused by small amounts of air trapped in the system, and will go away after running the engine for a few minutes and lightly pumping the brake pedal. Once the noise disappears, check the fluid level in the power steering pump and add fluid if necessary.

To bleed a Hydro-Boost power brake system:

1. Check and adjust the power steering fluid level.
2. Disable the vehicle ignition system to prevent the engine from starting. On diesel engines, disconnect the power supply wire to the fuel injection pump.
3. Crank the engine with the starter for several seconds.
4. Recheck the power steering fluid level, and top it up as necessary. Repeat Steps 3 and 4 until the fluid level remains constant.
5. Reconnect any wires removed in Step 2, and start the engine.
6. Turn the steering wheel slowly from lock to lock two times.
7. Shut off the engine, and discharge the accumulator.
8. Restart the engine and turn the steering wheel slowly from lock to lock two times.
9. Shut off the engine, then check and adjust the power steering fluid level.
10. If fluid foaming occurs when bleeding the Hydro-Boost system, shut off the engine and allow the car to sit for one hour. Then, check and adjust the power steering fluid level, and repeat the bleeding procedure.

POWERMASTER TESTING

The General Motors Powermaster brake system does not rely on hydraulic pressure from an outside source such as the power steering pump. Instead, the unit has a self-contained electro-hydraulic pump, and requires only a 12-volt power source to provide braking assist. As with other power boosters, a failure of the Powermaster system causes a high, hard brake pedal that makes it difficult for the driver to slow and stop the car. A Powermaster failure may also be indicated by illumination of the brake warning light on the instrument panel, or by unusual operation of the electro-hydraulic pump motor.

Before you suspect a faulty Powermaster unit, be sure the problem you are dealing with is booster related. For example, a faulty Powermaster cannot cause a low brake pedal; only about one-eighth inch (3 mm) of pedal pushrod travel is required to activate the booster valves. The Powermaster also cannot cause such common problems as brake pull, a pulsating or spongy pedal, brake squeal, or a failure in one half of the dual-circuit braking system.

If you suspect a problem with the Powermaster unit, check and adjust the fluid level in the reservoir as detailed in Chapter 3, then perform a function test. Depending on the results, perform the additional tests detailed below for external leaks, electrical problems, improper system pressures, and internal leaks. If the results of any test indicate that the Powermaster booster needs repair, the unit must be replaced. There are no rebuild kits available at this time.

Several Powermaster tests require that the electro-hydraulic pump be allowed to run in order to pressurize the system. When doing these tests, do not allow the pump to run for more than 20 seconds at a time, or it may overheat. To prevent pump damage, turn off the ignition and allow the pump to cool for two minutes between each operating cycle.

Powermaster Function Test

The Powermaster function test is a preliminary diagnostic procedure with several steps that check different aspects of the Powermaster system. To perform the test:

1. Inspect the outside of the Powermaster for signs of brake fluid leaks:
   a. If there are no signs of leakage, go to Step 2.
   b. If there is leakage, perform the external leak test below.
2. Apply the brake pedal with firm pressure and hold:
   a. If the pedal remains at a fixed height, go to Step 3.
   b. If the pedal drops toward the floor, the master cylinder section of the Powermaster is bypassing internally. Replace the Powermaster.
3. Pump the brake pedal at least 10 times with the ignition OFF to discharge the accumulator. The pedal feel will harden noticeably when the accumulator is discharged.
4. Release the parking brake.
5. Turn the ignition switch to RUN. The brake warning light should come on and the electric pump motor should begin to operate. After approximately 20 seconds, the pump motor should stop and the light should go out.
   a. If the pump motor and warning light operate as described, go to Step 6.
   b. If the pump motor does not run, perform the electrical checks below.

Figure 10-17. The special pressure test gauge installed on the Powermaster.

c. If the pump motor runs for longer than 20 seconds, perform the pressure tests below.
6. Leave the ignition switch in the RUN position, and do not apply the brake pedal for five minutes. If the pump motor begins to operate during that time, perform the internal leak test.

External Leak Test

External leakage from a Powermaster unit is usually easy to spot because the accumulator is pressurized to over 500 psi (3,450 kPa). As a result, even a small leak will cause fluid to seep or spray from the unit. Even when the leak is quite small, the Powermaster reservoir quickly runs dry of fluid.

To locate the source of a leak, wipe the Powermaster assembly and hoses clean with a rag soaked in brake cleaner or alcohol, and install a fender cover to protect the vehicle finish from possible brake fluid spray. Top up the reservoir with fresh brake fluid, and turn the ignition switch ON. As the pump motor runs to charge the accumulator, check for leaks at the reservoir cover, reservoir mounting grommets, pressure switch, accumulator, hose and pipe connections, and under the dash at the end of the pedal pushrod. Once you locate the leak, replace defective parts or tighten connections as needed. If fluid is leaking from around the brake pedal pushrod, replace the Powermaster unit.

If there is a great deal of fluid on the outside of the Powermaster reservoir, but the fluid level inside is normal, the fluid spill was probably caused by an overfilled reservoir and not by a leak. If the reservoir is filled when the accumulator is charged, excess fluid will be forced past the reservoir cover as fluid is returned to the reservoir from the accumulator during normal brake operation.

Electrical Checks

The electro-hydraulic pump of the Powermaster unit draws a great deal of current. In order for the booster to operate properly, it must have a constant 12-volt power supply, and the battery must be in good condition and fully charged. If the pump motor does not run during a function test, check the 30-amp Powermaster fuse, and make sure all electrical connections at the Powermaster unit are tight. If the fuse and connections are both okay, consult the factory shop manual of the vehicle being serviced for diagnosis information on the Powermaster electrical circuit.

Pressure Tests

The Powermaster pressure tests are used to determine why the electro-hydraulic pump motor runs too long or too often. There are three separate tests to check the upper limit pressure, the lower limit pressure, and the accumulator precharge pressure. Perform all three tests if the Powermaster function test or internal leak test instructs you to do so.

To perform the pressure tests requires a special pressure testing gauge (Kent-Moore No. J 35126 or equivalent). The tool consists of a high-pressure gauge, a bleeder valve, and a bleeder hose. To install the gauge, discharge the accumulator by applying the brake pedal at least ten times with medium pressure while the ignition is OFF. The pedal feel will become noticeably harder when the accumulator is discharged. Remove the pressure switch from the Powermaster unit, install the gauge its place, figure 10-17, then install the pressure switch into the opening in the gauge mounting boss. Route the bleeder hose from the gauge into the booster side of the fluid reservoir.

Upper limit pressure test

The upper limit test checks the pressure at which the electro-hydraulic pump shuts off. To perform the upper limit pressure test, close the bleeder valve on the special tool, then turn the ignition ON and allow the pump motor to run. The pump should shut off when the pressure

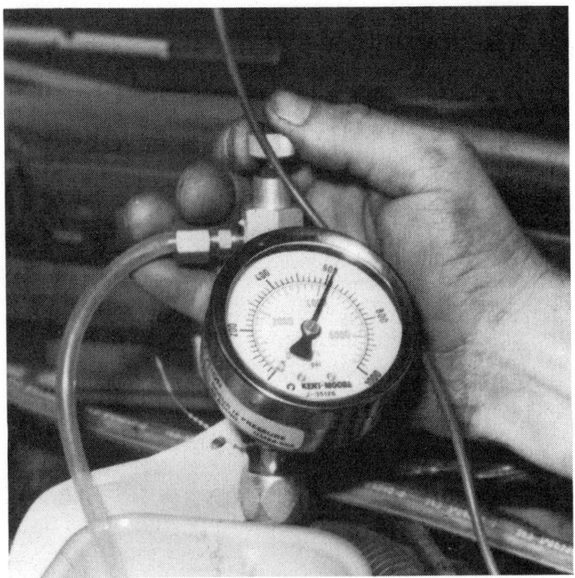

Figure 10-18. The Powermaster test gauge showing the correct reading for the upper limit pressure check.

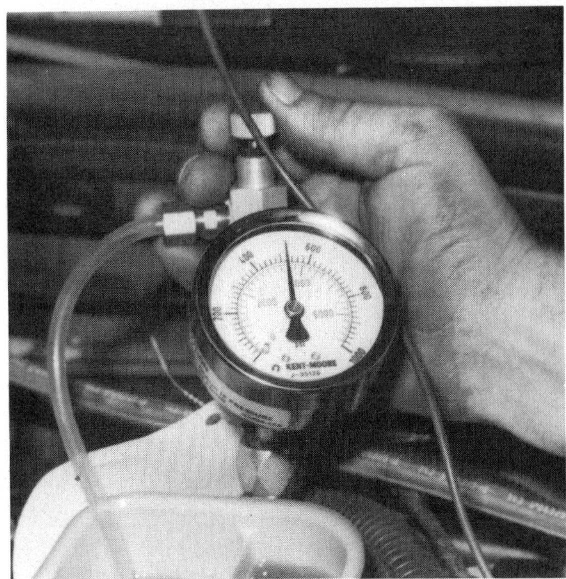

Figure 10-19. Open the bleeder valve on the test gauge to perform the Powermaster lower limit pressure check.

reading on the gauge reaches 635 to 735 psi (4,380 to 5,070 kPa), figure 10-18. If the pressure exceeds the upper limit, replace the pressure switch. If the pressure never reaches the upper limit and the pump motor runs longer than 20 seconds, replace the pump.

Lower limit pressure test

The lower limit test checks the pressure at which the electro-hydraulic pump turns on. To perform the lower limit pressure test, turn the ignition ON and wait until the pump motor shuts off, indicating that the accumulator is fully charged. Make sure the bleeder hose from the gauge is routed into the booster side of the fluid reservoir, then slowly open the bleeder valve, figure 10-19. Watch the reading on the gauge, and note the pressure at which the pump motor begins to run; this should be between 490 and 530 psi (3380 to 3655 kPa). If the pump starts to run at a higher pressure, or does not start to run at the lower limit, replace the pressure switch.

Accumulator precharge pressure test

The accumulator precharge pressure test checks the ability of the accumulator to hold a charge of hydraulic pressure. To perform the test, turn the ignition OFF, and have an assistant discharge the accumulator by repeatedly applying the brake pedal with medium force. As he does so, observe the gauge and note the pressure reading just before the brake pedal becomes hard. This is the accumulator precharge pressure, which should be between 200 and 300 psi (1,380 and 2,070 kPa).

Another way to test the accumulator precharge pressure is to discharge the accumulator as described above, then turn the ignition ON. The pressure reading on the gauge should immediately jump to the precharge pressure. If the Powermaster fails either of the precharge tests, replace the accumulator.

Internal Leak Test

If there is no evidence of an external leak, but the pump motor runs within a 5 minute period when the brake pedal is not applied and the ignition is ON, suspect an internal leak. To test for this problem:

1. Turn the ignition ON and pump the brake pedal until the electro-hydraulic pump motor begins to run.
2. Stop pumping the brake pedal, wait until the pump motor stops running, then turn the ignition OFF.
3. Hold a clear plastic hose, figure 10-20, over the fluid return port in the booster section of the Powermaster fluid reservoir, port B in figure 10-21:
   a. If brake fluid rises up the tube, fluid is leaking past the check valve. Replace the Powermaster.
   b. If fluid does not rise in the tube, go to Step 5.

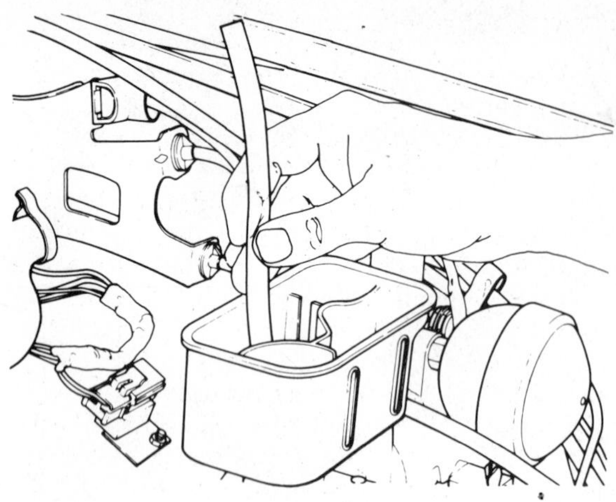

Figure 10-20. Use a clear plastic hose to check for internal leaks in the Powermaster.

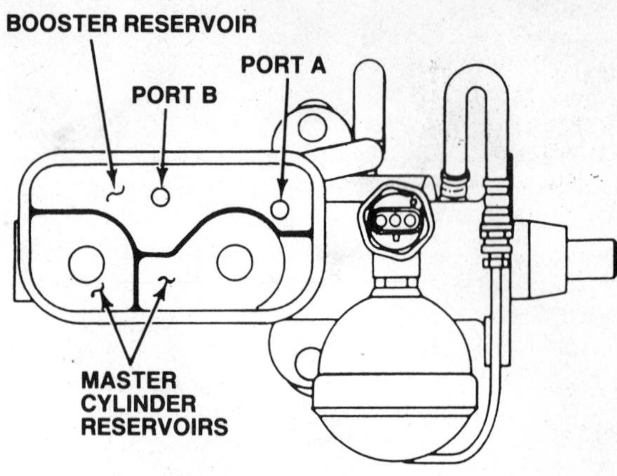

Figure 10-21. The Powermaster fluid return port (A) and supply port (B).

5. Hold a clear plastic hose over the pump supply port in the booster section of the Powermaster fluid reservoir, port A in figure 10-21:
   a. If fluid rises up the tube, there is a leak past the internal valves. Replace the Powermaster.
   b. If fluid does not rise in the tube, perform the pressure tests above.

POWERMASTER REPLACEMENT

The accumulator of a Powermaster unit holds over 500 psi (3,450 kPa) of pressure. As a result, it is *extremely important* to discharge the accumulator before you disconnect any of the lines or hoses attached to the booster. To discharge the accumulator, apply the brake pedal 10 or more times with medium force while the ignition is OFF. The pedal feel will harden noticeably when the accumulator is discharged.

To replace a Powermaster brake booster:

1. Remove or reposition any wires, brackets, hoses, or other components that obstruct access to the booster.
2. Detach the electrical connectors from the pressure switch and electro-hydraulic pump.
3. Disconnect the brake lines from the master cylinder portion of the Powermaster.
4. Disconnect the brake pedal pushrod from the pedal arm. To do this, it may be necessary to remove the stoplight switch.
5. Remove the retaining nuts that hold the booster to the firewall, then withdraw the booster from the vehicle.
6. Bench bleed the master cylinder portion of the Powermaster as described in Chapter 3.
7. Position the new booster on the firewall, then install and tighten the retaining nuts.
8. Connect the brake pedal pushrod to the pedal arm.
9. Install and adjust the stoplight switch if you removed it in Step 4. See Chapter 4 for information on switch adjustment.
10. Connect the brake lines to the master cylinder portion of the Powermaster.
11. Attach the electrical connectors to the pressure switch and electro-hydraulic pump.
12. Install any wires, brackets, hoses, or other components removed in Step 1.
13. Fill and bleed the Powermaster unit as described below.

POWERMASTER FILL AND BLEED

The master cylinder section of the Powermaster unit should be bench bled in a conventional manner before the booster is installed on the car. However, once the Powermaster unit is in place, you must use the procedure below to fill and bleed the booster section of the unit. During this process, do not allow the pump to run for longer than 20 seconds at a time.

To fill and bleed a Powermaster booster:

1. Remove the fluid reservoir cover and fill the booster side of the reservoir with fresh DOT 3 brake fluid.
2. Turn the ignition switch ON. As the pump motor runs, the fluid level in the booster side of the reservoir will drop as the accumulator is charged. Add fluid as needed to prevent the reservoir from running dry.
3. When the pump motor stops, adjust the fluid level so the ports in the bottom of the reservoir are just covered, then install the reservoir cover.
4. Turn the ignition OFF, and discharge the accumulator. Pump the brake pedal 10 or more

times with medium force. The pedal feel will harden noticeably when the accumulator is discharged.

5. Remove the reservoir cover, and top up the fluid level to the "maximum" mark on the reservoir.

6. Reinstall the reservoir cover and turn the ignition ON to charge the accumulator. Add fluid as needed to prevent the reservoir from running dry.

7. Repeat steps 4 through 7 a total of 10 to 15 times to purge all of the air from the booster. When you are completed, the fluid level in the booster side of the reservoir should always return to the "maximum" mark when the accumulator is discharged.

ANTI-LOCK BRAKE SERVICE

As mentioned at the outset of this chapter, anti-lock brake system (ABS) diagnosis is complex, so only very limited service information can be provided in this text. Any ABS diagnosis beyond the simple procedures below requires a factory shop manual for the vehicle being serviced, as well as special electrical test equipment. The following sections explain how to discharge the hydraulic accumulator of the Teves system, and make some basic diagnostic checks on both Teves and Bosch ABS.

Teves Accumulator Discharging

A fully charged hydraulic accumulator in a Teves integral ABS contains over 2,600 psi (18,000 kPa) of pressure. As a result, it is *extremely important* that you discharge the accumulator before disconnecting any lines or hoses attached to the unit. The accumulators on both Ford and General Motors Teves units are discharged the same way: apply the brake pedal a minimum of 20 times with approximately 50 lb (220 N) of force while the ignition is OFF. The pedal feel will harden noticeably when the accumulator is discharged.

Basic Diagnostic Checks

When an anti-lock brake system malfunctions, only the anti-lock capabilities are disabled; normal braking is not impaired in any way. Because of this, all anti-lock brake systems have an amber "anti-lock" warning light that indicates when the electronic control unit detects a problem in the system. The warning light also comes on when the car is first started while the computer is running its self-test sequence. If the light does not come on when the car is started, check for a burned out bulb.

Anti-lock system problems are signalled in a slightly different ways for Bosch and Teves systems. When a problem occurs in a Bosch ABS, the amber "anti-lock" warning light comes on and remains on constantly. However, when a problem occurs in a Teves ABS, the amber "anti-lock" warning light and/or the red "brake" warning light may come on. Either or both lights may stay on constantly, or flash on and off at certain times. A chart in the shop manual is used to interpret the light sequence, and direct you to more involved diagnostic procedures.

When the warning lights indicate that there is a problem in the anti-lock system, make the following basic checks:

- Check for blown fuses in the ABS circuits. Always replace a blown fuse with a new fuse of the same ampere rating.
- Make sure the battery is fully charged. Anti-lock units must have an uninterrupted power supply at a minimum of 12 volts or the control computer will disable the system.
- Check that the electrical connections at the control module, wheel speed sensors, and other ABS components are securely installed.
- Make a visual inspection of the wheel speed sensors to ensure they are not physically damaged.

Chapter

11

Bearing, Tire, Wheel, and Chassis Service

The wheel bearings, tires and wheels, and vehicle chassis all affect braking performance to one degree or another. The wheel bearings support the brake drums and rotors that help create the friction that stops the car. If the wheel bearings are not in good condition and adjusted properly, they can increase brake wear and vibration, reduce the stability of the vehicle while braking, and result in seal damage that can contaminate the brake linings with grease or axle lubricant. The tire and wheel assemblies form the friction link between the brake system and the road, so any problem in this area upsets traction, and therefore braking power and balance. The vehicle chassis ties all these elements together, and unless it is in good condition, the car may pull under braking.

This chapter is not a comprehensive guide to wheel bearing, tire and wheel, and chassis service; it deals only with those areas that are encountered in the course of diagnosing brake problems, or repairing the brake system. The bearing service section covers diagnosis, basic bearing service, and replacement procedures for the major types of wheel bearings and their seals. The tire and wheel service section covers installation, inspection, and checking procedures for radial and lateral runout. The chassis service section covers the basic inspection procedures for major chassis components.

TYPES OF WHEEL BEARINGS

As discussed in the *Classroom Manual,* three types of bearings are used as wheel bearings on automobiles: straight roller bearings, ball bearings, and tapered roller bearings. These bearings are used in three different ways to support the wheels, and each design requires somewhat different service procedures. The sections below describe the three basic styles of wheel bearings.

Adjustable Dual Wheel Bearings

The most common wheel bearing design is adjustable dual bearings, figure 11-1. This layout is used on non-driven wheels at both the front and rear axles. With this design, the inner bearing is always somewhat larger than the outer bearing because it supports the additional loads created by inertia and weight shift during cornering. All modern cars with adjustable dual wheel bearings use tapered roller bearings, however, some older cars have ball bearings instead.

Adjustable dual wheel bearings require inspection, adjustment and lubrication approximately every 30,000 miles (40,000 km) under

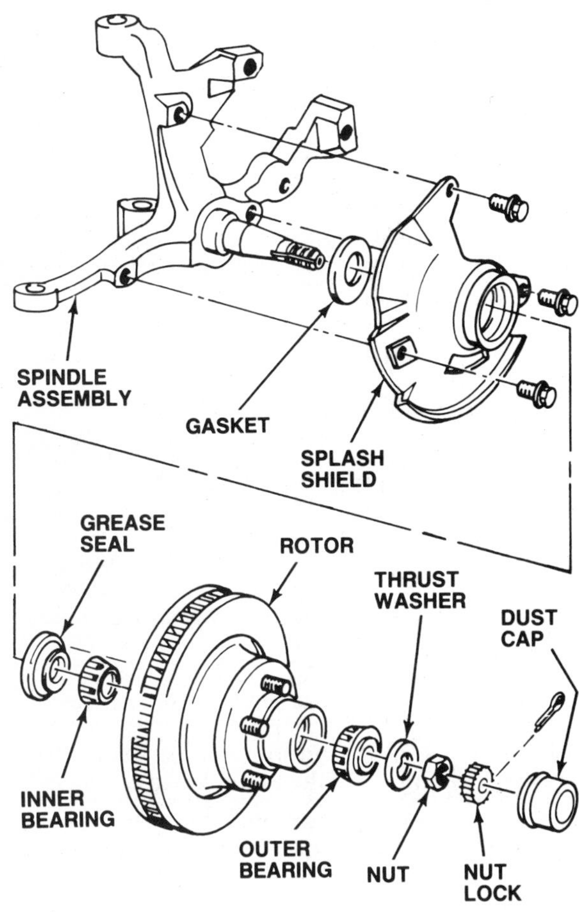

Figure 11-1. A typical adjustable dual wheel bearing assembly with tapered roller bearings.

normal driving conditions, and every 15,000 miles (20,000 km) in severe service. Consult the manufacturer's maintenance schedule to determine the exact intervals for the car you are servicing. These bearings should also be serviced if they are removed in the course of brake work or other vehicle repairs.

Sealed Wheel Bearings

The second type of wheel bearing is the sealed, non-adjustable, double-row bearing assembly found on the front wheels of most modern FWD cars, figure 11-2. A similar bearing assembly is used on some non-driven axles as well. Both bearings in this design have the same, fairly large, size. Large bearings are necessary because they are close together, and the loads created by supporting the axle are therefore somewhat greater. Originally, most sealed bearing units used dual ball bearings; however, some newer cars have sealed bearings with dual tapered roller bearings. Sealed bearings

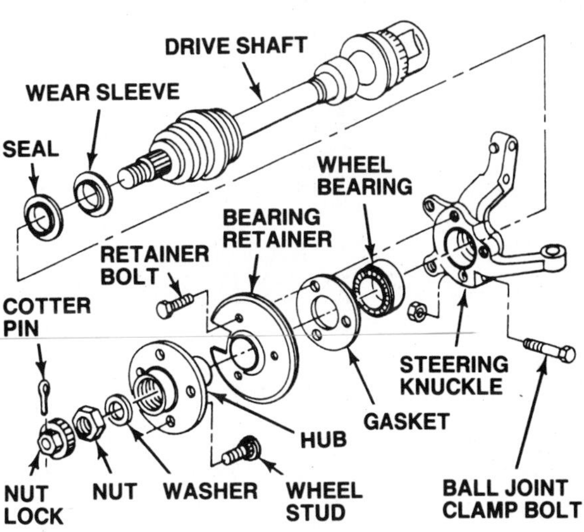

Figure 11-2. A typical sealed wheel bearing assembly.

require no periodic maintenance, and must be replaced when they become worn.

Solid Axle Wheel Bearings

The third type of wheel bearing is the single bearing used on cars with a solid rear axle. This type of wheel bearing, located near the outer end of the axle, can be either a straight roller bearing or a ball bearing. Straight roller bearings, figure 11-3, are pressed into the axle housing and use the axle shaft as the inner bearing race. The axle is retained in the housing by a C-lock that fits into a groove in the end of the shaft inside the differential. The axle lubricant also lubricates the wheel bearing, and a seal at the end of the housing keeps the lubricant inside the axle.

Ball bearings used on solid axles are generally pressed onto the axle shaft, figure 11-4, and the axle is held in place by a retainer plate that fits over the bearing and bolts to the axle housing. The retainer plate is usually secured by the same bolts that hold the brake backing plate in place. The ball bearings used on solid rear axles are either sealed units, or are unsealed and packed with bearing grease; rear axle ball bearings are not lubricated by the axle lubricant. An inner seal in the axle housing prevents axle lubricant from coming into contact with the bearing, and in some cases, an outer seal in the retainer plate prevents the bearing grease from escaping and contaminating the brake linings.

Like sealed wheel bearing assemblies, solid axle wheel bearings do not require periodic maintenance. They are only serviced when they become worn, or are removed in the course of brake work or other repairs.

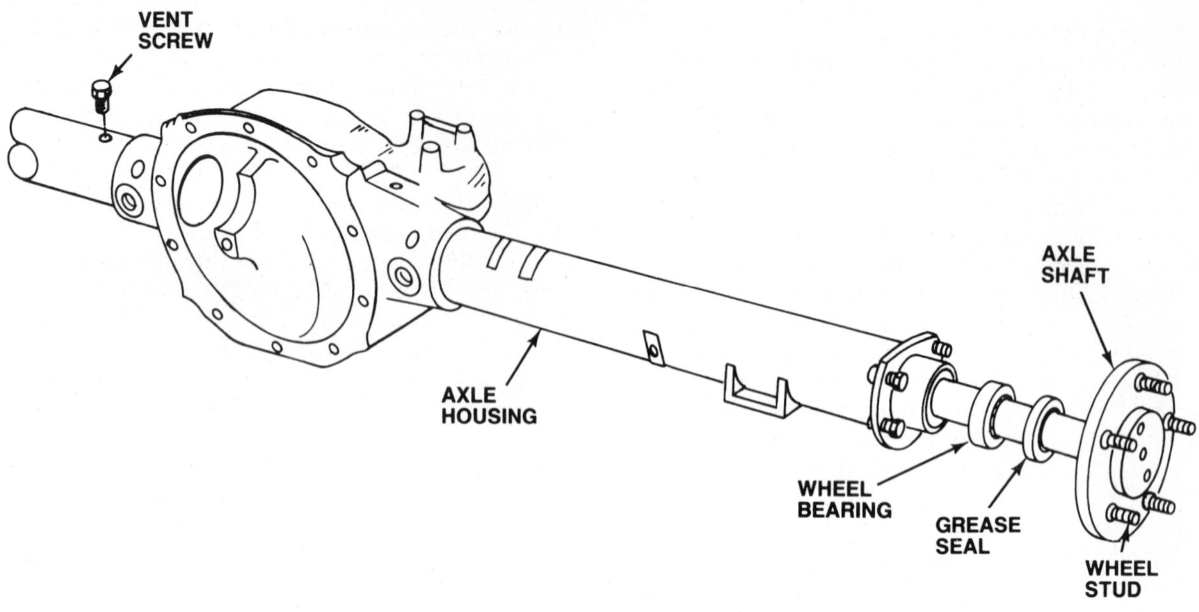

Figure 11-3. A solid rear axle with straight roller wheel bearings.

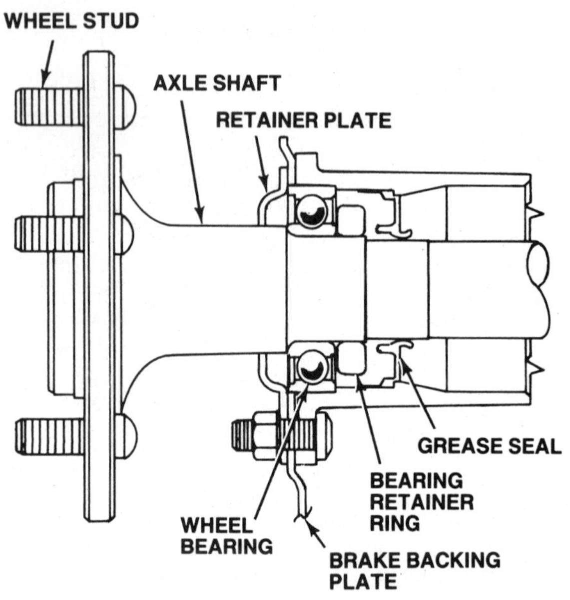

Figure 11-4. A solid rear axle with ball bearing wheel bearings.

WHEEL BEARING DIAGNOSIS

Worn or misadjusted wheel bearings can cause a number of brake problems. For example, loose or worn bearings allow brake drum and rotor runout that can lead to increased lining wear and excessive brake pedal travel. Loose bearings also make tire and wheel imbalance more noticeable, and at the front of the car, bearing problems allow the wheel alignment to vary; both these conditions cause instability during braking. At either axle, loose or worn bearings can lead to leaking wheel seals that allow grease or axle lubricant to contaminate the brake linings.

There are four signs that indicate a wheel bearing needs to be serviced: noise, roughness, grease seal leakage, and excessive axial play. These symptoms usually appear in various combinations depending on the severity of the problem and the type of bearing used. The sections below describe the symptoms in greater detail, and explain the procedures used to isolate bad wheel bearings.

Bearing Noise Tests

Wheel bearings rarely fail all at once; instead, the rolling surfaces deteriorate slowly and become rough. As the bearing rotates, this roughness causes a growling sound that is the most common symptom of a wheel bearing problem. As wear becomes worse, the sound grows louder and louder until the vehicle owner brings the car in for service.

It is usually not too difficult to tell whether a bearing noise is coming from the front or rear axle, although it is harder in some cases than in others. However, deciding which side of the car a bad bearing is located on can be very difficult. There are two tests that use noise to help isolate a bad bearing, one is done on the road, the other is done in the shop.

Bearing noise road test
The most common method used to isolate a bad wheel bearing is a road test. To road test a car with a bearing noise, find a wide street or large parking lot where there is plenty of room to safely manuever. Drive the car at approximately 20 mph (30 kph), and swerve back and forth to alternately load and unload the bearings on each side of the car. As you swerve to the right, the left-side bearings are put under greater load; as you swerve to the left, the right-side bearings are put under greater load. As you do this, listen for changes in the level of bearing noise. Unless there are bad bearings on both sides of the car, the noise is usually much louder when the car is turned in one direction than the other.

The results of this test are interpreted in different ways depending on the type of wheel bearings on the car. With adjustable dual wheel bearings, where the inner bearing carries a larger part of the total load and creates a louder noise when it fails, the bearing noise will increase when the side of the car with the defective bearing is loaded. For example, if the left-side inner bearing is bad, bearing noise will increase when the car is steered to the right. If the right side bearing is bad, bearing noise will increase when the car is steered to the left.

Double-row sealed bearing noises are interpreted the same way as those from adjustable dual bearings, providing one of the inner bearings is bad. However, if an outer bearing is defective the results will be reversed; the noise will increase when the side of the car opposite the bad bearing is loaded. For example, if the outer bearing on the right side is bad, the noise will increase when the car swerves to the right. This can create confusing test results, and one or more of the additional tests described below will be required to isolate a bad sealed bearing.

If a solid rear axle has ball bearings, the noise will increase when the side of the car with the bad bearing is more heavily loaded. The same is true of a solid axle with straight roller bearings, however, a road test may not be conclusive in these cases because straight roller bearings do not support axial loads, and therefore the noises they make are not as greatly affected by side-to-side weight shifts. One or more of the additional tests described below will be required to isolate a bad straight roller bearing.

Bearing noise shop test
If a suspect wheel bearing is on a driven axle, and the car does not have a limited-slip differential, you can use the engine to spin the wheels individually in the shop to help locate the problem. The bad bearing will generally make more noise when it is spun. This test can also be inconclusive, however, because the wheels are unloaded which reduces bearing noise somewhat. At the same time, the differential side gears are spinning which increases the amount of background noise.

This test requires that you run the vehicle in gear while it is raised off the ground in the shop. For safety reasons, this is best done with the car on a hoist, although jack stands are acceptable. In either case, use extreme care and make sure the car is properly supported so it cannot move in any way during the test. To check wheel bearing noise with the car in the shop:

1. Raise the vehicle so the wheels clear the floor by about two inches, then support the car so the suspension is in approximately the same position as when the car is resting on the ground. If the wheels are allowed to hang free, additional drivetrain noises will be created that interfere with bearing diagnosis. And, on FWD cars, the constant velocity joints may be damaged.
2. Block the wheels on one side of the driven axle so they cannot turn.
3. Have an assistant start the engine, place the car in gear, and accelerate the vehicle until approximately 20 mph (30 kph) shows on the speedometer with the car in high gear. With one driven wheel blocked, the opposite wheel will then rotate at a speed equivalent to about 40 mph (65 kph); do not exceed this limit or drivetrain damage may result.
4. Listen for bearing noise at the spinning wheel.
5. Stop the engine, block the wheel you just listened to so it cannot rotate, then repeat Steps 3 and 4 at the opposite wheel.

Bearing Feel Tests

The same wear that causes bearing noise can also be felt as a roughness if the wheel is rotated with the car supported off the ground. If the tests above are inconclusive, or the suspect bearing is not on a driven axle, raise and properly support the vehicle, then rotate the wheels by hand to check bearing condition. Feel for roughness in the bearing as it turns, and listen for a low-frequency rumbling noise. Compare the results on opposite sides of the car to determine where the problem lies.

Figure 11-5. A leaking grease seal can indicate a bad wheel bearing.

Bearing Grease Seal Tests

Another sign that a wheel bearing needs attention is leakage from the grease seal, figure 11-5. This is especially true of adjustable dual bearings and solid rear axle bearings. Grease or axle lubricant leaking from the seal can be a sign that excessive bearing play has caused seal damage. Also, once a seal begins to leak, brake dust and other contaminants can enter to harm the bearing. Always replace leaking grease seals to prevent contamination of the brake linings, and at the same time, service the wheel bearings.

Bearing Axial Play Tests

The final indication of wheel bearings that are worn or need adjustment is excessive axial play. If inward and outward movement of the bearing hub or axle exceeds specifications, the bearings must be serviced. Checking the axial play is the *only* means of determining the condition of sealed, double-row bearing assemblies; however, axial play *cannot* be used to check straight roller bearings in solid rear axles. Straight roller bearings do not support axial loads, so excessive side play does not indicate bearing condition, although it may indicate a worn C-lock or retaining groove in the axle.

The most common method used to check axial play on cars with adjustable dual wheel bearings is by hand. With the car raised and properly supported, grasp the top and bottom

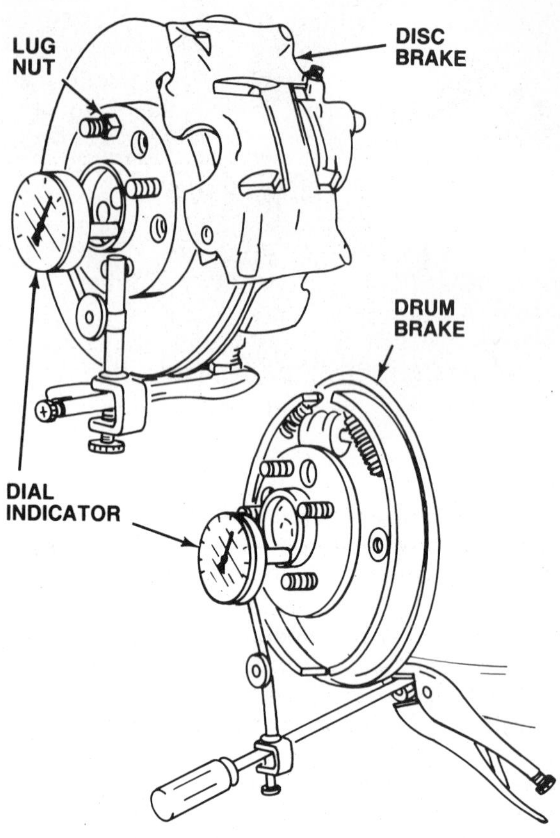

Figure 11-6. Using a dial indicator to check wheel bearing axial play.

of the tire and attempt to wobble it back and forth, and in and out. You should feel little or no axial play by hand. Experience will give you a better idea of how much play is normal. If play is excessive, service the bearings.

A more accurate way to check bearing axial play is to use a dial indicator. This method must be used to check sealed wheel bearing assemblies unless the bearing is so badly worn that movement is apparent in the hand test above. To check axial play with a dial indicator:

1. Raise and properly support the vehicle, then remove the wheel.
2. If the bearings to be checked are on a wheel equipped with a disc brake, use one of the methods described in Chapter 7 to push the caliper pistons into their bores just far enough that the brake pads do not drag against the rotor. It is not necessary to bottom the pistons in their bores.
3. Mount a dial indicator on the suspension, figure 11-6, and position the plunger against the edge of the bearing hub.
4. Push the hub inward toward the suspension until it will move no further, then zero the indicator dial.

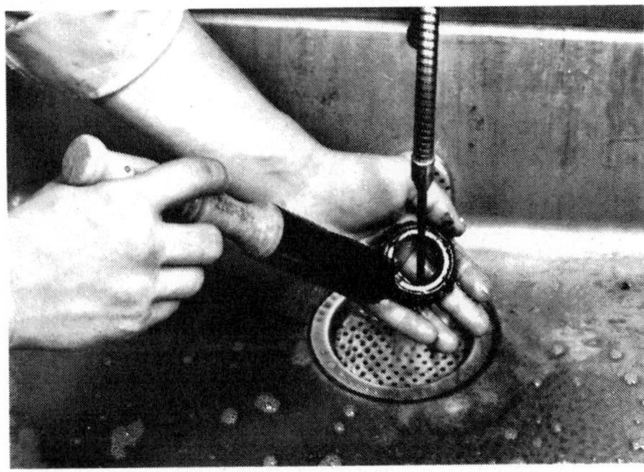

Figure 11-7. Wash unsealed wheel bearings in clean solvent.

5. Pull the hub outward away from the suspension until it will move no further. The reading on the indicator dial is the bearing axial play.

Adjustable dual wheel bearings (tapered roller bearings) are generally allowed between .001 and .005″ (.025 and .127 mm) axial play. A tapered roller bearing with less than .001″ (.025 mm) axial play will wear rapidly, and one with more than .005″ (.127 mm) axial play allows excessive drum or rotor runout. Though it is possible to only adjust loose bearings, the best policy is to always remove, clean, inspect, and repack the bearings unless you know they were recently serviced and are in good condition.

Sealed wheel bearing assemblies with ball bearings, or the single ball bearings used in solid rear axles, are allowed only a very small amount of axial play, usually a maximum of .002″ (.05 mm). Sealed bearing assemblies with tapered roller bearings are allowed up to .005″ (.125 mm) play like their adjustable counterparts. Always consult the factory shop manual to get the proper specifications for the vehicle being serviced. If the axial play of a sealed bearing, or a rear axle ball bearing, exceeds the amount allowed by the vehicle manufacturer, replace the bearing.

BASIC WHEEL BEARING SERVICE

Wheel bearings are removed for four reasons: they are diagnosed as being bad, they require repacking as part of routine maintenance, they must be removed so a leaking grease seal can be replaced, or they must be removed so other vehicle repairs can be made. The procedure used to remove and reinstall a wheel bearing varies with the type of bearing used, and detailed instructions for this job are given later in the chapter. However, in each case, the procedure is basically the same whether the bearing is being replaced or simply repacked.

Except for sealed bearings, all wheel bearings share a number of common service procedures that should be performed whenever the bearing is removed for any reason. The sections below provide some general bearing service tips, and describe how to clean wheel bearings, inspect them for wear and damage, and pack them with grease.

General Bearing Replacement Tips

When you replace an unsealed wheel bearing, leave the new part in its package until you are ready to install it; the special wrappings are designed to keep the bearing clean and rust free. Before you touch a new bearing, wash your hands and dip your fingers in clean motor oil. This ensures that the acids on your fingers will not harm the bearing.

Whenever you replace a tapered roller or ball bearing that has a separate outer race, you must replace the race as well. Even though an old outer race may appear to be in perfect condition, it has been subjected to the same stress and metal fatigue as the old bearing. An old race also has a certain wear pattern from running against the old bearing. If you install a new bearing in an old race, it will not fit properly, and premature bearing wear and failure will result.

Whenever you service any type of wheel bearing, install new grease seals. Seals wear with age just like bearings, and are often damaged during bearing removal. A new seal keeps grease or axle lubricant off the brake linings and in the bearing or axle where it belongs. At the same time, a new seal keeps out the dirt, water, and brake dust that can shorten a bearing or axle's service life.

Wheel Bearing Cleaning

When an unsealed wheel bearing is removed, wipe away as much old grease as possible using dry rags or paper towels; inspect the grease on the towels for metal chips or other indications of bearing wear or damage. Next, clean the bearing in a parts washer using *clean* petroleum-base solvent. A long-bristled brush is helpful to flush out hard to reach grit, figure 11-7. Wash each bearing individually, and keep bearings with detachable outer races separated so you can replace them in the same races from which they were removed. Clean unsealed rear axle ball bearings while they are still on the axle.

Once all of the old grease has been washed out, flush the bearings with a non-petroleum-base brake cleaning fluid; this removes any

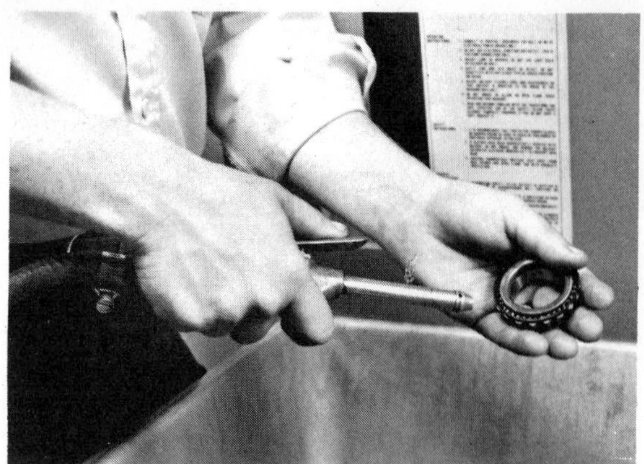

Figure 11-8. Use compressed air to dry wheel bearings.

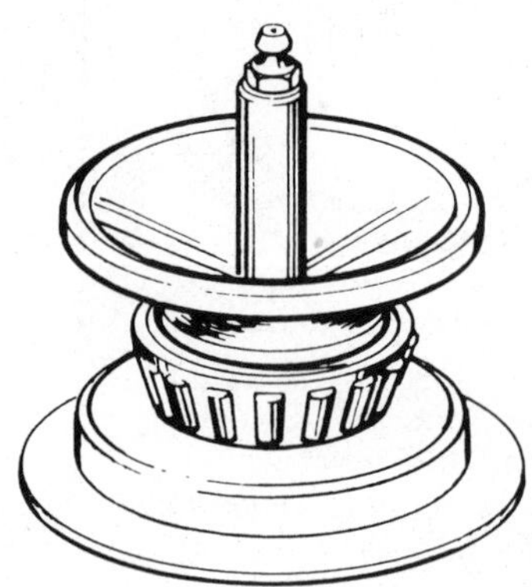

Figure 11-10. Bearing packers do an excellent job of lubricating wheel bearings.

traces of oil and solvent that can contaminate the new grease and lead to premature bearing failure. Finally, hold the bearings by their cages and dry them with clean, unlubricated compressed air. Direct the air through the bearing so it travels across the balls or rollers from side to side, figure 11-8. *Do not* spin a bearing with air while drying it. The bearing is not properly lubricated at this time, and spinning it at high speed can cause rapid wear and damage. In addition, if the air pressure or bearing speed is great enough, the balls or rollers may fly out of the cage, causing personal injury and property damage.

Figure 11-11. You can pack wheel bearings by hand.

Wheel Bearing Inspection

Once the bearings are clean and dry, inspect them for the problems shown in figure 11-9. Although the illustration depicts tapered roller bearings, most of the same problems occur with straight roller and ball bearings as well. To inspect a bearing, rotate it carefully in a good light so the complete surface of each ball, roller, and race can be fully checked. If any problems are apparent, or you have a question about its condition, replace the bearing. Never reuse a suspect bearing to save money. Like the brake system, the wheel bearings are a high liability area. An accident caused by a wheel bearing failure can result in personal injury, property damage, and a lawsuit.

Wheel Bearing Lubrication

Whenever you pack a new or used wheel bearing with grease, it is extremely important that you use the type of grease recommended by the vehicle manufacturer. All wheel bearing greases consist of oils that have been given a heavy consistency through the addition of a thickening agent. However, several different thickening agents are used, and most do not mix. *Always* clean away every trace of old grease before you repack a wheel bearing, and *never* add new grease to old, or bearing damage may result.

When you pack a wheel bearing, work the grease into the cage and races, and between the balls or rollers, so that no air spaces remain. The most effective and efficient way to do this is to use a bearing packer, figure 11-10. These devices use air or hydraulic pressure to force new grease through the entire bearing in one quick operation.

If a bearing packer is unavailable, you can pack wheel bearings by hand, although this method is slower and messier. To hand pack a tapered roller bearing, fill your palm with grease, place the large end of the bearing down, then draw the bearing across your palm, forcing grease into the cage and rollers until it oozes out the opposite side, figure 11-11. Repeat this process all around the bearing until it is completely filled with grease. Finish by spreading a medium coating of grease around

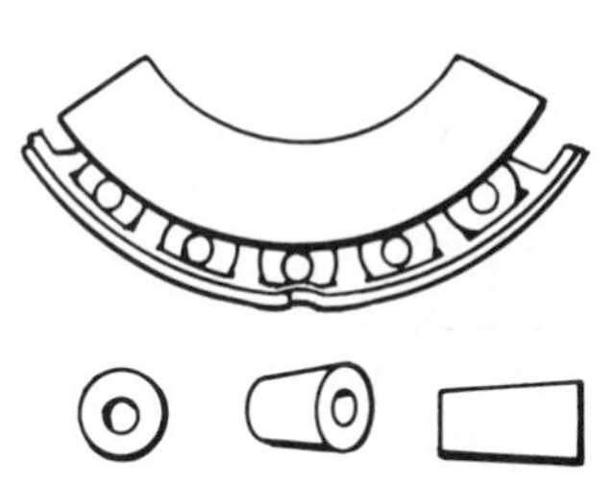

BENT CAGE

CAGE DAMAGE CAUSED BY IMPROPER HANDLING OR TOOL USE

GALLING

METAL SMEARS ON ROLLER ENDS CAUSED BY OVERHEATING, OVERLOADING, OR INADEQUATE LUBRICATION

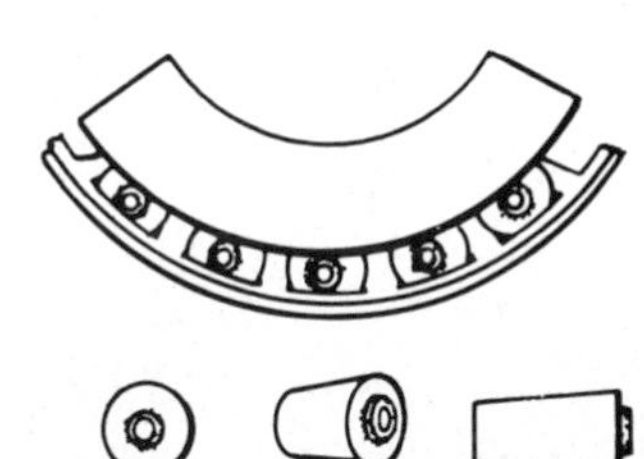

STEP WEAR

NOTCHED WEAR PATTERN ON ROLLER ENDS CAUSED BY ABRASIVES IN THE LUBRICANT

ETCHING AND CORROSION

EATEN AWAY BEARING SURFACE WITH GRAY OR GRAY-BLACK COLOR CAUSED BY MOISTURE CONTAMINATION OF THE LUBRICANT

PITTING AND BRUISING

PITS, DEPRESSIONS, AND GROOVES IN THE BEARING SURFACES CAUSED BY PARTICULATE CONTAMINATION OF THE LUBRICANT

SPALLING

FLAKING AWAY OF THE BEARING SURFACE METAL CAUSED BY FATIGUE

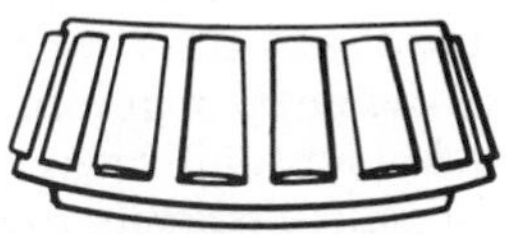

MISALIGNMENT

SKEWED WEAR PATTERN CAUSED BY BENT SPINDLE OR IMPROPER BEARING INSTALLATION

HEAT DISCOLORATION

FAINT YELLOW TO DARK BLUE DISCOLORATION FROM OVERHEATING CAUSED BY OVERLOADING OR INADEQUATE LUBRICATION

BRINELLING

INDENTATIONS IN THE RACES CAUSED BY IMPACT LOADS OR VIBRATION WHEN THE BEARING IS NOT TURNING

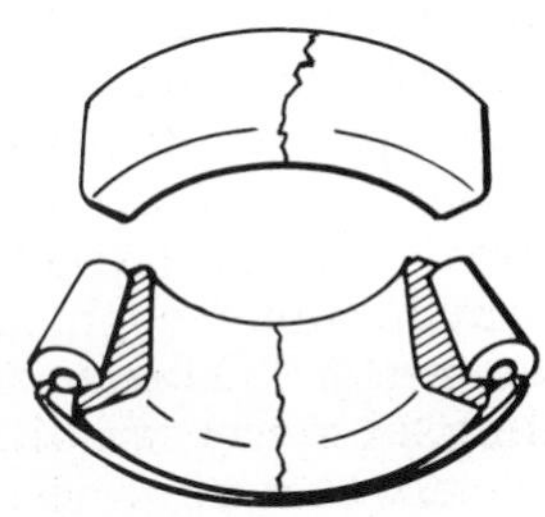

CRACKED RACE

CRACKING OF THE RACE CAUSED BY EXCESSIVE PRESS FIT, IMPROPER INSTALLATION, OR DAMAGED BEARING SEATS

SMEARING

SMEARED METAL FROM SLIPPAGE CAUSED BY POOR FIT, POOR LUBRICATION, OVERLOADING, OVERHEATING, OR HANDLING DAMAGE

FRETTAGE

ETCHING OR CORROSION CAUSED BY SMALL RELATIVE MOVEMENTS BETWEEN PARTS WITH NO LUBRICATION

Figure 11-9. A wheel bearing inspection chart.

Figure 11-12. To prevent damage, use a special tool to remove the dust cap.

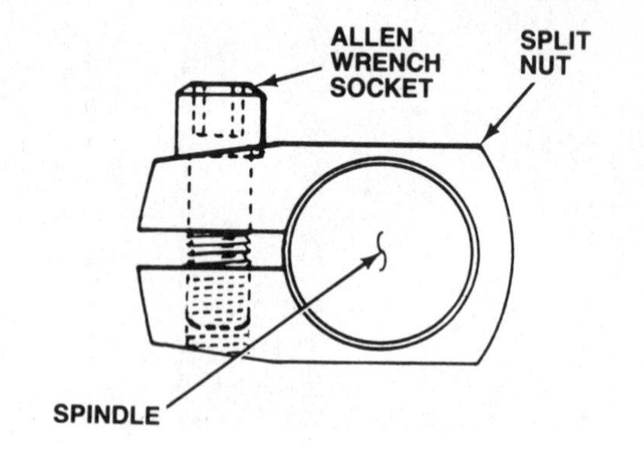

Figure 11-13. A split wheel bearing adjusting nut is used on some imported cars.

the outside of the bearing. When you pack a ball bearing by hand, work the grease in from both sides until the space between the inner and outer races is completely filled with grease.

ADJUSTABLE DUAL WHEEL BEARING SERVICE

Virtually all modern cars with adjustable dual wheel bearings use tapered roller bearings, so only that type is dealt with in the procedure below. Except for removing and installing the bearing races, the procedure is basically the same whether tapered roller bearings are being replaced or simply repacked. The procedure is also the same whether the bearings are fitted in a brake drum hub or a brake rotor hub.

To service a set of adjustable dual wheel bearings:

1. Raise and properly support the vehicle so the wheels with the bearings to be serviced hang free, then remove the wheels.
2. If the axle is equipped with disc brakes, remove the brake caliper and anchor plate as needed so the rotor can be removed from the axle (see Chapter 7).
3. Pull the dust cap from the center of the hub to expose the adjusting nut, figure 11-12.
4. Loosen the adjusting nut and back it off approximately ½ inch (13 mm). On most cars, the nut is secured by a castellated nut lock and cotter pin; remove the pin and nut lock, then unscrew the nut. Some imported cars have a split nut with a pinch bolt, figure 11-13; loosen the bolt, then unscrew the nut.

Figure 11-14. Remove the outer wheel bearing and related parts.

5. Pull the drum or rotor outward to free the thrust washer and outer wheel bearing, then push the drum or rotor inward on the spindle.
6. Remove the adjusting nut, thrust washer, and outer wheel bearing from the hub, figure 11-14, and set them aside.
7. Pull the drum or rotor outward and slide it off the spindle taking care not to drag the inner wheel bearing across the adjusting nut threads. To remove some brake drums, you may have to loosen the brake adjustment as described in Chapter 6.
8. If the inner wheel bearing sticks on the spindle and makes the drum or rotor difficult to remove, use a puller or slide hammer to remove the drum or rotor, figure 11-15. Once the drum or rotor is off the axle, use a puller or a pair of pry bars to carefully remove the inner bearing and grease seal.

Figure 11-15. A puller may be required to remove a drum or rotor.

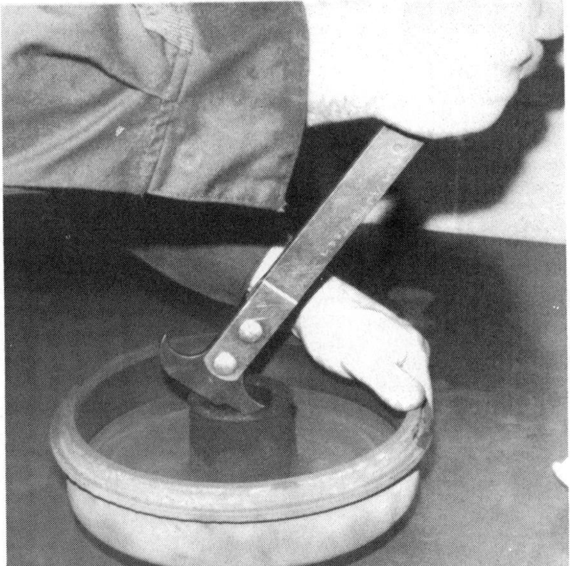

Figure 11-16. A seal puller speeds seal removal.

9. Support the drum or rotor on the workbench and remove the grease seal. This can be done in two ways:
   a. Hook the claw of a seal puller under the metal retaining ring of the seal, and lever the seal out, figure 11-16; a large screwdriver can also be used to pry out the seal.
   b. Position a non-metallic drift about three-quarters of an inch (19 mm) in diameter through the outer bearing opening so it contacts the inner race of the inner bearing, figure 11-17. Strike the drift with a hammer until the bearing and seal are driven from the hub.
10. Clean and inspect the bearings and bearing races as described earlier. Also clean the inside of the drum or rotor hub.

Figure 11-17. Using a drift to remove an inner wheel bearing and grease seal.

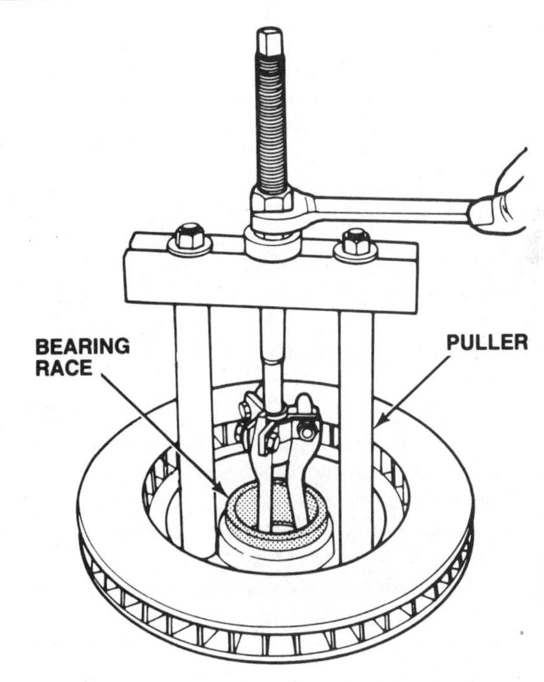

Figure 11-18. A wheel bearing race puller.

11. If a new bearing must be installed, remove the old race from the drum or rotor hub. This can be done in two ways:
   a. Use a special bearing race puller, figure 11-18.
   b. Insert a metal drift through the hub and position it against the backside of the race; use a drift that is softer than the hub metal to prevent damage. Strike the drift with a hammer while moving it around the race to drive the race from the hub.
12. Use a bearing race driver or a suitably sized socket to install the new race in the drum or

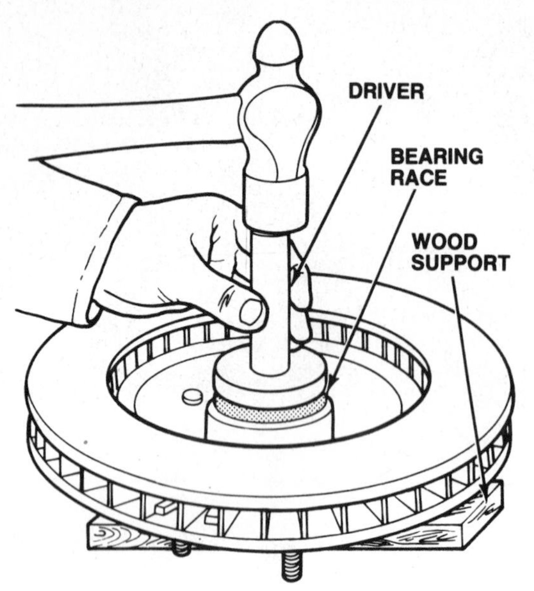

Figure 11-19. Installing a bearing race with a driver.

Figure 11-20. Install the grease seal into the hub over the inner bearing.

rotor hub, figure 11-19. The tool should only contact the outer edge of the race. Drive the race in until it is fully seated; the pitch of the sound made when striking the driver will change when the bearing is seated.

13. Clean and inspect the spindle for rust, nicks, and scratches. Remove any blemishes with fine sand paper, and make sure the inner races of both bearings slip easily over the spindle. If the spindle is badly scored, or cracked in any way, it must be replaced.

14. Lightly coat the spindle with grease. This prevents rust and allows the inner bearing races to creep slightly so wear is distributed evenly around the bearings.

15. Pack the wheel bearings with grease as described earlier.

Figure 11-21. Install the drum or rotor on the spindle.

Figure 11-22. Install the outer bearing into the hub.

16. Place the drum or rotor outer side down on the workbench, then coat the inside of the hub with grease to prevent rust.

17. Put a medium coating of grease on the inner bearing race, place the inner bearing in the race.

18. Install the grease seal with a seal driver, figure 11-20, then put a light coating of grease on the seal lip.

19. Turn the drum or rotor over and put a medium coating of grease on the outer bearing race.

20. Install the drum or rotor onto the spindle, figure 11-21, taking care to keep it centered so the inner bearing or outer bearing race does not drag on the adjusting nut threads.

21. Install the outer bearing into the hub, figure 11-22, place the thrust washer over it, and install the adjusting nut finger tight.

22. Adjust the wheel bearings as described below.

23. If the axle is equipped with disc brakes, install the anchor plate and brake caliper as described in Chapter 7. If the axle has drum brakes, and the brake adjustment was loosened

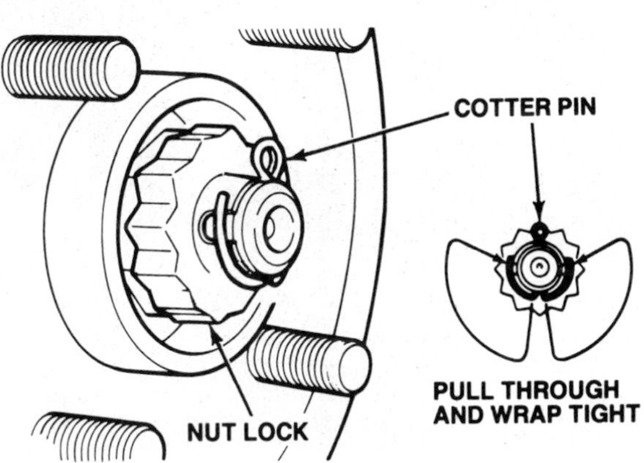

Figure 11-23. A properly secured wheel bearing adjusting nut.

Figure 11-24. A dial indicator allows precise wheel bearing adjustment.

in Step 7 or 22, adjust the brakes as described in Chapter 6.
24. Install the wheel and tighten the lug nuts to the correct torque in the proper sequence.

Tapered Roller Bearing Adjustment

Tapered roller wheel bearings require adjustment whenever you service them. Before you adjust the bearings, make sure the adjusting nut turns freely on the spindle threads. If the nut does not turn easily by hand, remove any nicks or burrs from the spindle threads with a thread file. If the adjusting nut still binds, run a tap through it or replace the nut. On cars with drum brakes, back off the brake adjustment as needed so the brakes do not drag.

There are three ways to adjust tapered roller bearings: by hand, with a torque wrench, or using a dial indicator. *Always use the adjusting method recommended by the vehicle manufacturer.* Once you set the proper axial play, lock the adjusting nut in place, and install the dust cap with a plastic hammer. If the adjusting nut is secured with a cotter pin, place the nut lock over the adjusting nut so the slots in the lock align with the cotter pin hole in the spindle. Insert a new cotter pin through the hole, and wrap the tabs around the nut lock, figure 11-23. If the car is equipped with a split adjusting nut, hold it in position with a suitably sized wrench, and tighten the pinch bolt to the torque specified by the vehicle manufacturer.

Hand adjustment
To adjust the wheel bearings by hand, rotate the wheel and snug up the adjusting nut with a wrench to seat the bearings. As you continue to rotate the wheel, back off the adjusting nut ¼ to ½ turn, or until it is just loose, then tighten the nut with your fingers to a snug fit. Lock the adjusting nut in place.

Torque wrench adjustment
To adjust the wheel bearings with a torque wrench, rotate the wheel and torque the adjusting nut to the value specified by the vehicle manufacturer, typically 12 to 25 ft-lb (15 to 35 Nm). Back off the adjusting nut ⅓ turn, then retighten it to the value specified by the vehicle manufacturer, typically 10 to 15 in-lb (1 to 1.5 Nm). Lock the adjusting nut in place.

Dial indicator adjustment
To adjust the wheel bearings with a dial indicator, rotate the wheel and torque the adjusting nut to 12 to 25 ft-lb (15 to 35 Nm). Back off the adjusting nut ¼ to ½ turn, or until it is just loose, then tighten the nut with your fingers to a snug fit. Mount a dial indicator on the drum or rotor and position the plunger against the end of the spindle, figure 11-24. Move the drum or rotor in and out to measure the axial play, and turn the adjusting nut as needed to obtain the clearance specified by the vehicle manufacturer, typically between .001 and .005 inch (.025 to .127 mm) end play. Lock the adjusting nut in place.

SEALED WHEEL BEARING REPLACEMENT

As described earlier, the sealed, double-row wheel bearing assemblies used on the front wheels of all FWD cars, and the non-driven rear wheels of a few cars, are serviced by replacing them when their axial play exceeds the manufacturer's specifications. The replacement procedure varies depending on whether the bearing is on a front or rear axle. Procedures for both are given below.

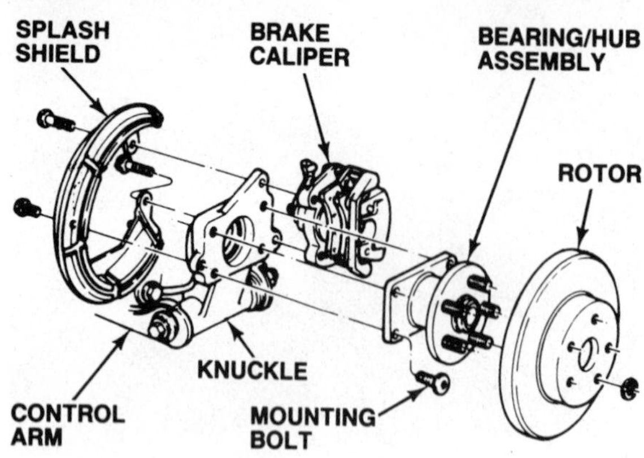

Figure 11-25. A rear-wheel sealed bearing/hub assembly.

Figure 11-26. A wheel bearing retaining nut with a crimped collar.

Rear Axle Sealed Bearing Replacement

General Motors is the only domestic automaker that uses sealed rear wheel bearings at this time. Their design combines the bearings with the wheel hub assembly, figure 11-25, which makes replacement a relatively simple job. To replace one of these bearing/hub assemblies:

1. Raise and properly support the vehicle, then remove the wheel at the corner of the car with the defective bearing.
2. If the axle is equipped with disc brakes, remove the caliper as described in Chapter 7.
3. Remove and discard any speed nuts, or retaining bolts or screws, then remove the brake drum or rotor from the hub.
4. Remove the four bearing/hub retaining bolts, and remove the assembly.
5. Install the new bearing/hub assembly, and tighten the four retaining bolts to the torque specified by the vehicle manufacturer.
6. Position the brake drum or rotor over the wheel studs.
7. If the axle is equipped with disc brakes, install the brake caliper as described in Chapter 7.

Front Axle Sealed Bearing Replacement

Replacing a sealed wheel bearing assembly on a driven front axle is much more involved than the same job on a non-driven rear axle. On Chrysler and Ford cars you must remove the steering knuckle from the vehicle before you can replace the wheel bearing. General Motors bearings can be replaced in this manner as well, however, special tools are available that allow the combined bearing/hub assembly to be replaced without removing the steering knuckle.

The bearing replacement procedure below applies to domestic FWD cars; the same basic steps apply to most imports as well.

To replace a sealed wheel bearing assembly on a driven front axle:

1. On Chrysler cars, remove the cotter pin and nut lock from the hub retaining nut. The crimped collar on Ford retaining nuts, figure 11-26, does not have to be straightened prior to nut removal. General Motors cars have self-locking nuts.
2. Have an assistant apply the brakes while you loosen the hub nut with a socket and breaker bar. Do not use an impact wrench for this job because the shocks it delivers can damage constant-velocity joints.
3. Raise and properly support the vehicle, then remove the wheel.
4. Remove the brake caliper and anchor plate as needed (see Chapter 7) so the rotor can be removed from the axle.
5. Remove any speed nuts, or retaining bolts or screws, then remove the brake rotor.
6. Remove the hub nut and washer. On General Motors cars go to Step 13.
7. Remove the nut securing the tie rod end, then use a puller to separate the tie rod end from the steering arm, figure 11-27.
8. Loosen the clamp bolt that holds the ball joint stud in the steering knuckle, figure 11-28, then disconnect the lower control arm from the steering knuckle.
9. Loosen the retaining nuts on the bolts that attach the shock strut to the steering knuckle. If the bolts are used to set the wheel alignment, mark the position of the adjusting cams before you loosen the nuts, figure 11-29.
10. Push the axle shaft out of the steering knuckle using a puller if necessary.

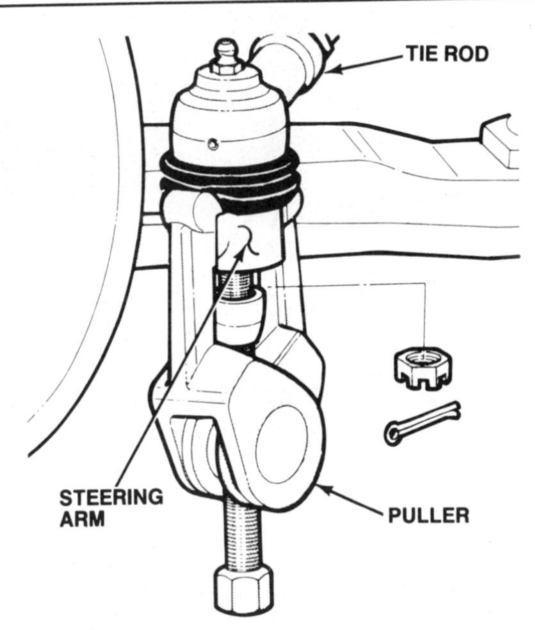

Figure 11-27. Use a puller to separate the tie rod from the steering arm.

Figure 11-28. Loosen the ball joint stud clamp bolt to detach the lower control arm.

Figure 11-29. Mark the adjusting cam position before you loosen the strut attachment bolts.

Figure 11-30. Removing the hub with a drift punch.

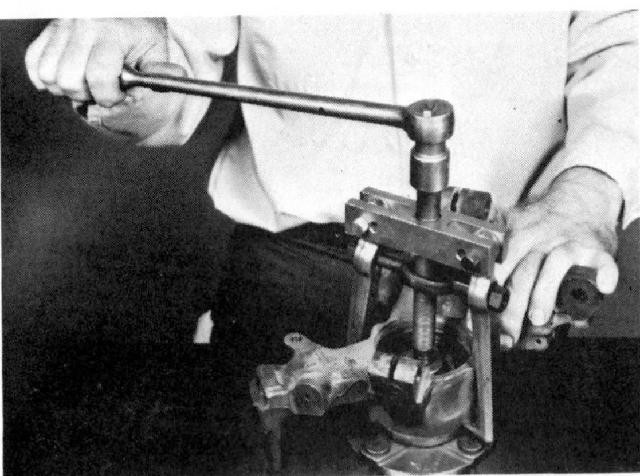

Figure 11-31. Removing the hub with a puller.

11. Remove the bolts that attach the steering knuckle to the shock strut, and remove the steering knuckle from the vehicle.
12. Remove the hub from the center of the sealed bearing assembly:
   a. On Chrysler cars, use a drift punch to remove the hub, figure 11-30. If an inner bearing race comes out with the hub, use a puller to remove it.
   b. On Ford cars, use a puller to remove the hub, figure 11-31.
13. Remove the bearing from the steering knuckle:
   a. On Chrysler cars, remove the three bolts securing the bearing retainer, then press the bearing out of the steering knuckle, figure 11-32.
   b. On Ford cars, use a screwdriver to remove the snap ring that retains the bearing, then press the bearing out of the steering knuckle.

Figure 11-32. Using a press to remove the bearing assembly from the steering knuckle.

Figure 11-34. Using a puller to remove the bearing/hub assembly from a GM steering knuckle.

Figure 11-33. Removing the bearing/hub assembly retainer bolts on a GM steering knuckle.

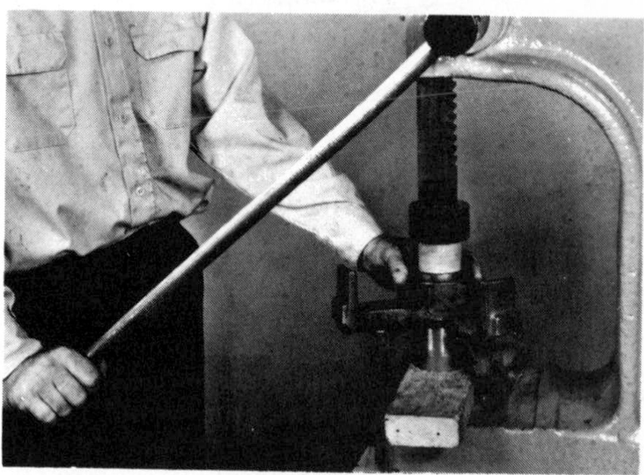
Figure 11-35. Use a press to install the new bearing assembly.

c. On General Motors cars, remove the bearing/hub retaining bolts and splash shield, figure 11-33. Then, use a puller to remove the bearing/hub assembly, figure 11-34.

14. Remove the seal:
   a. On Chrysler cars, use a screwdriver to pry the seal out of the steering knuckle.
   b. On Ford cars, remove the halfshaft and place it in a vise. Use a screwdriver and hammer to remove the seal from the axle.
   c. On General Motors cars, use a punch to drive the seal out of the steering knuckle toward the engine, then cut the seal off the drive shaft.
15. Clean the bore of the steering knuckle and, on Ford vehicles, the seal mounting surface of the driveshaft. Inspect the bore and mounting surface, and smooth away any nicks or burrs with sandpaper. On General Motors cars go to Step 24.
16. Apply a light film of grease to the steering knuckle bore, then press the new wheel bearing into place, figure 11-35. Apply pressure only on the outer bearing race.
   a. On Chrysler cars, install the bearing retainer and tighten the three bolts to the torque specified by the vehicle manufacturer.
   b. On Ford cars, install the snap ring that retains the bearing into its groove in the steering knuckle.
17. Apply a light film of grease to the inner bearing bore, then press the hub into the center of the bearing assembly, figure 11-36. Support or press *only* on the inner race of the bearing to prevent bearing damage.
18. Install the new seal with a seal driver, then coat the lip of the seal with wheel bearing grease:
   a. On Chrysler cars, install the seal in the steering knuckle.
   b. On Ford cars, install the seal onto the driveshaft, figure 11-37.

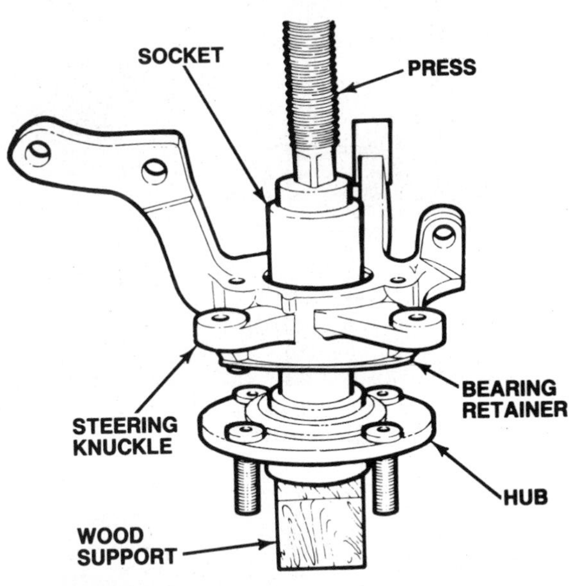

Figure 11-36. Use a press to install the hub into the wheel bearing assembly.

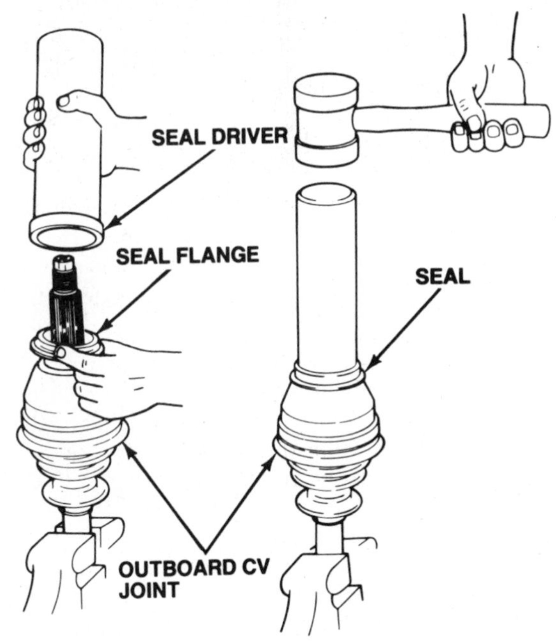

Figure 11-37. The seal on Ford wheel bearings installs on the axle.

19. Align the axle splines with those in the hub, then press the hub and steering knuckle onto the axle. If necessary, use a special tool to fully seat the hub on the driveshaft.
20. Attach the steering knuckle to the shock strut, but do not tighten the retaining nuts.
21. Position the steering knuckle on the ball

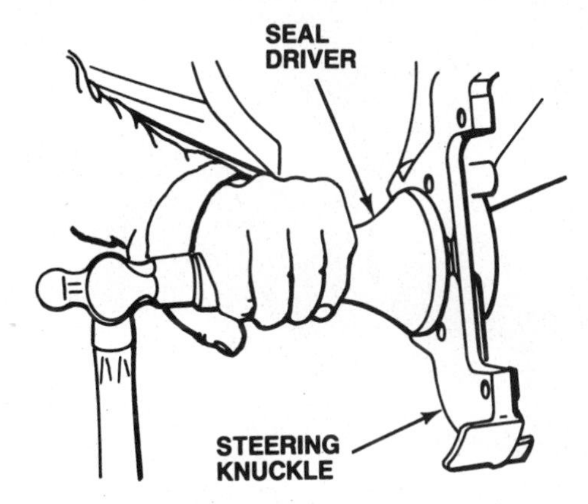

Figure 11-38. Use a driver to install a new dust seal.

Figure 11-39. Installing the GM bearing/hub assembly onto the axle.

joint stud, and tighten the clamp bolt to the torque specified by the vehicle manufacturer.
22. If necessary, align the adjusting cams on the bolts that attach the steering knuckle to the shock strut with the marks made in Step 9. Tighten the retaining nuts to the torque specified by the vehicle manufacturer.
23. Connect the tie rod end to the steering arm and tighten the retaining nut to the torque specified by the vehicle manufacturer.
Go to Step 27.
24. Apply a light film of grease to the steering knuckle bore, then use a seal driver to install the new seal, figure 11-38.
25. Install the bearing/hub assembly onto the axle, figure 11-39, and install a new washer and hub retaining nut. Tighten the nut until the bearing/hub assembly is fully seated.
26. Install the splash shield and bearing/hub retaining bolts. Tighten the bolts to the torque specified by the vehicle manufacturer.
27. Install the brake rotor over the wheel studs.
28. Install the brake caliper as described in Chapter 7.

Figure 11-40. Torque the hub nut before you lower the car.

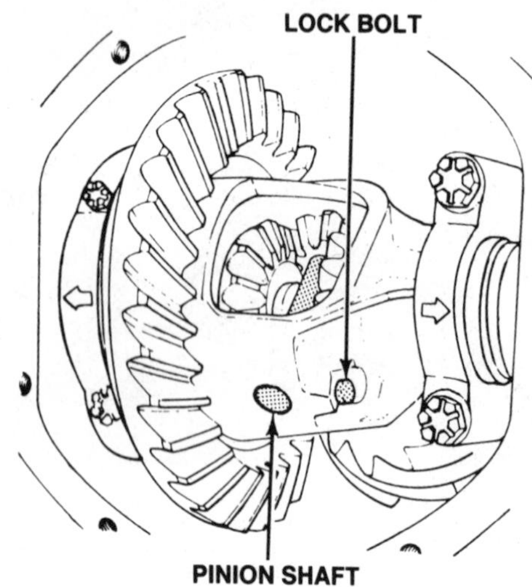

Figure 11-41. To remove a C-lock, first remove the pinion shaft and lock bolt.

29. Have an assistant apply the brakes as you tighten the hub retaining nut to the torque specified by the vehicle manufacturer, figure 11-40.
    a. On Chrysler cars, install the nut lock and a new cotter pin.
    b. On Ford cars, use a blunt chisel or similar tool to crimp the retaining nut collar into the groove on the axle.
30. Install the wheel and lower the car.

SOLID AXLE WHEEL BEARING SERVICE

One of the most common reasons for removing the wheel bearings from a solid rear axle is to replace leaking grease seals that have allowed axle lubricant to contaminate the brake linings. However, before you undertake this job, first check the axle vent on top of the axle housing, figure 11-3. If the vent is plugged, heat can cause a pressure build up inside the housing that will force lubricant past the seals even if they are in perfect condition. If the vent is not obstructed, replace the leaking seals.

As discussed at the beginning of the chapter, there are two ways axle shafts are retained in solid rear axle housings: axles that ride on straight roller bearings have C-locks; axles that ride on ball bearings have retainer plates. Because the two types require different service procedures, determine which type of axle you are servicing before you begin work.

To ready a car with either type of axle for bearing service, raise and properly support the vehicle, then remove the wheels. Next, remove the brake drums, or brake calipers and rotors. Drum removal information is covered in Chapter 6, and brake caliper removal is covered in Chapter 7.

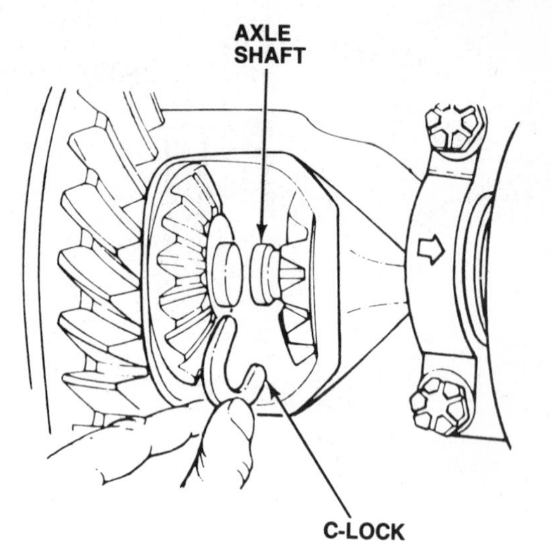

Figure 11-42. Remove the C-lock from its groove in the axle.

Bearing and Seal Service — C-Lock Axles

To service the wheel bearing and grease seal on a solid rear axle retained by a C-lock:

1. Clean any dirt from around the differential cover, then remove the cover and drain the axle lubricant into a suitable drain pan.
2. Rotate the differential carrier until the pinion shaft lock bolt is accessible, then remove the lock bolt and pinion shaft, figure 11-41.
3. Without rotating the axle, push it toward the center of the car until the C-lock clears the counterbore in the differential side gear, then remove the C-lock from its groove, figure 11-42.

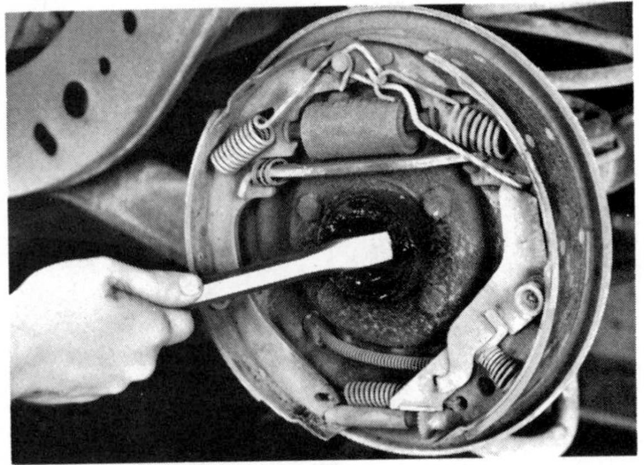

Figure 11-43. Removing the seal from the axle housing.

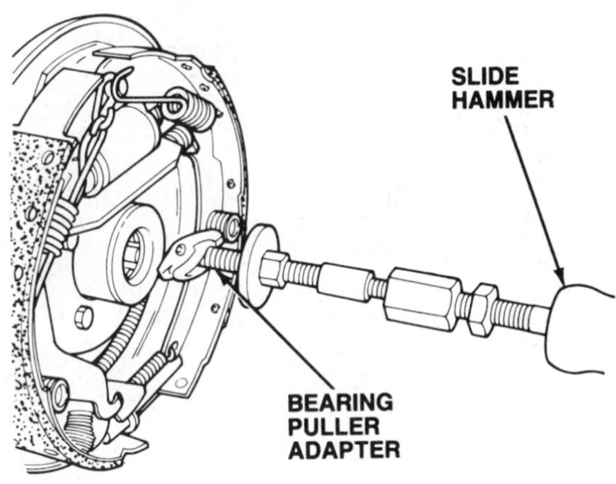

Figure 11-44. Use a slide hammer to pull the bearing out of the axle housing.

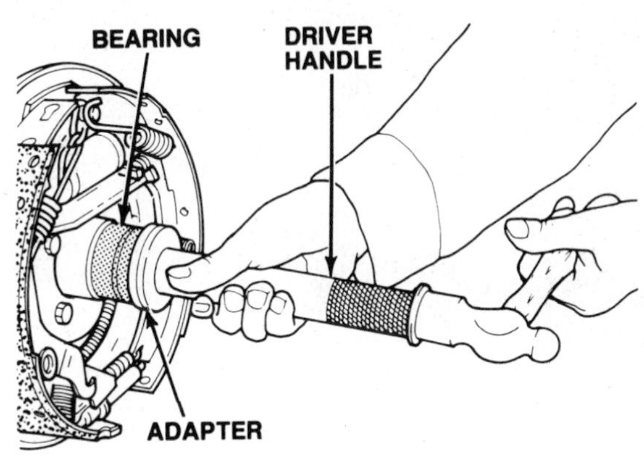

Figure 11-45. Install the new bearing with a bearing driver.

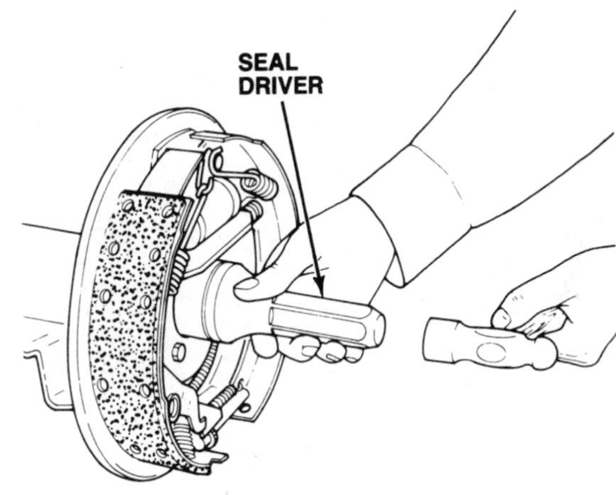

Figure 11-46. Drive the grease seal into the axle housing.

4. Return the axle to its original position, and replace the pinion shaft and lock bolt to hold the differential carrier pinion gears in position.
5. Pull the axle straight out of the housing by hand, making sure it does not drag on the wheel bearing and cause damage.
6. Pry the seal from the axle housing with a seal puller or large screwdriver, figure 11-43.
7. Use a slide hammer to remove the bearing from the axle housing, figure 11-44.
8. Clean the axle and inspect the surface where the bearing and seal ride. It should be smooth, dull gray in color, and free of any pitting, indentations, scratches, or flaking metal. If any damage is present, replace both the axle and bearing.
9. Clean and inspect the bearing as described earlier. Replace the bearing if any problems are present.
10. Clean the axle housing bore where the bearing and seal fit, and check for damage.
11. Lubricate the outside of the bearing with wheel bearing grease, then align the bearing squarely with the opening in the axle housing. Drive the bearing into place with a suitably sized socket or bearing driver, figure 11-45, that contacts only the outer race of the bearing. The pitch of the sound made when striking the driver will change when the bearing is seated.
12. Install a new grease seal into the axle housing with a seal driver, figure 11-46.
13. Lubricate the bearing rollers, seal lip, and axle sealing surface with axle lubricant, then carefully slide the axle into the housing. Do not allow the splines on the axle to damage the seal or bearing during installation. It may be necessary to rotate the axle slightly to align its splines with those in the differential side gear.
14. Remove the lock bolt and pinion shaft from the differential carrier, and push the axle inward until the C-lock groove clears the counterbore in the differential side gear. Do not rotate

Figure 11-47. Use a new gasket when you install the differential cover.

Figure 11-49. Pull the axle from the housing with a slide hammer.

Figure 11-48. You can unscrew retainer plate nuts through a hole in the axle flange.

Figure 11-50. One way to free a bearing retaining ring is to notch it with a chisel.

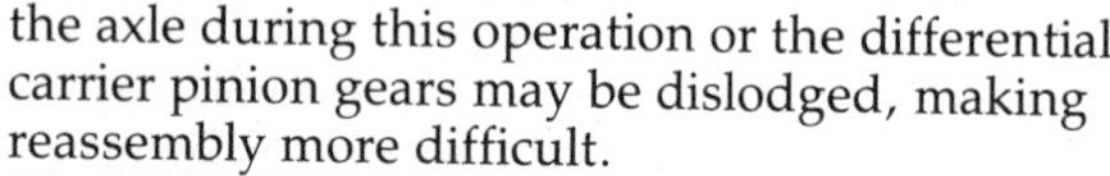

the axle during this operation or the differential carrier pinion gears may be dislodged, making reassembly more difficult.

15. Insert the C-lock into the groove in the inner end of the axle, then pull the axle outward to seat the C-lock in the counterbore of the differential side gear.
16. Install the pinion shaft and lock bolt, then tighten the lock bolt to the torque specified by the manufacturer.
17. Install the differential cover using a new gasket or silicone sealer as specified by the vehicle manufacturer, figure 11-47.
18. With the axle in its normal operating position, fill the differential to the bottom edge of the fill hole with the lubricant specified by the vehicle manufacturer.

Bearing and Seal Service — Retainer Plate Axles

On most solid rear axles where the wheel bearing and grease seal are secured by a retainer plate, service the bearing and seal as follows:

1. Remove the nuts that hold the retainer plate to the backing plate, figure 11-48.
2. Slide the retainer plate off of the studs, then reinstall two of the nuts finger tight to keep the brake backing plate in place.
3. Use a puller or slide hammer to remove the axle shaft and bearing from the axle housing, figure 11-49.
4. Remove the seal from the axle housing.
5. Clean the axle housing bore where the bearing and seal fit, and check for damage.
6. If the wheel bearing is unsealed, clean and inspect it as described earlier. If the bearing is sealed, turn it by hand to feel for roughness, listen for noise, and make sure the metal side plates are not dented. Replace the bearing if a problem is apparent, or you have any question about the bearing's condition.
7. If the bearing must be replaced, remove the retaining ring that secures it on the axle. Depending on the hardness of the ring and how

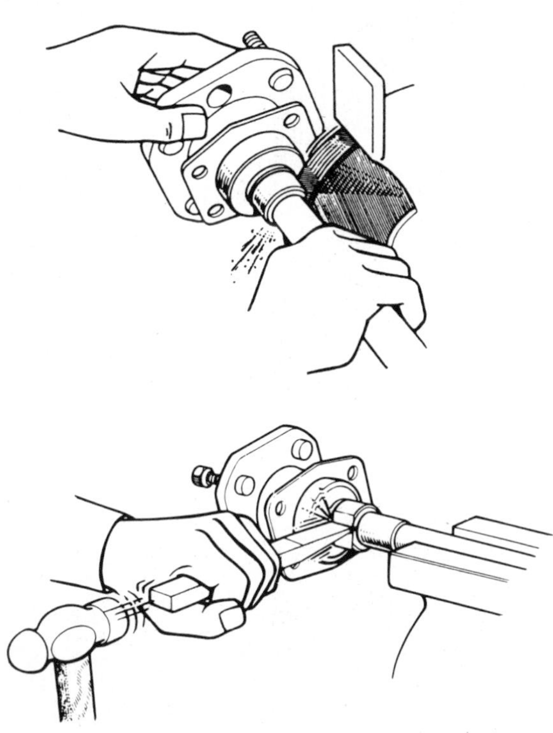

Figure 11-51. Stubborn bearing retaining rings must be ground and cut away.

tightly it fits, use one of these methods, being sure to wear safety glasses to protect your eyes:

a. Mount the axle in a vise, and use a hammer and chisel to notch the retaining ring in several places around its outer edge, figure 11-50. The ring may loosen enough to remove it from the axle.

b. Use a grinder to cut away the ring almost to the axle, figure 11-51, then use a hammer and chisel to split the ring and remove it from the axle. Take care not to damage the axle in any way.

8. Using a bearing splitter or other adapter that contacts the *inner* race, press the old bearing off the axle. Never attempt to remove a wheel bearing by pressing against the outer race; the bearing may come apart and cause an injury.
9. Remove the retainer plate and inspect the area on the axle where the bearing and seal make contact; it should be smooth and have a dull gray appearance. Replace the axle if it is rough or worn.
10. Clean all traces of the old gasket from the retainer plate and the brake backing plate. If the retainer plate contains an outer seal, install a new seal and lubricate the seal lip with wheel bearing grease. Replace the retainer plate and a new gasket on the axle.

Figure 11-52. Press the new bearing into position on the axle.

11. If the new wheel bearing is unsealed, pack it with grease as described earlier.
12. Press the new wheel bearing onto the axle, figure 11-52. Apply pressure only to the inner race of the bearing.
13. Press a new bearing retaining ring onto the axle. Never attempt to press a bearing *and* retaining ring onto the axle at the same time.
14. Install a new seal into the axle housing with a seal driver, and lightly lubricate the lip of the seal with axle lubricant.
15. Carefully slide the axle into the housing so as not to damage the seal. Rotate the axle as necessary to align its splines with those in the differential side gear.
16. Gently tap the end of the axle with a hammer until the bearing is fully seated in the axle housing.
17. Remove the two nuts holding the backing plate in place, then slide the retainer plate onto the axle housing studs.
18. Install the retainer plate nuts, and tighten them to the torque specified by the vehicle manufacturer.

TIRE AND WHEEL SERVICE

The effectiveness of even the best brake repairs can be reduced if there are tire and wheel problems that reduce traction. In some cases, problems associated with tire and wheel service can directly affect the friction developed between the brake drum or rotor, and the brake linings. The sections below describe: how to properly install the tire and wheel assemblies, how to visually inspect tires and wheels for problems that may affect braking, and how to measure tire and wheel runout.

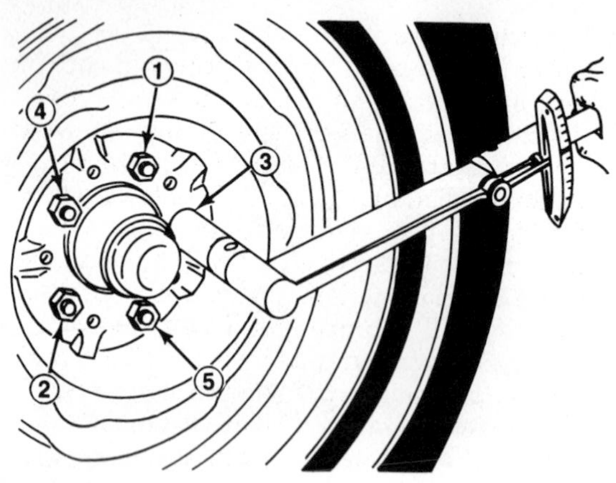

Figure 11-53. Use a torque wrench to tighten lug nuts or bolts.

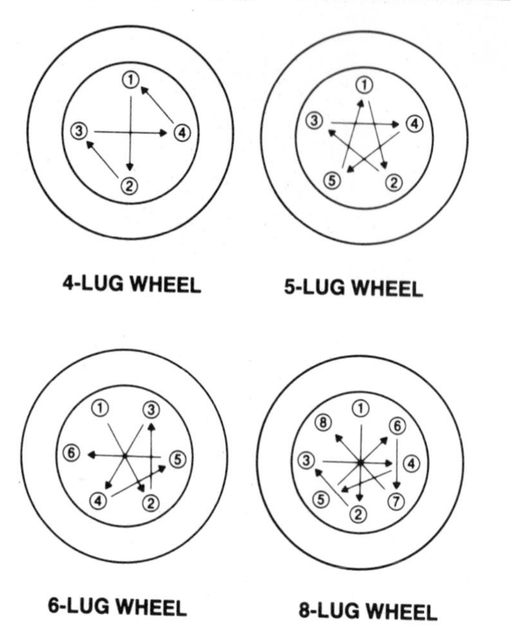

Figure 11-54. Proper lug nut tightening sequences.

Tire and Wheel Installation

Although the job of installing a tire and wheel assembly on the car may seem relatively simple, unless certain procedures are strictly observed, braking problems may result. The two main areas of concern are the lug nut or bolt torque, and the sequence in which the lug nuts or bolts are tightened. If you overtighten lug nuts and bolts, tighten them unevenly, or tighten them in the wrong sequence, you can distort the wheel, the hub, and the brake drum or rotor. This distortion causes runout and results in a pulsating brake pedal, reduced traction, uneven drum or rotor wear, and poor contact between the drum or rotor and the brake linings. In addition, if you overtighten the lug nuts or bolts too much, the car owner may be unable to remove them with the vehicle lug wrench in the event of a flat tire.

Always use an accurate torque wrench to tighten the lug nuts and bolts to the torque specified by the vehicle manufacturer, figure 11-53. This figure can be found in the owner's manual, the shop manual, and in publications from tire companies and other aftermarket firms that provide specifications to the service industry. Although an impact wrench is useful for *removing* lug nuts and bolts, never use such a wrench to install a wheel because the torque it applies cannot be accurately controlled.

In addition to the proper tightening torque, it is also important that you tighten lug nuts or bolts in the proper sequence, figure 11-54. As with the proper torque, this helps prevent distortion. All sequences are essentially criss-cross patterns in which you tighten the nuts or bolts alternating from one side of the wheel to the other.

To properly install a tire and wheel assembly, make sure the mounting surfaces on both the wheel and vehicle are clean; remove any dirt, rust, or corrosion with a wire brush. Install the wheel straight onto the car, taking care not to drag it across the wheel stud threads, then install all of the lug nuts or bolts. Do not use any type of lubricant on the wheel studs or lug bolts unless specifically instructed to do so by the vehicle manufacturer. Tighten the nuts or bolts by hand with a wrench until they are snug, then tighten each nut or bolt in the proper sequence to approximately half the specified torque. Go through the sequence once more, and bring the nuts or bolts to their final tightness.

Tire and Wheel Inspection

As discussed in the *Classroom Manual,* tires that are improperly inflated, excessively worn, or mismatched in some manner are all possible causes of braking problems. Bent, cracked, or otherwise damaged wheels can also affect braking performance. The paragraphs below describe the checks and inspections used to identify these problems.

Tire inflation

Improper inflation is a common tire problem, and many studies show that over half the vehicles on the road have incorrectly inflated tires at any given time. Inflation pressures higher or lower than those specified by the vehicle manufacturer decrease the tire contact patch with the road and reduce braking traction and stability.

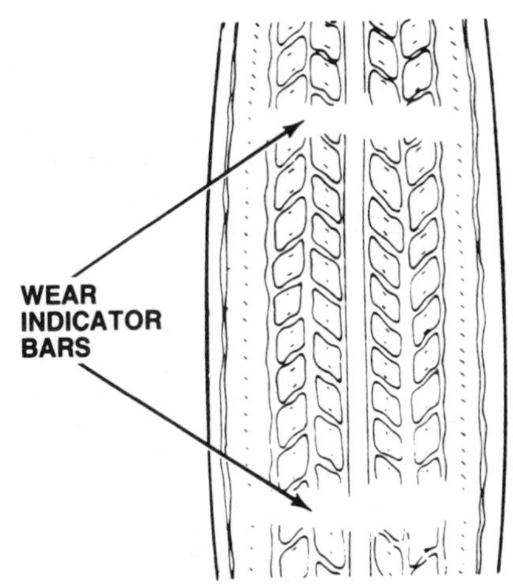

Figure 11-55. The wear indicators on this tire are showing.

Use an accurate, high-quality gauge to check the pressure in each tire. If necessary, adjust the pressure to the figure specified by the vehicle manufacturer. Most cars have a sticker in the glove box or on the driver's door jamb that gives the recommended tire pressures for the size of tires that came on the vehicle. These pressures generally apply to similarly sized replacement tires as well. However, you should never inflate a tire beyond the maximum pressure molded into its sidewall.

Tire pressures often vary depending on the load the vehicle carries and speed at which the car is driven; heavier loads and faster speeds require higher tire inflation pressures. Be sure to set the pressures to a level appropriate for your customer's driving conditions.

Tire wear

Tire wear can affect braking in two ways. On dry roads, worn tires actually offer greater braking traction because tread squirm is reduced and there may be more rubber on the road. However, on wet pavement, worn tires have reduced braking traction because they have a much greater tendency to hydroplane. When you check tire wear to trace a brake pull, look primarily for uneven amounts of wear from side to side on an axle.

All modern tires are equipped with tread wear indicator bars that show when only about 1/16 inch (1.5 mm) of tread is left. The bars appear as bald strips approximately ½ inch (13 mm) wide that run across the tire, figure 11-55. Replace a tire when the wear indicators appear in two or more adjacent tread grooves, if there is localized balding, or if cord is visible at any point.

Mismatched tires

In order to have even braking power at each wheel, all four tires should be the same type and size unless otherwise specified by the vehicle manufacturer. If the tires differ in their construction (radial, bias-belted, bias-ply), section width (155, 165, etc.), aspect ratio (60, 70, etc.), or diameter (13, 14, etc.), they will have different traction characteristics. This upsets brake balance if the difference is between the front and rear axles, or can cause a brake pull if the difference is between tires on the same axle.

If you have eliminated all other possible causes of a brake balance or pull problem, and suspect that the tires may be contributing to the difficulty, check the size designations molded into the tire sidewalls. If tires of different types or sizes have been mixed, replace them with tires that match the vehicle manufacturer's recommendations.

Wheel damage

Mild forms of wheel damage, such as a slightly bent rim, may only cause the tire to lose air pressure. More dramatic damage reduces braking traction by causing runout that makes the wheel difficult or impossible to balance. In addition, an impact that causes such damage will usually affect wheel alignment as well. Cracked or badly rusted wheels are especially dangerous because they can fail entirely under the stress of heavy braking.

Inspect all wheels for bent or distorted rims, cracks, and rust or corrosion. If the amount of distortion is minor, you may have to use a dial indicator to check for wheel runout as described in the next section. Replace any wheel that is cracked or badly rusted. Bent or distorted steel wheels can sometimes be straightened by a wheel repair specialist if the damage is not too severe. Alloy wheels, however, cannot be repaired and must be replaced if they are damaged in any way.

TIRE AND WHEEL RUNOUT

Tire and wheel assemblies suffer two kinds of runout, radial and lateral. Radial runout is when the assembly is out of round; this reveals itself as an irregular up and down motion. Lateral runout is when the assembly does not run true from side to side; this reveals itself as an oscillation or wobble. Because it cannot be balanced properly, a tire and wheel assembly with excessive runout does not maintain smooth contact with the road, which upsets braking traction. Tire and wheel runout is most often felt as a vibration in the car body, or a shimmy in the steering wheel.

Figure 11-56. A dial indicator designed to check tire runout.

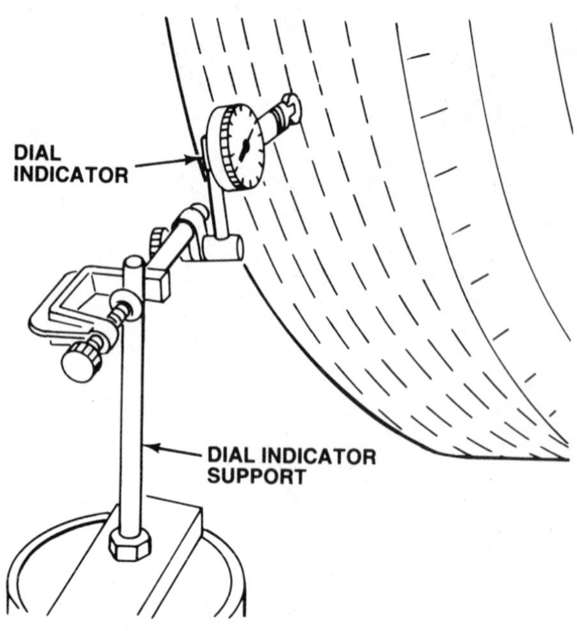

Figure 11-57. Check tire radial runout with the dial indicator against a center tread.

Checking Tire and Wheel Runout

Runout of the tire and wheel assembly is usually caused by the tire only, the wheel only, or a combination of the two. In rare instances, the vehicle hub may have runout as well. When you check for runout, check at the tire first because this gives you the combined runout of the entire assembly. If runout at the tire is acceptable, there is no need to check further. However, if runout at the tire is excessive, you should then check the wheel runout to determine which part of the assembly is causing the problem.

Before you measure tire and wheel runout, take the following steps to prevent false readings. First, make sure the wheel bearings are properly adjusted; loose bearings increase the amount of runout. Some technicians prefer to check runout with the tire and wheel assembly mounted on a computer wheel balancer because this eliminates wheel bearing play as a factor in the measurement.

Second, make sure the lug nuts or bolts are properly torqued. As discussed earlier in the chapter, overtightened lug nuts or bolts distort the hubs and wheels, and this makes it impossible to get an accurate runout reading.

Finally, drive the car for several miles to warm the tires to operating temperature. When a vehicle is parked for any length of time, the cords in the flattened portion of the tire where it contacts the road take a set. Unless the tire is warmed up, the cord set will cause runout during the test. Immediately after you drive the car, raise and properly support it to prevent the tires from taking a new set.

It is easiest to check tire and wheel runout with a special dial indicator made for this purpose, figure 11-56. This tool has a small roller on the end of its plunger that makes it less likely to drag or catch on the tire or wheel. Check for radial runout of the tire and wheel assembly first because it is a more common source of problems, then check for lateral runout.

Radial runout test

1. Raise and properly support the vehicle so the tire and wheel to be checked spin freely.
2. Position the plunger of the dial indicator against a center tire tread, figure 11-57.
3. Rotate the tire until the lowest reading shows on the indicator dial, then zero the dial. Ignore any sudden changes in reading caused by irregularities in the tread.
4. Rotate the tire until the highest reading shows on the indicator dial; this is the total radial runout of the tire and wheel assembly.
   a. If the total radial runout is less than .060″ (1.5 mm), check the lateral runout as described in the next section.
   b. If the total radial runout is greater than .060″ (1.5 mm), continue with Step 5 to check the wheel runout.
5. Position the plunger of the dial indicator against the inside of the wheel rim, figure 11-58.
6. Rotate the wheel until the lowest reading shows on the indicator dial, then zero the dial. Ignore any sudden changes in reading caused by irregularities in the wheel surface.

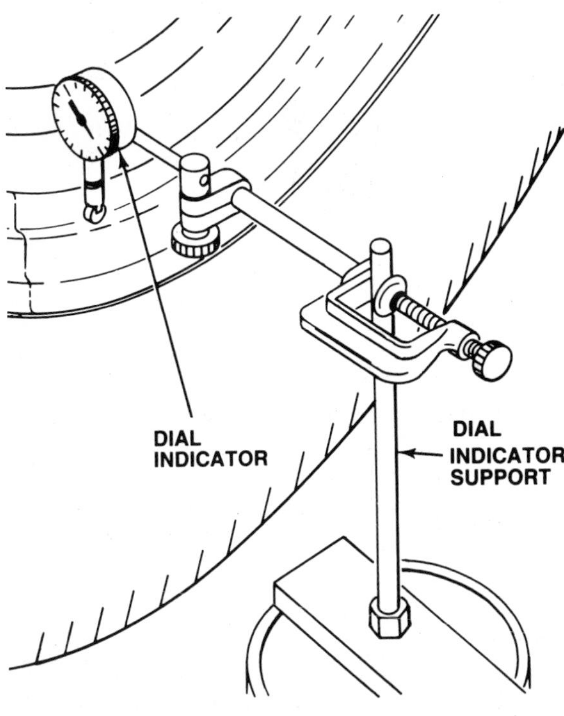

Figure 11-58. Check wheel radial runout with the dial indicator against a horizontal surface.

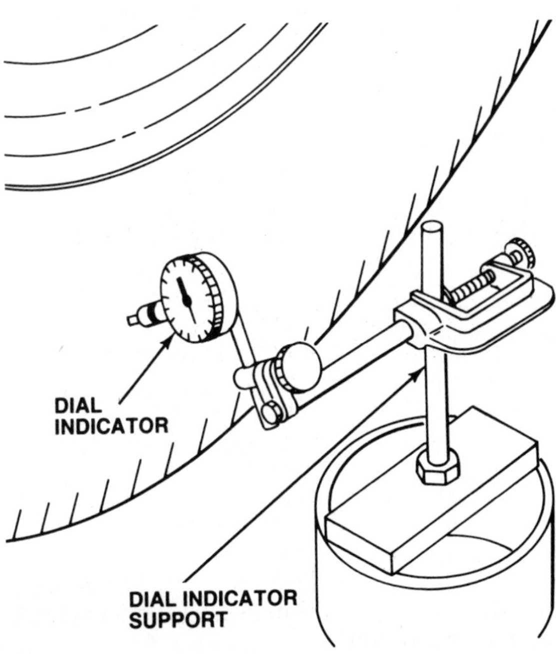

Figure 11-59. Check tire lateral runout with the dial indicator against the sidewall.

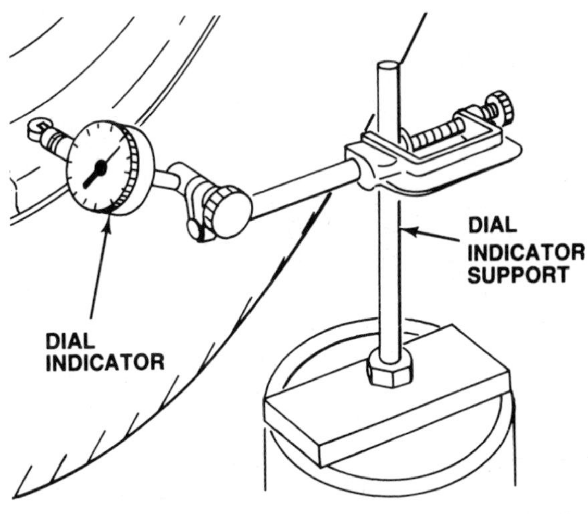

Figure 11-60. Check wheel lateral runout with the dial indicator against a vertical surface.

7. Rotate the wheel until the highest reading shows on the indicator dial; this is the total radial runout of the wheel. The runout should be less than .040" (1.00 mm) for a steel wheel or .030" (.75 mm) for an alloy wheel.
   a. If the wheel radial runout is within specifications, attempt to correct the total radial runout by match mounting the tire on the rim as described below. If this is unsuccessful, replace the tire.
   b. If the wheel radial runout is outside specifications, attempt to correct it by repositioning the wheel on the hub as described below. If this is unsuccessful, replace the wheel.

Lateral runout test

1. Position the plunger of the dial indicator against a smooth portion of the tire sidewall, figure 11-59.
2. Rotate the wheel until the lowest reading shows on the indicator dial, then zero the dial. Ignore any sudden changes in reading caused by irregularities in the sidewall.
3. Rotate the wheel until the highest reading shows on the indicator dial; this is the total lateral runout of the tire and wheel assembly.
   a. If the total lateral runout is less than .080" (2.00 mm), tire and wheel runout is not a problem.
   b. If the total lateral runout is greater than .080" (2.00 mm), continue with Step 4 to check the wheel runout.
4. Position the plunger of the dial indicator against the side of the wheel rim, figure 11-60.
5. Rotate the wheel until the lowest reading shows on the indicator dial, then zero the dial. Ignore any sudden changes in reading caused by irregularities in the wheel surface.

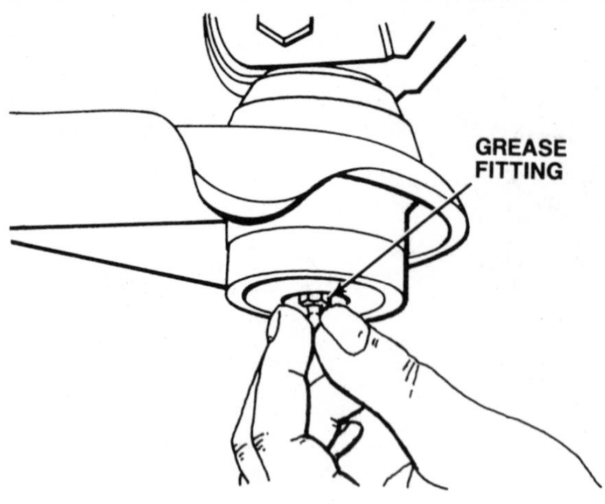

Figure 11-62. Checking a Chrysler wear indicator ball joint.

6. Rotate the wheel until the highest reading shows on the indicator dial; this is the total lateral runout of the wheel. The runout should be less than .045" (1.15 mm) for a steel wheel or .030" (.75 mm) for an alloy wheel.
   a. If the wheel lateral runout is within specifications, attempt to correct the total lateral runout by match mounting the tire on the rim as described below. If this is unsuccessful, replace the tire.
   b. If the wheel lateral runout is outside specifications, attempt to correct it by repositioning the wheel on the hub as described below. If this is unsuccessful, replace the wheel.

Correcting Tire and Wheel Runout

If the runout of a tire and wheel assembly is not within specifications, there are four ways you can correct it: reposition the wheel on the hub, match mount the tire on the wheel, replace the tire, or replace the wheel. As instructed in the procedures above, you should always attempt to reposition the wheel or match mount the tire before you replace the tire or wheel.

Repositioning the wheel on the hub can correct runout if a stack up of manufacturing tolerances is contributing to the problem. If you install the wheel in a new position, the tolerances may offset one another and reduce the runout to an acceptable figure. To do this, unbolt the wheel, rotate it two studs over from its original position, and remount it making sure you tighten the lug nuts or bolts to the correct torque in the proper sequence. Recheck for radial and lateral runout as described above.

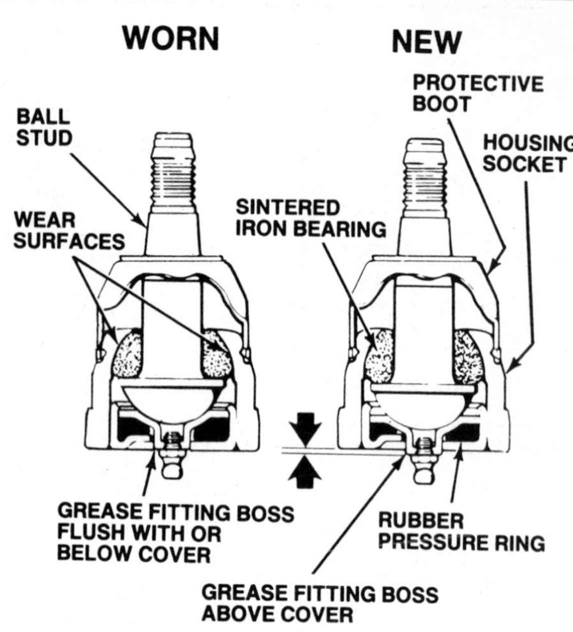

Figure 11-63. Grease fitting position indicates the condition of Ford and GM wear indicator ball joints.

Match mounting the tire on the wheel can also correct for runout if the high spots of the tire and wheel are aligned. To do this, break the bead of the tire loose from the rim, and reposition the tire 180 degrees from its original position. Recheck for radial and lateral runout as described above.

CHASSIS SERVICE

Chassis problems that affect braking usually result in a pull toward one side when the brakes are applied. This can be caused by damaged or worn suspension components, or by incorrect wheel alignment. If the brakes, tires, and wheels, are all in good condition, but a pull under braking persists, inspect the suspension and replace any defective parts you may find. Then, put the car on an alignment rack and adjust the suspension to the vehicle manufacturer's specifications.

Suspension Inspection

You can make a partial check of the suspension while the car is still on the ground. Examine the car at rest, and look for any obvious leanings that may indicate a broken spring or spring mount, or a weak and sagging spring. If you suspect a problem, check the vehicle ride height against the specifications in the shop manual to make sure the car is level.

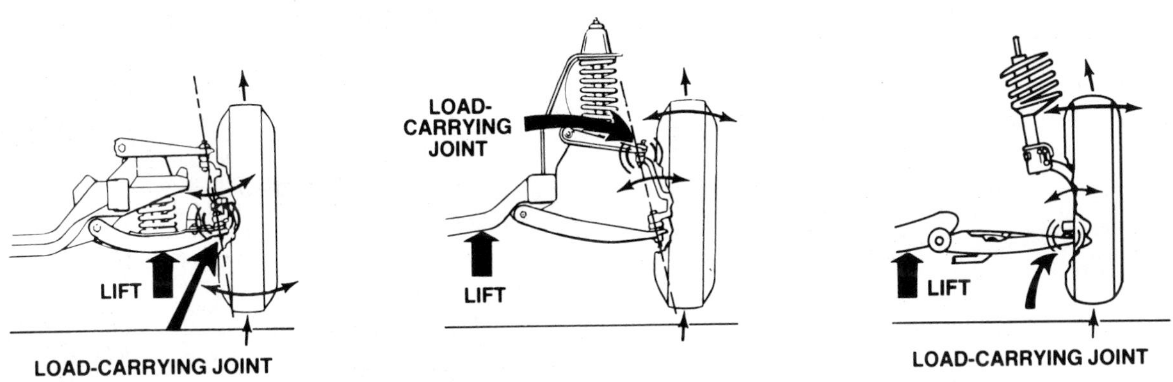

Figure 11-64. Support the suspension as shown to relieve tension on the load-carrying ball joint.

Next, raise and properly support the vehicle, then inspect the shock absorbers, tie-rod ends, and ball joints. Check the shock absorbers or struts for large amounts of oil leakage that indicate a blown seal; a small amount of seepage is considered normal. Also check for broken or loose shock absorbers or strut mounts.

Check the tie-rod ends for looseness and wear that results in radial and axial play. To test a tie-rod end for radial play, grasp the tire and turn it sharply from side to side while watching for play. To test a tie-rod end for axial play, grasp the tie rod next to the tie-rod end, and attempt to move the rod up and down while watching for play. Replace any tie-rod end that has detectable play.

Ball joints suffer the most wear of any suspension component, and are therefore the most likely to contribute to a braking problem. A worn ball joint allows the spindle to continually change position as the load on the wheel changes. This directly affects wheel alignment, and the stability of the car under braking. The wear limits for ball joints on most cars are shown in figure 11-61 on the next page.

Wear indicator ball joints

As noted in figure 11-61, many newer cars are equipped with wear indicator ball joints. You check the wear on these ball joints while the weight of the vehicle is resting on the wheels. On Chrysler cars, grasp the ball joint grease fitting and attempt to wiggle it, figure 11-62; replace the joint if any movement occurs. On Ford Motor Company and General Motors cars, check the position of the ball joint grease fitting, figure 11-63. Replace the joint if the fitting boss is flush with, or extends below, the ball joint cover.

Ball joints without wear indicators

On cars that do not have wear indicator ball joints, you check the wear of the joints with the weight of the vehicle off the wheels. Typically, the wear will be concentrated in the load-carrying ball joint, figure 11-64, that supports the weight of the car. If necessary, adjust the wheel bearings before you check the ball joints so you do not mistake bearing play for ball joint wear.

To check the ball joints, raise the car with a floor jack at the appropriate point shown in figure 11-64; this removes tension from the load-carrying ball joint so it will be free to move. Grasp the top and bottom of the tire, and rock it in and out while watching for radial play in the ball joints. Next, place a pry bar under the tire and lever it up and down while looking for axial play in the ball joints. Replace the load-carrying ball joint if its radial or axial play exceeds the specifications given in figure 11-61. Replace nonload-carrying ball joints if any play is detectable.

Alignment Inspection

While you inspect the suspension as described above, you should also look for any signs of accident damage that may have bent the chassis and upset the wheel alignment. Once you have replaced any damaged or worn parts, put the vehicle on an alignment rack and check the alignment. Pay special attention to the caster variation from one side of the car to the other.

A brake pull is most likely to occur if the caster is negative at one front wheel and positive on the other. This can occur even with new parts set to specifications. For example, if the manufacturer specifies a maximum 1 degree caster variation between the wheels, you could have ½ degree negative caster at one wheel and ½ degree positive caster at the other. Never align a front end so there is positive caster at one wheel and negative caster at the other.

Make & Model	Year	Max. Axial Tolerance
DOMESTIC CARS		
American Motors		
All ex. Alliance, Encore, Pacer	1987-78	.080" Note 1
Alliance, Encore, GTA	1987-83	Note 2
Pacer	1980-78	Note 3
Chrysler Corp.		
All FWD	1987-81	Note 4
All RWD	1987-78	.030"
All FWD	1980-78	.050"
Ford Motor Company		
All FWD	1987-81	Note 2
All RWD	1987-81	Note 5
All ex. Pinto, Bobcat, Granada, Monarch, Versailles	1980	Note 5
Pinto, Bobcat	1980	Note 6
Granada, Monarch, Versailles	1980	Note 2
All ex. Mustang, Capri, Granada, Monarch, Versailles	1979	Note 6
Mustang, Capri	1979	Note 5
All ex. Granada, Monarch, Versailles	1978	Note 6
General Motors Corp.		
Buick		
All FWD ex. 1985-79 Riviera	1987-80	Note 2
Riviera	1985-79	.125"
All RWD	1987-78	Note 5
Cadillac		
All FWD ex.1985-78, Eldorado, 1985-80 Seville	1987-82	Note 2
All RWD	1987-78	Note 5
Seville	1985-80	.125"
Eldorado	1985-78	.125"
Chevrolet		
All FWD	1987-80	Note 2
All RWD ex. Camaro	1987-83	Note 5
Camaro	1987-82	Note 2
Corvette	1982-78	.0625"
Nova	1979-78	.0625"
All others	1982-78	Note 5
Oldsmobile		
All FWD ex. 1985-77 Toronado	1987-80	Note 2
All RWD	1987-78	Note 5
Toronado	1985-78	.125"
Pontiac		
All RWD ex. Firebird	1987-82	Note 5
Firebird	1987-82	Note 2
All FWD	1987-80	Note 2
All RWD	1981-78	Note 5
IMPORTED CARS/ LIGHT TRUCKS		
Chrysler Imports		
All	1978	4-7 Note 12
All ex. Pickup	1979-80	2.2-4.3 Note 12
Pickup	1979-87	.020"
Challenger, Sapporo	1981-83	3.6-5.8 Note 12
Colt, Champ	1981-84	2.2-4.3 Note 12
Vista	1984	5.5 Note 12
Conquest	1984-87	3.5-5.8 Note 12
Vista	1985-87	2.2-7.2 Note 12
Colt	1985-87	1.7-7.2 Note 12
Ford		
Courier	1978-82	.031"
Fiesta, Capri	1978-80	Note 11
General Motors		
Luv	1978-82	.060"
Nova	1985-87	0.8-2.5 Note 12
All others	1985-87	Note 11
Honda		
All models	1978-84	.020"
All models	1985-87	Note 11
Mazda		
All models ex. RX-7	1978-80	.040" Note 14
RX-7	1979-80	Note 11
RX-7, 626, GLC wagon	1981-82	0.9 Note 12
B2000	1981-83	.040" Note 14
GLC	1981-83	Note 11
RX-7, GLC wagon	1983-85	0.9 Note 12
626	1983-85	1.4-2.5 Note 12
All others	1984-87	Note 11
Nissan/Datsun		
810	1978	3 min. Note 12
510	1978	1.5 min. Note 12
200SX, B210, 280Z	1978	3.5 min. Note 12
Pickup	1978	0.7 min. Note 12
210, 510	1979-82	1.5 min. Note 12
Pickup	1979-83 early	.004"-.040"
200 SX	1979	3.5 min. Note 12
810	1979-80	1.0 min. Note 12
280ZX	1979-84	1.0 min. Note 12
310, Sentra, Pulsar	1979-86	0.6 min. Note 12
200SX	1980-83	1.5-5.5 Note 12
810, Maxima	1981-84	3.5 min. Note 12
Stanza	1982-85	0.3 min. Note 12
Pickup	1983 late 1985 early	0.6 min. Note 12
Micra	1984-87	0
200SX	1984-87	0
300ZX	1984-87	.004"-.040"
Maxima	1985-87	.004"-.040"
Stanza	1986-87	.004"-.040"
Pickup 2WD	Late 1986-87	.004"-.043"
4WD	Late 1986-87	0-.008"
Toyota		
Celica, Corolla, Corona	1978-82	0.10"
Pickup, 4-Runner	1978-87	.090"
Tercel, Starlet	1980-87	Note 2
Cressida	1980-87	0.10"
Corolla RWD	1983-87	0.10 Note 15
Camry	1983-87	Note 2
Celica, Supra 2.8	1983-87	0.10"
Corolla FWD	1983-84	Note 2
Van	1984-85	.090"
MR2	1985-87	0.6-2.0 Note 12
Corolla FWD	1985-87	0.7-2.5 Note 12
Supra 3.0	1986-87	Note 2
Volkswagen		
Type 1	1978-79	0.10
Type 2	1978-79	.080"
All others	1978-87	.100"
Volvo		
All models	1978-87	.120"
DOMESTIC TRUCKS		
Chrysler Corp.		
All 2WD	1987-78	.020"
All 4WD	1987-78	Note 2
Ford Motor Company		
All 2WD	1987-78	.030"
All 4WD	1987-78	Note 11
General Motors Corp.		
All 2WD: Upper	1987-78	Note 8
Lower	1987-78	Note 9
All 4WD	1987-78	Note 10
Jeep		
All	1987-78	Note 5

Note 1: Maximum radial tolerance, .160 inch.

Note 2: Replace the ball joint if any measurable play is evident.

Note 3: Place the car on a level surface. DO NOT LIFT. Insert a rod into the lubrication hole until it contacts the ball stud. Mark the rod at the outer edge of the hole. Withdraw the rod and measure the distance from the end of the rod to the mark. Replace the ball joint if the distance exceeds 7/16 inch.

Note 4: Place the car on a level surface. DO NOT LIFT. Grasp the grease fitting of the ball joint and attempt to move it. Replace the ball joint if any movement occurs.

Note 5: Place the car on a level surface. DO NOT LIFT. Inspect the position of the roundboss into which the grease fitting is threaded. Replace the ball joint if the boss is flush with, or receded below, the surface of the ball joint cover.

Note 6: Replace the ball joint if radial play exceeds .250 inch.

Note 7: Remove the ball joint stud from the spindle, install the locknut, and use a torque wrench to measure the amount of force required to turn the ball stud. A new part should require 2-10 ft-lb (2.7-13.6 Nm) to turn the ball stud. Replace the ball joint if zero torque is needed to turn the ball stud.

Note 8: Replace if any lateral shake is noted or if the ball stud can be rotated by fingers.

Note 9: Support lower control arm at wheel hub and measure distance from tip of ball stud to the tip of the grease fitting. Remove support and remeasure distance. Replace if distance exceeds .094".

Note 10: Ball joint is adjustable, refer to factory shop manual for procedure.

Note 11: Not specified by manufacturer.

Note 12: Ft-lb of torque required to rotate ball joint stud when removed from king pin.

Note 13: Ball joints are adjustable by adding shims. See service manual for detailed procedure.

Note 14: Maximum radial tolerance.

Note 15: On solid, non-spring loaded ball joints, replace if any measureable play is evident.

Figure 11-61. Ball joint testing specifications.

Index

* See individual topic entry for more detailed information

* See individual topic entry for more detailed information

* See individual topic entry for more detailed information

* See individual topic entry for more detailed information